PUBLIC HEALTH ENGINEERING PRACTICE
FOURTH EDITION

Volume II

SEWERAGE AND SEWAGE DISPOSAL

L. B. Escritt

MACDONALD & EVANS LTD
8 JOHN STREET, LONDON WC1N 2HY
1972

First published in 1935
Second edition rewritten and reset to Royal 8vo. June 1949
Reprinted March 1951
Reprinted February 1954
Third edition February 1959
Fourth edition reset in two volumes March 1972

S.B.N.: 7121 1641 9

©

MACDONALD AND EVANS LIMITED
1972

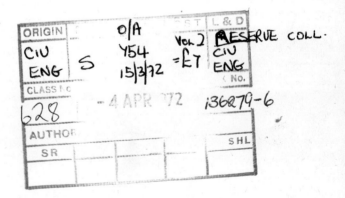

Printed in Great Britain by Fletcher & Son Ltd, Norwich
and bound by Richard Clay (The Chaucer Press) Ltd, Bungay, Suffolk

PUBLIC HEALTH ENGINEERING PRACTICE

Volume II
SEWERAGE AND SEWAGE DISPOSAL

Other books by L. B. Escritt

SEWERAGE ENGINEERING
REGIONAL PLANNING
SURFACE DRAINAGE
SEWERAGE DESIGN AND SPECIFICATION
THE MUNICIPAL ENGINEER
SEWAGE TREATMENT: DESIGN AND SPECIFICATION
A CODE FOR SEWERAGE PRACTICE
SURFACE-WATER SEWERAGE
BUILDING SANITATION
RIFLE AND GUN
RIFLEMAN AND PISTOLMAN
WATER SUPPLY AND BUILDING SANITATION
THE SMALL CELLAR
PUMPING-STATION EQUIPMENT AND DESIGN
DESIGN OF SURFACE-WATER SEWERS
SEWERS AND SEWAGE WORKS WITH METRIC CALCULATIONS AND FORMULAE
ESCRITTS' TABLES OF METRIC HYDRAULIC FLOW

PREFACE TO THE FOURTH EDITION

FROM the first revision by the present author of *The Work of the Public Health Engineer*, the book has sold steadily with no sign of diminution for twenty years and it is now known to be a standard work, adopted by several universities and colleges at home and abroad. The third edition appeared in 1959, since when new knowledge and practice call for a new publication. This in turn has necessitated considerable expansion of the content of the previous edition and, for ease of reference and general convenience to the reader, two separate volumes have now been created entitled respectively: *Water Supply and Building Sanitation* and *Sewerage and Sewage Disposal*.

There are, however, some changes of circumstance which call for modification other than enlarging and bringing the work up-to-date. First, the rate at which statutes and regulations are now re-enacted or modified makes it impracticable to include legal material other than in specifically legal works intended for revision or republication at very short intervals. It is therefore with regret that the part prepared by Dr. Sidney F. Rich, O.B.E., J.P., LL.B., has had to be omitted, as has also the very short section on public cleansing because that subject, while a public health service, is seldom part of a public health engineer's work.

Another change, setting a problem to authors and publishers alike, is the proposed general adoption of the metric system in lieu of imperial units. Engineering, particularly public health engineering, differs from other matters such as coinage and pure science in that the advantages of adoption of the metric system will be much outweighed by the disadvantages. The use of metric units will not make the majority of calculations easier, for all hydraulic-flow calculations involve time and the force of gravity which introduce decimal fractions. All the metric system does for public health engineering is to introduce a new system of arbitrary units which eventually will become familiar but with the result that the time-honoured imperial units will become forgotten by the majority of students and practising engineers. The difficulty will then be that, when engineers have to refer (as they frequently do) to the great wealth of engineering literature in textbooks and the proceedings of institutions in the various English-speaking countries, it will be as if they were reading a foreign language.

The problem which the author has to solve is to decide to what extent the text of this book should be converted to the metric system. There is the temptation to be thoroughly "modern" and turn completely metric. But having sought the advice of universities and given careful consideration to the matter he has decided that to do this would be to render the book of reduced value during at least the life of the new edition.

In spite of there being a target date by which it is intended the change to the metric system shall be completed, it should be remembered that the first move towards the adoption of the metric system in England was made over a century ago and this was followed by further attempts, and the system was certainly made legal in the Weights and Measures (Metric System) Act, 1897. Thus it is possible that the target date may not be achieved and, even if it is, it is certain to be followed by a further period of some years during which the mistakes made by the committees are being ironed out.

For these reasons imperial units have been retained in many of the formulae and most of the example calculations in this edition. However, metric equivalents of formulae have been added where these appear to be desirable, important calculations have been made in terms of both imperial and metric units and appendixes have been prepared, giving the metric definitions in force at the time of writing, on imperial and metric equivalents and on the nominal equivalent diameters of pipes.

Shortly after revising this book, the author retired from practice and it is for this reason that reference to his membership of various institutions does not appear on the title-page. Retirement has had one distinct advantage: it has given him freedom to express frank opinions on matters of interest to the public and the public health engineering profession.

In recent years there has been much activity of technical committees but not all of this could be described as helpful. The publications which appeared early in the century, many of which are still referred to, suggested that their authors were the technical experts of the time. Now the position is different: committee members seem to be, in the main or entirely, representatives of (often financially) interested bodies, selected with little regard to their knowledge of the subject. Thus there have been occasions when an "expert" committee has had a very small minority of members, or even none at all, with first-hand experience of the subject in question. It follows that it is not uncommon for one to find, in proposed codes or recommendations, suggestions that not only offend against the interests of public health but also against Statutory Law and common sense, or involve mathematical absurdities.

There is no short cut to expertness. It can be achieved by study and long experience only. But once one has become an expert, one has no need for guidance by codes prepared by those less qualified. Then why are codes written?

The danger of a code is that it can act as an ignis fatuus for those who would practise as advisers where they are not qualified to practise. And codes prepared by official or semi-official bodies are particularly dangerous because they are liable to be accepted as gospel no matter now inaccurate or even ridiculous their contents.

To be able to prepare written material for the guidance of members of the profession one must have more knowledge of the subject than the majority of the readers. One must also be prepared to spend weeks, months or years in research and to give careful consideration to everything that is put into print. But how many members of committees, even when they have reasonable knowledge of their subject, are willing to give a large amount of their time outside the committee room? In the words of one public servant whose business is committees, "Committees will agree anything provided they do not have to do the work themselves."

Even when initially good, stereotyped codes have their dangers. For the short time that they are up-to-date they can be of use to engineering designers in that they save time that would be otherwise lost in searching through literature or arriving at decisions. But they seldom do remain up-to-date for very long and then they become a serious bar to progress.

A fair example is the notorious "Ministry of Health 'Requirements,' etc." These "Requirements" are understood to have originated in an internal circular of the Local Government Board many years prior to the creation of the Ministry of Health in 1919. The circular was, at the time, a fair summary of some of the information given in the Reports of the Royal Commission on Sewage Disposal. The contents of this circular became known outside the Board and were later privately published in a pamphlet which began with the remark: "The following are *believed** to be the requirements of the Local Government Board." In 1921 the same material, with some moderate additions, was published in a small book in which it was stated that the contents were *believed* to be the requirements of the Ministry of Health.

Thereafter the main part of this book was copied in various publications without regard to copyright until many engineers believed that this obsolescent code was, in fact, official, and this belief was so widely held that questions on the "Ministry of Health Requirements" appeared in the examination papers of professional bodies and at least one university college.

The effect of this code was to hold back progress in design and stifle innovations in Great Britain for nearly twenty years, for many engineers, including even specialist consultants, would not deviate from it for fear of Ministry disapproval. The author started a campaign in the Press against the "Requirements" in 1938, but it was not until during the Second World War that it could be said that the belief in them had generally died out.

The literal meaning of the word "practice" is repeated exercise in an art or science. It therefore would not seem right that any new idea proposed in committee, but never used or even tried experimentally, should be included in a code of practice: to do this would be to mislead the public. But this has been done.

Another activity which is not an unmixed blessing is official research. Such research is necessary and, on the whole, desirable; and it is sometimes very good work. The danger lies in that too much is expected of the average scientific worker and too much is blindly accepted if emanating from an official department. Thus new discoveries based on too brief researches,

* Author's italics.

even when contrary to common experience or mathematics, or unsupported by data or reason, are liable to become embodied in new advisory codes.

From the foregoing it will be appreciated that the contents of this book do not always conform to the expressed opinions of the latest committees even when these have the backing of professional institutions or government departments. For the opinions herein are based on forty-five years' experience in the design and construction of public health engineering works costing tens of millions of pounds, supplemented by the study of international literature and years of both official and individual research. In nearly all instances the methods described or innovations made have been used and found satisfactory in practice and all the new methods have been proposed because there has been a need for them.

Moreover, the writer has not retained any of his own theories when further experience or study have shown that there is room for improvement, as will be seen by comparing this edition of *Public Health Engineering Practice* with previous editions.

It is only in prefaces to new editions that authors have the opportunity of replying to reviewers. All reviewers are entitled to their honest opinions but some comment not on the contents of a book but on what it does not contain, without considering that there may be good reason for the omission.

A textbook should provide, in the least space practicable, the information which will be most useful to the greatest number of readers. It follows that not everything can or should be included. The material that must not be omitted from a public health engineering textbook is the "bread and butter" of engineering practice; on the other hand, the purely topical or intriguing oddity is best left out. On one occasion the author was criticised for not describing a method of "sanitation" that had had much notice in the Press. As this was a proprietary process he could hardly reply that the reasons for the omission were that the method had been very little used indeed, that in his opinion it was insanitary and that its use was prohibited by law!

Similarly, careful consideration has to be given to which mathematical procedures of design are recommended or even described. Several new theories are eagerly taken up by enthusiastic young engineers who waste much time at their employers' expense in applying them. There seems to be an appeal in complexity for its own sake, whereas the simple theory not only saves time but is often the more accurate.

Mathematics is the simplification of the facts of nature so as to bring them within the limited compass of the human mind. The engineer has to think in symbols, usually visual, such as the straight line, the square, circle, pyramid and sphere; things that in fact exist in approximate form only. Of necessity, some mathematics is complicated and some theories difficult to grasp. But the delight in being brilliantly obscure should not be allowed to militate against the principle that the good method and the good design are essentially simple. This is one reason why some complicated procedures are not included in this book. Another more important one is that no negligible proportion of new methods, on examination, proves to be based on unsound mathematics or reasoning.

The omission of description of any materials, of construction, of proprietary methods or of theories, however, must not necessarily be taken to mean that the author does not favour them.

August 1971 L. B. Escritt

CONTENTS

	PAGE
PREFACE TO THE FOURTH EDITION	v
LIST OF ILLUSTRATIONS	xiii
LIST OF TABLES	xvii

PART I
SEWERAGE AND LAND DRAINAGE

CHAPTER		
I.	GENERAL PRINCIPLES OF SEWERAGE	3
II.	SOIL SEWERAGE	16
III.	SURFACE-WATER SEWERAGE: THE LLOYD-DAVIES METHOD AND THE RATIONAL METHOD	24
IV.	SURFACE-WATER SEWERAGE: ELABORATIONS OF THEORY .	49
V.	STORM-RELIEF WORKS	74
VI.	SOAKAWAY SYSTEMS FOR SURFACE WATER . . .	80
VII.	SEWERS ACROSS VALLEYS	83
VIII.	VENTILATION AND FLUSHING OF SEWERS . . .	92
IX.	DRAINAGE OF ROAD SURFACES	96
X.	THE CONSTRUCTION OF SEWERS OF SMALL OR MEDIUM DIAMETER	104
XI.	THE CONSTRUCTION OF SEWERS OF LARGE DIAMETER .	115
XII.	THE CONSTRUCTION OF MANHOLES AND CHAMBERS .	127
XIII.	SEWAGE PUMPING	138
XIV.	PNEUMATIC EJECTORS	165
XV.	PROBLEMS PECULIAR TO THE SEWERAGE OF COASTAL TOWNS	173
XVI.	CONSTRUCTION OF STORAGE TANKS AND SEA OUTFALLS .	182
XVII.	UNDER-DRAINAGE	189
XVIII.	RIVER MAINTENANCE	195
XIX.	SAFETY OF SEWERAGE OPERATIVES	209

PART II
SEWAGE TREATMENT AND DISPOSAL

XX.	RATES OF FLOW AND CHEMISTRY OF SEWAGE . . .	217
XXI.	DISPOSAL OF SEWAGE BY DILUTION	231
XXII.	TREATMENT OF SEWAGE ON LAND	235

CHAPTER		PAGE
XXIII.	SITING AND LAY-OUT OF SEWAGE-TREATMENT WORKS	238
XXIV.	PRELIMINARY TREATMENT: SCREENING AND GRIT AND GREASE REMOVAL	249
XXV.	SEPARATION AND TREATMENT OF STORM WATER	279
XXVI.	THEORIES OF CONTINUOUS-FLOW SEDIMENTATION	291
XXVII.	SEDIMENTATION-TANK DETAILS	307
XXVIII.	TREATMENT AND DISPOSAL OF SEWAGE SLUDGE	324
XXIX.	PERCOLATING-FILTER TREATMENT	365
XXX.	ACTIVATED-SLUDGE SYSTEMS	390
XXXI.	METHODS OF SEWAGE TREATMENT OF HISTORICAL INTEREST	417
XXXII.	TREATMENT OF TRADE WASTES	424
XXXIII.	SEWAGE DISPOSAL FOR ISOLATED BUILDINGS	433
	APPENDIX I. DEFINITIONS AND EQUIVALENTS OF BRITISH AND METRIC UNITS	471
	APPENDIX II. NOMINAL EQUIVALENT IMPERIAL AND METRIC PIPE DIAMETERS	474
	APPENDIX III. FLOW IN PARTLY-FILLED SEWERS	476
	BIBLIOGRAPHY	477
	INDEX	488

LIST OF ILLUSTRATIONS

FIGURE		PAGE
1.	Typical arrangement of individual house connections to sewer laid in road	17
2.	Typical sewer plan and section	18
3.	Hourly variations in flow and strength of sewage	20
4.	Drainage diagram illustrating soil-sewerage calculation	21
5.	Run-off curves for storms shorter than, equal to and longer than time of concentration	35
6.	Drainage area	37
7.	Change of impermeability during rainfall	41
8.	Drainage area: irregular distribution	50
9.	Diagram illustrating time-area graph and the Ormsby and Hart method	51
10.	Tangent method	54
11.	"Tangent curves"	56
12.	Application of tangent curves to time-area graph	57
13.	Storm curve for use with Ormsby and Hart method	61
14.	Diagram illustrating basis of storm-water storage calculation	67
15.	Diagram illustrating basis of storm-water storage calculation	68
16.	Simple storm-overflow weir	78
17.	Soakaway constructed of pre-cast concrete segments and suitable for use in limestone formations	82
18.	Steel sewer supported on piers above ground-level	86
19.	Single inverted siphon	87
20.	Double inverted siphon	88
21.	Typical flushing siphon	93
22.	Typical flushing-tank	94
23.	Longitudinal section of road showing falls of channel and kerb line	97
24.	Cross-section of cambered road showing crown off centre	98
25.	Plan showing levels on the surface of a cambered road as required to secure adequate drainage	99
26.	Intersection of two roads with cross-falls in the same direction	102
27.	Intersection of two roads with cross-falls resulting in a channel line	103
28.	Intersection of two roads with cross-falls resulting in a ridge or crown	103
29.	Cement mortar joint on vitrified clay pipes	104
30.	Hot-pour joint on vitrified clay pipes	104
31.	Polyester mouldings and rubber "O"-ring on vitrified clay pipes	106
32.	Plasticised P.V.C. mouldings on vitrified clay pipes	106
33.	Benching, haunching and concrete surround to stoneware or concrete pipe	108
34.	Section of cast-iron segmental sewer	118
35.	Details of cast-iron tubbing	118
36.	Reinforced-concrete sewer	121
37.	Mass concrete culvert	121
38.	Horseshoe-shaped sewer	121
39.	Cross-section of U-shaped sewer	122

LIST OF ILLUSTRATIONS

FIGURE		PAGE
40.	"Normal" egg-shaped sewer	122
41.	Culvert with low headroom	122
42.	Large concrete pipe sewer	122
43.	Steel sewer being laid in trench	125
44.	Deep side-entrance manhole constructed of pre-cast concrete pipes	128
45.	Pre-cast concrete manhole with backdrop	128
46.	Shallow manhole showing good and bad designs of benching	129
47.	Thickness of manhole walls	131
48.	Side-entrance manhole on large sewer	132
49.	Brick manhole with ramp or inclined backdrop	134
50.	Manhole designed to withstand internal pressure due to surcharge	136
51.	"Amphistoma" pump showing impeller	144
52.	Loss of head through reflux valves	154
53.	"Dry-well" pumping station	160
54.	"Barrington" self-priming sewage pump	161
55.	"Barrington" self-priming sewage pump	163
56.	"Shone" ejector	165
57.	"Lift and force" ejectors	166
58.	Three-point problem	176
59.	Steel outfall being floated into position	183
60.	Plan of sea outfall	185
61.	Automatic sewage-effluent sampler, Crossness sewage-treatment works	227
62.	Principle of automatic sampler	227
63.	Control room, Crossness sewage-treatment works	228
64.	General plan of Crossness sewage-treatment works	243
65.	Faulty design of tank	244
66.	Diagram illustrating the development of embankments in various circumstances	245
67.	Example of a good lay-out for small sewage-treatment works but with uneconomical arrangement of pipework	246
68.	Example of a well-conceived small sewage-works lay-out with economical arrangement of pipework	247
69.	Dorr type "T" screen and disintegrator	252
70.	Mechanical screens in chamber	254
71.	"Comminutor"	261
72.	Dorr "Detritor" type "A"	264
73.	Dorr "Detritor"	266
74.	Proportional-flow plate weir	267
75.	Cross-section of constant-velocity detritus channel controlled by specially shaped weir plate	268
76.	Constant-velocity detritus channel installation	269
77.	Ames Crosta Mills grit extractor installed at Troqueer sewage-purification works, Dumfries	270
78.	Cross-section of parabolic constant-velocity detritus channel	271
79.	Cross-section of W-shaped constant-velocity detritus channel	271
80.	Constant-velocity detritus channels, Crossness sewage-treatment works	272
81.	East elevation of Crossness detritus channels	272

LIST OF ILLUSTRATIONS

FIGURE		PAGE
82.	Constant-velocity detritus channels for small sewage works: plan	273
83.	Constant-velocity detritus channels for small sewage works: cross-section	273
84.	Storm overflow weir chamber for large sewage-treatment works	280
85.	Balancing tank	289
86.	Relation of sedimentation to surface area of tank, according to Hazen's figures	292
87.	Relation of sedimentation to capacity of tank, according to Hazen's figures	292
88.	Degree of settlement produced at various detention periods	299
89.	Degree of settlement produced at various detention periods (plotted to logarithmic scales)	299
90.	The effect of detention period on settlement of river mud	300
91.	The effect of detention period on settlement of activated sludge	300
92.	Dortmund upward-flow sedimentation tank	302
93.	Cross-section of pyramidal sedimentation tank	310
94.	Bridge-type sedimentation tank for activated sludge	311
95.	Sludging mechanism for circular tank	312
96.	Dorr "Clarifier" type "A"	315
97.	Dorr "Clarifier" type "SC"	317
98.	Rectangular sedimentation tanks with travelling scraper and transporter carriage	318
99.	Hartley sludging mechanism, Crossness sewage-treatment works	319
100.	"Mieder" scraper	320
101.	Diagram illustrating the system of progressive bifurcation of feed-pipes of equal length	321
102.	Pyramidal sedimentation tank constructed above ground-level in mass concrete	322
103.	Pyramidal sedimentation tank constructed above ground-level in reinforced concrete and resting on reinforced-concrete raft	322
104.	Pyramidal sedimentation tank constructed above ground-level in reinforced concrete to an economical design	322
105.	Diagram illustrating how the moisture content of sludge affects the volume	325
106.	Cross-section of part of a filter-press	331
107.	Sludge cake from filter-press	332
108.	Flow diagram of a Coilfilter installation	333
109.	Surface of digesting sludge	334
110.	Dorr digester	336
111.	Heater and circulating chambers of Dorr-Oliver "B.G.H." digesters, Crossness sewage-treatment works	341
112.	View between Dorr-Oliver "B.G.H." digestion tanks, Crossness sewage-treatment works	342
113.	Dorr primary digester type "B.G.H."	344
114.	Dorr type "B" sludge heating and circulating unit	346
115.	Ames Crosta Mills "Simplex" digestion tank with floating roof and suspended circulating pump	351
116.	Gasholder of the Dorr-Oliver "B.G.H." digester, Crossness sewage-treatment works	352

FIGURE		PAGE
117.	Excess gas burner and primary sludge dewatering tanks, Beckton sewage-treatment works	352
118.	Secondary digestion tanks, Crossness sewage-treatment works	354
119.	Dorr secondary digester type "S"	355
120.	Ames Crosta Mills primary and secondary digestion tanks	357
121.	Diagram showing utilisation of sludge gas and recovery of waste heat from engine exhaust and coolant	359
122.	Flow diagram of Porteous plant	362
123.	Sedimentation tank and percolating filter	376
124.	Section of dosing tank and percolating filter	377
125.	Illustrating loss of head from sedimentation tank to percolating filter	378
126.	Travelling distributor for rectangular percolating filter	382
127.	Flow diagrams of Dorr "Biofiltration" process	385
128.	Flow diagram of typical diffused-air plant	391
129.	Dome diffusers mounted on air main	393
130.	Flat-bottomed aeration tank	394
131.	Arrangement of dome diffusers in flat-bottomed aeration tank	394
132.	Aerial view of Maple Lodge sewage-treatment works: a diffused-air plant	399
133.	Lay-out of activated-sludge plant, diffused-air system	403
134.	Diagram illustrating approximate calculation of quantity of surplus activated sludge	406
135.	Lay-out of aeration tanks and final sedimentation tanks of a Simplex surface aeration plant	408
136.	Primary sedimentation tanks, Crossness, feeding to aeration tanks (Fig. 137)	410
137.	"Simplex" aeration tanks, Crossness, fed from primary sedimentation tanks (Fig. 136)	411
138.	Simplex high-intensity aerating cone	411
139.	Cross-section of tank illustrating the principle of the Kessener process	412
140.	Lime-mixing plant	427
141.	Labyrinth for chemical treatment of sewage	428
142.	"Elsan" chemical closet	437
143.	Cesspool emptier	438
144.	Chain pump	441
145.	Typical septic tank	445
146.	Septic tank and percolating filter	448
147.	Septic tank, automatic pump and percolating filter	450
148.	Cast-iron dip-pipes for septic tank	452
149.	Small rectangular percolating filter with Tuke and Bell tipping-tray mechanism	456
150.	Small circular percolating filter with Ames Crosta Mills rotating distributor	458
151.	Small sewage-treatment works suitable for an institution, village or camp	459
152.	Tipping-tray mechanism and one of the distribution channels to which tipping-tray discharges	460
153.	Wire-netting cover to percolating filter to keep out falling leaves	460
154.	Simplex aeration plant for an institution	463
155.	Simplex aeration plant for an institution	465

LIST OF TABLES

TABLE		PAGE
1.	Proportional depths, areas, velocities and discharges for circular sewers running partly full	11
2.	Minimum gradients for sewers and optimum gradients for rising mains	12
3.	Calculations for soil sewers	22
4.	Impermeability factor calculation	26
5.	Typical impermeability factors	27
6.	Intensities of rainfall	32
7.	Example calculation	38
8.	Example calculation	38
9.	Example calculation for surface-water sewerage scheme	39
10.	Probable run-off to sewers	42
11.	Run-off and rainfall coefficients for rational method	45
12.	Example calculation for surface-water sewerage scheme	46
13.	Example calculation for surface-water sewerage scheme	46
14.	Example calculation for surface-water sewerage scheme	47
15.	Example storm for use in Coleman and Johnson method	58
16.	Coleman and Johnson calculation	59
17.	Coleman and Johnson calculation	59
18.	Addition of run-off from subsidiary areas in Coleman and Johnson method	60
19.	Hypothetical storm as used by Ormsby and Hart	62
20.	Ormsby and Hart calculation	63
21.	Time-area graph data	64
22.	Hydrograph method calculation	65
23.	Hydrograph method calculation	66
24.	Maximum theoretical gully spacing	97
25.	Difference of road channel levels	101
26.	Proportions of cast-iron segmental sewers	117
27.	Thickness of circular sewers	126
28.	Starting times of pump sets	153
29.	Frictional resistance of fittings	155
30.	Example of cut-in and cut-out levels of automatic pumps	156
31.	Pressure and volume of air and power required for compression	169
32.	Loss of pressure in air mains	170
33.	Loss of pressure in air mains	171
34.	Example of survey book for float experiments	176
35.	Value of $3.575 D^{2\frac{1}{8}}$ for circular pipes	179
36.	Example stage-by-stage calculation for discharge of storage tank	181
37.	Minimum gradients for land drains	191
38.	Calculation for unit hydrograph	201
39.	Application of unit hydrograph	202
40.	B.O.D. and strength of sewage	223
41.	Areas of land required for land treatment of sewage	237

TABLE		PAGE
42.	Quantities of screenings	255
43.	Falling speeds of siliceous particles in water	265
44.	Mechanical analysis of detritus	274
45.	Effect of velocity on quantity and density of sludge	296
46.	Specific heat of sludge	343
47.	Thermal conductivities of building materials and subsoils	345
48.	Percentage of gas produced	353
49.	Discharge of gas through orifices	356
50.	Minimum gradients for sludge mains	364
51.	Loading of percolating filters	372
52.	Gallons retained in a cubic yard of medium	373
53.	Effect of size of percolating-filter medium	375
54.	Recommended velocities in air mains	397
55.	Efficiencies of low-pressure air-blowers	397
56.	Summary of operating data, Mogden sewage-works	404
57.	Dimensions of typical Simplex aeration tanks	409
58.	Selection of products manufactured on a trading estate, effluents produced and treatment methods applied before discharge to sewer	430
59.	Head of population served by septic tanks	451
60.	Proportions of septic tanks	451
61.	Delivery of chain pumps	454
62.	Capacities of small percolating filters	461

PART I
SEWERAGE AND LAND DRAINAGE

CHAPTER I
GENERAL PRINCIPLES OF SEWERAGE

THE earliest sewers were either open or culverted watercourses and ditches, and when in time special sewers had to be constructed they were built to various shapes and sizes and generally of brickwork. Most of them were made large enough for men to go through them and clear them periodically, for the flow was generally too sluggish to carry on the solids, which formed offensive deposits in their inverts. The condition of affairs that prevailed in London is well described in the following passage from the "Main Drainage of London," by Sir George W. Humphreys[115]:

". . . the Act (19 Chas. II., cap 3) 'for rebuilding the City of London', passed in 1667, contained much of great importance to the health of the City, inasmuch as in this Act seems to lie the origin of the Commissioners of Sewers for the City of London, a body to whose hands the sanitary well-being of the City was practically entrusted for two hundred years. The above-mentioned Act conferred upon them large powers in connection with sewerage and paving, and these powers were extended from time to time by further Acts, dealing, *inter alia*, with cleansing, lighting, dust removal, etc.

Outside the City the inhabitants do not seem to have been troubled with numerous or exacting sanitary regulations. Each parish managed its own affairs, appointing from among its own body, its various unpaid officers, including the surveyor and the scavenger, and for a long period this somewhat primitive organisation sufficed to meet, according to the not very exacting standards of the time, the sanitary needs of these rural communities.

When, however, at varying dates from the beginning of the 17th century, these outlying districts commenced to 'develop', and the population went up by leaps and bounds, the existing arrangements were found unequal to the strain.

Paid officials or contractors were appointed to perform the sanitary work, and scores of local Acts made their appearance in the Statute Book, dealing with the cleansing of streets, lighting, dust removal, etc., applying sometimes to the whole of a parish, but in numerous instances only to a much more limited area. The first Act, however, on these lines applying to an area* comparable with the present County was the Metropolis General Paving Act, commonly known as Michael Angelo Taylor's Act, 1817 (57 Geo., III, cap. 29), which consolidated and reduced into a kind of system many of the provisions relating to paving and the regulation and improvements of streets, which were contained in the special Acts above referred to. It also provided for the cleansing of drains, cesspools, etc., and street watering and regulated certain offensive trades.

A few years before the passing of this Act, an invention had been introduced which was to have a very important effect on sanitation. This was the water-closet,

---

* It applied to London, Westminster, Southwark, the districts included within the **Weekly Bills of Mortality,** St. Pancras and St. Marylebone.

which offered facilities, before unknown, for the entire removal of sewage. At first the water-closet was made to discharge not into the sewers, but into a cesspool, the ancient receptacle for offensive household refuse, the contents of which were removed from time to time.

The large addition thus caused to the contents of the cesspools, however, made it necessary to introduce overflow drains running from them into the street sewers; and other reasons also gave inducement to discharge the sewage, with the aid of a sufficient water supply, direct from the water-closet to the street sewers. Originally, the discharge of offensive matter into the sewers was a penal offence, and so continued up to about the year 1815. Afterwards it became permissive to drain houses into sewers, and in 1847 it was made compulsory.

These sewers were originally banked-up open watercourses, intended solely for the purpose of carrying off the surface drainage. They were under the control of commissioners constituted under the general Act 23 Henry VIII, cap. 5 (The Bill of Sewers) as amended by later legislation, and in 1847 the sewerage of the larger portion of the districts now forming the County of London (apart from the City) was governed by seven distinct commissions, viz., the Westminster, the Holborn and Finsbury, the Tower Hamlets, the St. Katharine's, the Poplar and Blackwall, the Greenwich, and the Surrey and Kent. Each of these Commissions had its own mode of conducting business, and its own regulations as to sizes of drains, their rates of inclination, mode of execution and cost; for long before the supersession of these bodies the open watercourses of the earlier period had to a very large extent been culverted and placed underground. The sewerage of the City was under the control of the City Commissioners of Sewers, mentioned above. It was not to be expected that under such divided management any advantageous general principles of drainage could be acted upon."

The following letter which was printed in *The Times* for Thursday, 5 July 1849 gives an intimate picture of conditions which must have been by no means uncommon.

"To the Editur of the Times Paper.
Sur,—May we beg and beseach your proteckshion and power. We are Sur, as it may be, livin in a Wilderniss, so far as the rest of London knows anything of us, or as the rich and great people care about. We live in muck and filthe. We ain't got no priviz, no dust bins, no drains, no watersplies, and no drain or suer in the hole place. The Suer Company, in Greek St. Soho Square, all great, rich and powerfool men, take no notice whatsomedever of our cumplaints. The Stenche of a Gullyhole is disgustin. We all of us suffer, and numbers are ill, and if the Colera comes Lord help us.

Some gentlemans comed yesterday, and we thought they was comishoners from the Suer Company, but they was complaining of the noosance and stenche our lanes and corts was to them in New Oxforde Street. They was much surpised to see the seller in No. 12, Carrier St., in our lane, where a child was dying from fever, and would not believe that Sixty persons sleep in it every night. This here seller you couldent swing a cat in, and the rent is five shillings a week; but there are greate many sich deare sellars. Sur, we hope you will let us have our cumplaints put into your hinfluenshall paper, and make these landlords of our houses and these comishoners (the friends we spose of the landlords) make our houses decent for Christians to live in.

Praye Sir com and see us, for we are living like piggs, and it aint faire we shoulde be so ill treted.

We are your respeckfull servents in Church Lane, Carrier St., and other corts.

Teusday, July 3, 1849."

In 1856 the Metropolitan Board of Works superseded the Metropolitan Commission of Sewers and it was not long before large-scale works were put in hand. These were a series of intercepting sewers designed by the Board's engineer, Sir Joseph William Bazalgette, to take all flows of sewage other than storm water to two "Outfall Works" at Beckton and Crossness respectively on the north and south of the Thames outside London. Storm water was permitted to overflow and discharge to the Thames locally as it does to this day. The Southern Outfall Works were opened by the Prince of Wales (King Edward VII) in April 1865: the Northern Outfall Works had been completed in 1864.

At that time very little was known of sewage treatment. The outfall works consisted merely of storage tanks arranged so that sewage should be discharged on the ebb tide only, in the hope that it would be carried well away from London. This did not prove satisfactory and, on the recommendations of a Royal Commission of 1882, "that the solid matter in the sewage should be separated from the liquid by some process of deposition or precipitation..." and "that the liquid proportion of sewage remaining after precipitation might, as a temporary measure, be then allowed to pass into the river." Precipitation works were completed in 1891 shortly after the creation of the London County Council, who then became the responsible body. The settled sludge from these precipitation tanks was shipped to sea.

Following the recommendations of the aforementioned Royal Commission, researches into means of sewage treatment were made. These included the treatment of settled effluent on bacteria beds which were described in "The Experimental Bacterial Treatment of London Sewage" by F. Clowes and H. C. Houston, which was published by the London County Council in 1904. But in spite of encouraging results nothing effective was done for many years and, with the increasing population and industry of London, the polluted state of the River Thames became progressively worse.

During a period of several years it appeared that the policy of the officers of the Council was to economise in cost by omitting the use of chemical precipitants. It was not until 1921 (thirty years after the completion of the "temporary measure") that small-scale experimental plants of the Diffused Air, Sheffield and Simplex activated sludge types were installed at the Southern Outfall Works. A somewhat larger experimental plant was completed at Beckton in 1932. This was later enlarged to six times its original size in 1939 but eventually was acknowledged a failure.

These experiments were carried out at a time when other large local authorities had constructed or were constructing their sewage works according to well-established principles,* for in 1915 the Royal Commission on Sewage Disposal, having considered a vast body of evidence since 1898, published their Final (tenth) Report, being a General Summary of Conclusions and Recommendations, in which it was stated: "It is practicable to purify the sewage of towns to any degree required, either by land treatment or by artificial filters, and there is no essential difference between the two processes." Also the activated sludge method had been invented prior to the First World War and this was used in the largest sewage works in Europe, the Mogden Works, West Middlesex, which were completed in 1934.

The first serious attempt at treating London sewage was the installation at the Northern Outfall Works, Beckton, of a Diffused Air activated sludge plant to treat approximately one-third of the dry-weather flow. These works were opened by H.R.H. the Duke of Edinburgh in October 1959.

In 1954 the author was instructed to prepare a scheme for reconstruction of the old "sand pit" (a crude detritus tank) and sedimentation channels and the construction of an aeration plant and sludge-digestion plant to deal with two-thirds of the flow at the Southern Outfall Works, Crossness, but he pointed out that a complete sewage-treatment works could be constructed for about £10,600,000—a not much greater cost than that of the original proposal. This scheme was eventually approved and the works opened in May 1964, about ninety-nine years after the opening of the original outfall works. (At the same time he had in hand the basic design of the Northern Outfall Works, which, at the time of writing, the Greater London Council is completing.)

The early history of the development of sewerage has left its mark on practice in more than one particular. First, the drainage of surface water together with foul sewage is known as the "combined system." Second, the necessity for maintaining velocities of flow so as to prevent silting of the open watercourses and culverts originated the practice of designing sewerage schemes so that all the sewers are laid to "self-cleansing" gradients throughout. Third, whereas in these days the vast majority of sewers are circular pipes or culverts entirely enclosed and buried in the ground, they are still designed as if they were open watercourses: the invert and sides retain the water, and are laid at gradients from top end to bottom, consistently falling in the direction of the flow, while the crown serves the purpose of holding out the superincumbent earth. The sewer is assumed to be running only just full, and it is not under internal pressure, as is a water-main. For this reason the hydraulic gradient of a sewer is for all practical purposes the gradient of the invert and/or the crown, and for design purposes it is usual to assume that the slope of the crown is the hydraulic gradient when the sewer is run-

---

* This extraordinary circumstance is explained by the fact that, *circa* 1930, there was no member of L.C.C. Main Drainage staff, above the level of temporary junior assistant, with any previous experience of sewage-works design.

ning full, and the fall of the invert the hydraulic gradient at the minimum rate of flow.

A legacy of the early days which is not so beneficial is the storm overflow. At the time when Bazalgette was deciding to spill over into the Thames all storm flows exceeding about one-quarter inch per day of rainfall on the north side of the Thames and one-tenth inch on the south side it was not appreciated how great storm flows could be and how polluting to the watercourses to which they were discharged. Thus the principle of providing storm overflows to discharge crude sewage to streams was established and, although this is now contrary to Sections 30 and 31 of the *Public Health Act*, 1936, there are engineers who are still installing them (unnecessarily, in the author's opinion).

Sewerage systems. The earlier sewers were all on the "combined system"—i.e. soil or foul sewage from premises was discharged to the same sewers as the run-off of rain-water from roofs and roads. Later a new system —the "separate system"—was devised in which soil sewage was discharged to soil sewers from which surface water was excluded, and surface water from roofs and roads discharged to surface-water sewers. There was also the "partially separate system" in which, while separate sewers were provided, some surface water, usually from the back roofs and yards, was discharged to the soil sewers by permission of the local authority.

The combined system was in early years most commonly preferred both in practice and by the authorities, and it was complained that the separate system was being applied without good reason. The Fifth Report of the Royal Commission on Sewage Disposal contains the following paragraph in this connection:

"In our opinion the cases in which the provision of separate systems can properly be justified are rare, and it appears to us that the relative merits of the separate and combined systems have not always been sufficiently considered."

It must be admitted that the choice of sewerage systems has been, and still is, largely a matter of fashion. For some time after the separate system had been introduced, it was permitted but by no means favoured by the Ministry of Health. Later, fashion tended towards the separate system almost to the extent of totally excluding the combined system except in connection with extensions of existing works where the sewers were already combined. The reasons given for favouring the separate system were:

1. Combined sewers, being large enough to take storm flows, were excessively large and liable to silt during dry weather, unless laid at gradients involving excessive depth or necessitating pumping.

2. The discharge of large quantities of storm water at the outfall works upsets treatment processes.

Neither of these arguments is strictly true. In the majority of instances, particularly when the cost of drains as well as sewers is taken into account, *the combined system involves considerably less cost than the separate system.* This has been proved by several comparative estimates. When one takes into account the improvement and enlargement of watercourses made necessary by the increased discharges to them it becomes evident that the separate system is more expensive than the combined system costing nearly twice as much.

If the combined sewers are laid at the same gradients as would be required by separate soil sewers, it is generally found that they are self-cleansing during dry weather. Where combined sewers have become foul as a result of inadequate gradients, the fault lies with the engineers who designed them and not with the system. A not uncommon fault has been to lay combined sewers at the gradients that will produce self-cleansing conditions only when the sewers are running full, instead of at the gradients required to produce self-cleansing conditions when they are flowing partly full in dry weather.

When storm water is discharged to sewage treatment works, all flow above that which can be given full treatment is passed to storm tanks. The effect on the treatment works is thus not at all serious, in fact, improvement of quality of final effluent during wet weather is a not uncommon observation.

The choice of sewerage system should depend on circumstances. Generally existing systems should be extended without alteration. Where new housing estates are being developed at the cost of the local authority, the combined system is almost certain to prove most economical unless surface water can be discharged to soakaways. This system is also to be preferred for densely built-up towns and industrial areas where the run-off from roads is liable to be foul, and all places where there is a risk of illicit soil connections being made to the surface-water sewers.

The separate system should always be used where it it will not prove unduly expensive or result in the possible fouling of watercourses. In this connection it should be mentioned that the rainfall run-off from modern housing estates is not polluted to any great extent. The partially-separate system is of limited value, and it is doubtful if it can be applied to any new sewerage schemes that are not extensions of existing systems.

Designers of sewerage schemes have often overlooked the fact that the *Public Health Act*, 1936, empowers owners and occupiers of property to discharge their soil sewage and surface water to the public sewers and that the only restrictions are that, where a separate system *has been provided*, the foul sewage and surface water must be discharged to the appropriate sewers, also that discharge may not be made to a storm-relief sewer. A sewerage system is not a separate system unless separate soil and surface water sewers have, in fact, been provided. It is therefore not possible to design and construct a sewer intended to take soil sewage only, as many engineers have done, for unless a surface-water sewer is also provided, owners and occupiers of property may discharge surface water to the so-called soil sewer which, legally, would appear to be a combined sewer.

GENERAL PRINCIPLES OF SEWERAGE

"Self-cleansing" gradients. Sewers are designed on two main principles: first, that the capacities shall be adequate but not excessive for accommodating the rate of flow and second, that the size of sewer and the fall shall be so arranged that, once a day, at the peak rate of flow, the velocity of flow is sufficient to cleanse the sewer of settled solids.

The subject of self-cleansing gradients deserves discussion at length, because too much has been taken for granted in the past, and gross errors have been perpetuated by repetition and the acceptance of "authority" in the face of all evidence to the contrary.

A case in point is as follows: Maguire's rule, devised long ago, was that the gradient of a drain or sewer should be 1 in 10 times the diameter in inches—i.e. a 6-inch diameter drain should be laid at 1 in 60, a 9-inch diameter at 1 in 90, a 12-inch-diameter at 1 in 120, and so on to absurdity. This rule was suggested at a time when very little was known about self-cleansing velocities, and while it may have served its purpose to some extent, it was found to be quite impracticable for sewerage design and, for that purpose, it was superseded over a hundred years ago. Nevertheless, it has been retained by many authorities as a rule for the design of *drains* to premises, or even private sewers and, as recently as just after the Second World War, it was included in a code of practice! But there is no excuse whatsoever for the perpetuation of this ancient, inaccurate and extravagant rule of thumb, in view of the fact that there is now considerable knowledge of what are effective self-cleansing gradients under normal conditions.

The practice of the majority of present-day engineers is to lay sewers to gradients sufficient to ensure velocities of $2\frac{1}{2}$ feet per second when they are flowing full or half-full, which condition also ensures that (in the case of circular pipes) when the sewer is flowing between one-quarter and one-fifth of its total capacity the velocity is 2 feet per second. The velocity, for all practical purposes, may be considered a self-cleansing velocity for a small or medium-size sewer. Thus, when designing sewers other than of large diameter, it is generally safe to allow for gradients sufficient to produce a velocity of $2\frac{1}{2}$ feet per second when the sewers are running full, even when it is known that the sewers in question will seldom run full.

The conditions are complex and not easily approached by theory. Observation of settlement and scour in rivers suggests that fine sediments such as alluvium are just prevented from settlement by velocities that vary as the square root of the hydraulic mean depth. But if Nixon's formula as given in Chapter XVIII were applied to sewers, very low velocities, much lower than those known to be satisfactory in practice, might be calculated as satisfactory for sewers of the smallest diameter.

The solids found in sewers are not, however, fine sands or silts only but include stones and light materials of large size. Stones are not taken into suspension except at very high velocities but they easily slide along the invert of the sewer if the invert is smooth. The velocity required to move stones theoretically increases directly as the diameter of the stone, being something in the region of 2 feet per second for the 1-inch stone. But the velocity close to the invert of a sewer is only about 60 per cent of the average in the sewer. Thus, small gravel would probably tend to remain in the invert of a large sewer in spite of the average velocity in the sewer being theoretically sufficient to move it, whereas a large stone occupying half the depth of a 4-inch drain would be moved by the theoretical velocity applicable to it.

The large organic solids which are kept in suspension by the turbulence of flow can have a cleansing effect in that they will pick up and sweep away any settled detritus provided the quantity of detritus is not so great for deposition to exceed the sweeping effect. On the other hand heavy deposition of detritus can in exceptional circumstances lead to fouling of sewers which theoretically have self-cleansing gradients, organic solids being held in position with the inorganic deposit, In one instance the author found a 3-inch deposit of sand in a 15-inch diameter sewer that was running at a self-cleansing velocity and nearly full at the time of observation. The reason for this unusual circumstance was that large quantities of moulding sand were finding their way to the sewer from a foundry just upstream of the point of observation.

The organic solids themselves tend to form a greasy scum at the water-line and this can become a heavy deposit if the velocity is inadequate. Such a deposit is also found in the crown of a rising main or inverted siphon in which the normal velocity is inadequate. This is usually associated with a deposit of detritus in the invert.

Apart from theoretical consideration it remains a fact that the rule, discovered in the 19th century, that a sewer having a flow velocity of $2\frac{1}{2}$ feet per second will remain clean, can be used with satisfaction. In parts of America soil sewers are designed to a minimum velocity of 2 feet per second but surface-water sewers to $2\frac{1}{2}$ feet per second. But while many sewers will keep clean at the lower velocity it is always better to design to the higher velocity as this makes some allowance for the times when flow must be below normal.

At a velocity of 1 foot per second detritus will settle in sewers of almost any diameter and this is so regular that detritus channels from less than 2 to 10 or more feet in depth have been designed to settle detritus at a constant velocity of 1 foot per second (see Chapter XXIV). At a velocity of 8 inches per second almost anything will settle.

Where sewers have to be constructed of large diameter so as to accommodate maximum storm flows, but it is known that most of the time the actual flows of sewage will be very small, special precautions have to be taken to avoid silting. In these cases the velocity at the peak daily rate of flow should be considered, and this should not be less than 2 feet per second, and greater if possible. Table 1 gives the proportional values for areas, discharges, and velocities in circular pipes when partly full.

GENERAL PRINCIPLES OF SEWERAGE

At one time, on the basis of theoretical calculations, it was thought that the maximum velocity in a circular sewer occurred when the sewer was running 0·81 of its depth full and that the maximum discharge occurred when it was running 0·94 of its depth full. Careful experiments in the laboratory and observations in the field suggest that this theoretical consideration is fallacious and that the maximum discharge occurs when the sewer is running full. The proportional velocities and discharges given in Table 1 are the *maximum*

TABLE 1
PROPORTIONAL DEPTHS, AREAS, VELOCITIES AND DISCHARGES FOR CIRCULAR SEWERS RUNNING PARTLY FULL

Proportional depth	Proportional area	Proportional velocity*	Proportional discharge*
0·05	0·0187	0·2406	0·0045
0·10	0·0520	0·3769	0·0196
0·15	0·0941	0·4835	0·0455
0·20	0·1424	0·5766	0·0821
0·25	0·1955	0·6568	0·1284
0·30	0·2523	0·7273	0·1835
0·35	0·3119	0·7897	0·2463
0·40	0·3735	0·8455	0·3158
0·45	0·4364	0·8944	0·3903
0·50	0·5000	0·9370	0·4685
0·55	0·5636	0·9737	0·5488
0·60	0·6265	1·0048	0·6295
0·65	0·6881	1·0307	0·7087
0·70	0·7477	1·0491	0·7844
0·75	0·8045	1·0621	0·8545
0·80	0·8576	1·0679	0·9159
0·85	0·9059	1·0658	0·9655
0·90	0·9480	1·0534	0·9986
0·95	0·9813	1·0268	1·0076
1·00	1·0000	1·0000	1·0000

* Maximum empiric figures.

empiric figures. The normal average velocities and discharges in sewers of average construction are probably about 10 per cent less than as given in this table except when the sewers are running full. (See also Appendix III.)

Sewer designers often develop the bad habit of putting in a larger-size sewer than required by the flow, so as to give the *appearance on paper* of a self-cleansing gradient. The self-cleansing gradient depends mostly on the flow, not merely on the size of pipe installed, and at a small flow a small-diameter pipe will keep cleaner than a larger pipe laid to the same gradient.

Generally the designer should work to the minimum gradients given in Table 2, which give a velocity of $2\frac{1}{2}$ feet per second when flowing full. Table 2 is based on Scobey's formula.

TABLE 2

MINIMUM GRADIENTS FOR SEWERS AND OPTIMUM GRADIENTS FOR RISING MAINS

Diameter in inches	Minimum self-cleansing gradient 1 in:	Discharge in cubic feet per minute	Optimum hydraulic gradient for rising mains 1 in:	Discharge in cubic feet per minute
4	110	13·0	89	14·4
5	145	20·3	120	22·3
6	180	29·4	150	32·2
7	220	39·9	180	44·0
8	260	52·1	215	57·2
9	300	66·0	245	73·0
10	340	81·8	280	90·1
12	425	118·0	355	129·1
15	560	185·0	465	202·7
18	710	265·0	590	290·4
21	860	360·0	710	397·0
24	1000	475·0	840	518·0
27	1200	590·0	970	656·0
30	1350	734·0	1100	812·0
33	1500	894·0	1250	979·0
36	1700	1055·0	1400	1163·0
39	1850	1248·0	1550	1364·0
42	2050	1440·0	1700	1582·0
45	2250	1648·0	1850	1817·0
48	2400	1890·0	2000	2070·0
51	2600	2129·0	2150	2341·0
54	2800	2384·0	2300	2630·0
57	3000	2654·0	2500	2907·0
60	3200	2940·0	2650	3231·0
63	3400	3242·0	2800	3572·0
66	3600	3560·0	3000	3900·0
69	3800	3894·0	3150	4277·0
72	4000	4244·0	3300	4672·0
75	4200	4610·0	3500	5050·0
78	4400	4992·0	3650	5481·0
81	4650	5362·0	3850	5893·0
84	4850	5776·0	4000	6360·0
87	5100	6176·0	4200	6806·0
90	5300	6623·0	4350	7310·0
93	5500	7085·0	4550	7790·0
96	5700	7565·0	4750	8287·0
99	6000	7994·0	4950	8801·0
102	6200	8505·0	5100	9377·0
105	6400	9032·0	5300	9926·0
108	6600	9577·0	5500	10491·0

Maximum gradients. Opinion varies considerably as to the maximum gradients desirable in sewers. There are two reasons given for avoiding excessive gradients, one being that too high a velocity causes scour, and the other that if the gradient is steep and there is little flow, the depth of flow is insufficient to submerge and transport the larger solids. The practice of the past has been for some engineers to avoid velocities greater than 6 feet per second in all cases, while others have worked to a maximum velocity of 10 feet per second. American practice has been to allow a maximum velocity of 8 feet per second, and this is now being adopted for rising mains and inverted siphons in Great Britain.

In efforts to avoid high velocities in sewers, too much money has been spent on excessive excavation and construction of backdrops. At one time it was not uncommon practice to lay sewers at flat gradients and absorb the surplus fall in backdrop manholes. Now in similar circumstances most engineers lay *small* sewers to such gradients as the contours of the land determine, without troubling whether the gradients are inclined to be steep, for, particularly where small-diameter sewers carrying small flows are concerned, it is not economical to make special provision to avoid high velocities in gravitating sewers. It is, however, desirable to avoid velocities of more than about 8 feet per second in *rising mains* and *inverted siphons*.

The type of scour that occurs in large sewers is mainly the formation of deep grooves in the inverts (even of cast-iron pipes) due to the movement of heavy solids which are not in suspension but slide along. Some engineers consider that increased velocity would lift these solids from the invert and reduce scour.

The hanging up of solids in the inverts of sewers with little depth of flow has from time to time been observed. Nevertheless, this is not important, for although particles of faecal matter, etc., may lodge for a time, further accumulations have the effect of damming the flow until the pressure of water behind the solid material breaks it away and clears the sewer.

Normal requirements in design. Soil sewers are designed to accommodate from four to six times the average dry-weather flow when flowing full; and surface-water sewers and combined sewers, when flowing full, should take (as is hereinafter described) the run-off from a storm liable to occur once in three years. Any excess on these flows is accommodated by surcharge of the sewers—that is, by the hydraulic gradient rising above the crown of the pipe and pressure developing in the sewer. Thus, in the design of sewers, the hydraulic gradient is taken as being the line of the crown of the sewer when the flow is the maximum the sewer is intended to accommodate; and for this reason wherever there are intersections of sewers or changes of diameter, as far as is practicable the crowns of the different diameter sewers are kept level one with another, and the change of diameter causes a step or slope-down in the invert at the point of connection. Nevertheless, in practice it is the invert, not the crown, levels that are shown on the drawings and the calculations of gradients are made according to the differences of invert level over the lengths concerned.

To facilitate cleansing with the aid of sewer rods and other devices, sewers up to and including 30 inches diameter must be laid in straight lines between manholes or other points of access, both in the horizontal and vertical planes. Manholes must be placed at all changes of direction or gradient: in this it is implied that there shall be a manhole at every sewer intersection. Manholes are also desirable at all changes of diameter. On the other hand, *drains* normally connect to sewers by simple junctions, and private sewers (other than for the sewerage of new streets) may connect to other sewers, private or public, with an oblique junction, provided there is a manhole on the incoming private sewer within 40 feet of the junction.

Ministry of Housing and Local Government requirements as regards manholes and concrete protection of glazed-ware and concrete sewer-pipes. The following requirements are quoted from the Ministry of Housing and Local Government Form K29:

"*A—Sewer pipes with flexible joints which, together with their bedding, are designed to carry specific loads*

1. In the absence of complete computations of the total external design loads the values given in the "Simplified Tables of External Loads on Buried Pipelines," published by the Building Research Station (National Building Studies—Special Report No. 32)* may be used for *single* pipes in trenches, pending the publication by the British Standards Institution of revised Codes of Practice for Sewerage and Drainage.
2. Where the pipes are laid in conditions outside the scope of the tables, e.g. less than the minimum cover, two or more pipes in the same trench, excessive trench width, under embankments or in heading, etc., special provision should be made to suit local loading.
3. Pipes under roads and verges should normally be laid with at least 4 ft. of cover, and under fields and gardens with at least 3 ft. of cover.
4. The backfilling should be so selected, placed and compacted as to give proper support to the road or other structures concerned including existing pipelines.

"*B—Concrete support for glazed vitrified clay ware and concrete sewer pipes NOT designed to carry specific loads*

1. Pipes of 30 ins. diameter and under with 20 ft. and more of cover in trenches to be surrounded with at least 6 ins. of concrete. Pipes of more than 30 ins. diameter under the above conditions may require additional concrete protection.
2. Subject to (1) all pipes with over 14 ft. of cover to be bedded on and haunched with at least 6 ins. of concrete to at least the horizontal diameter of the pipe. The splaying of the concrete above that level to be tangential to the pipe.
3. Subject to (1) all pipes of 18 ins. diameter and over to be bedded and haunched with at least 6 ins. of concrete to at least the horizontal diameter of the pipe. The splaying of the concrete above that level to be tangential to the pipe.

* At the time of writing it is understood that this publication is *considered* obsolete (author).

4. Subject to (5) all pipes under 18 ins. in diameter and with less than 14 ft. of cover may be laid without concrete, if the joints are of the socket or collar type.
5. All pipes and tubes with less than 4 ft. of cover under roads (except roads formed of concrete), or 3 ft. not under roads, to be surrounded with at least 6 ins. of concrete.
6. In all cases the filling or concrete support must be well rammed and consolidated at the sides and haunches of the pipe. Selected filling should be used up to at least one foot above the top of the pipe."

Minimum and maximum diameters. For many years the minimum permissible diameter for sewers had been 6 inches, but the 1952 edition of *Model Byelaws Series IV Buildings** permitted private sewers of 4 inches internal diameter. It is, however, unlikely that local authorities will adopt this diameter when constructing public sewers. There is no maximum diameter to which sewers are limited. It has been suggested that diameters much greater than 16 feet tend to become uneconomical where work is constructed in tunnel. This is questionable: however, the problem rarely arises.

Location of sewers. In most schemes sewers are laid as near as possible in the centre of the carriage-way, but deviating from side to side where necessary because, while sewers must be laid to dead straight lines between manholes, roads curve, and unless excessive numbers of manholes are inserted at considerable cost, both sewers and manholes must at times be laid near the sides of the roads. Otherwise the sides of the carriage-way are occupied by gas- and water-mains which, not being designed according to the same practice as sewers, do not need to take the central position.

In rural districts and usually in residential areas the centre of the road is the most practical and economical position for sewers. Where, however, wide or trunk roads constructed of expensive materials are flanked by properties it may be advisable to lay a main sewer at one side of the road clear of the carriage-way and provide parallel lateral sewers on the other side, connecting across the carriage-way at intervals whenever it becomes necessary to increase the diameters of the parallel sewers above the minimum.

* The *Public Health Act*, 1961, did away with building byelaws.

CHAPTER II
SOIL SEWERAGE

THE methods used by different engineers when designing sewerage schemes vary considerably in detail and in efficiency, and naturally on such a matter there is room for difference of opinion. All experienced designers attack problems of this kind according to habits developed over years of practice: they know what they are doing, but might not be able clearly to describe their procedure to a layman. A novice attempting to design a sewerage scheme without proper guidance must have some difficulty in deciding how to set about his work in the first place. It is for the benefit of such that this chapter is written, but it should be remembered that the method described herein is one of many.

Preliminaries. When it has been decided that a scheme shall be prepared, the outlines are roughly sketched on plan. To do this the assistant responsible for design takes a large-scale Ordnance map of the municipality and, noting all existing and proposed future roads and properties to be sewered, marks each road in, say, blue pencil on a sheet of tracing-paper laid over the map. Working from the information made available by spot levels and contours, he indicates with arrows the direction of fall of each blue-pencil line. Then he joins all the lines together, using as little extra length as possible, so as to form one complete sewerage system discharging to the outfall works, the position of which must have been decided. In joining the branch sewers together in this manner it is as well for the assistant to use a different coloured pencil in order that he may see at a glance the proportion of useful sewers serving properties as compared with main sewers, which are only collecting sewage from other sewers and delivering to the outfall. The cost of main sewers should be kept as low as local conditions permit if the scheme is not to be unduly expensive. The assistant should not be content with the first scheme prepared in this manner; he should try out a number of alternatives and measure up the total lengths, selecting the most economical scheme.

With this tentative proposal before him, the assistant may carry out a chain-and-level survey in the field, taking accurate levels along all proposed lines of sewers, which will by now have been marked in pencil on large-scale Ordnance sheets. As he traverses the ground with his instruments, he should decide the actual lines to be taken by the sewers according to the terrain, or, alternatively, take levels on more than one line to permit adjustment when the work is on the drawing-board. Not only should he take chainage and levels and decide on the best location for each sewer, but he should make notes on all relevant points, observe when houses lie far back from the road or at low levels, and be sure that all properties within a reasonable distance of the public highway, and certainly all in which the buildings are within 100 feet, can be connected. He should note the bed-levels of streams, top-water levels and flood-levels, collect information on places liable to floods, and record the materials of which roads are constructed. This will save searching through records or further expeditions into the field. The site for the sewage-treatment works should be spot levelled all over at at least 100-foot intervals, and closer where irregularities of the ground make it advisable.

Preparation of sections. Work in the drawing-office commences with preparation of sections along the lines of the scheme as so far envisaged, together with determination of positions of manholes. It is in this part of the work that there is perhaps the most variation of procedure. Some engineers work to consistent systems, some are extraordinarily haphazard, others work to systems which cannot be applied consistently. The following system is one which has been used extensively and found to be capable of consistent application. It has also been compared with others experimentally, and proved to save time.

First, the main trunk sewer of the scheme is chosen and other sewers considered to be branch sewers. The main sewer is that which extends farthest from the outfall works in large diameter. Whenever two sewers meet at a point, the main sewer is the larger incoming sewer. The main sewer can usually be decided by inspection before the sizes are accurately determined. The manholes on this line of sewer are all designated manhole "A," and numbered consecutively M.H.A.1, M.H.A.2, M.H.A.3, etc., commencing at the lowest end of the line of sewer and finishing at the top end. The subsidiary or branch sewers are designated B, C, etc., the B line being that which connects to the main line nearest the outfall.

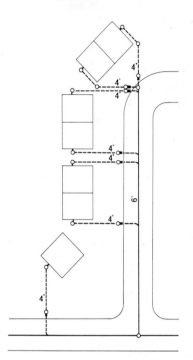

Fig. 1—Typical arrangement of individual house connections to sewer laid in road

Circles indicate manholes, thick lines in contact with circles indicate intercepting traps, full lines indicate sewers and broken lines indicate drains.

When all the sewers connecting with the main line have been given their distinctive letters, the branches to the branches are similarly given distinctive letters, commencing with the one which connects with the lowest part of the B line, and so on. It is thus possible for anyone referring to the plan to know at once where to look on his section sheets for any particular manhole, or, if scrutinising his section sheets, similarly to know where to look on the general plan: there is no time wasted in searching for manholes, as there would be if they were indiscriminately numbered.

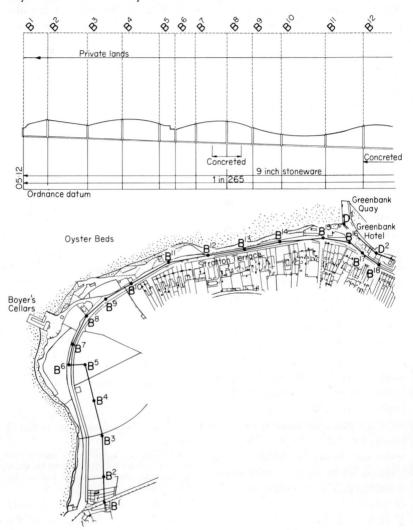

FIG. 2—Typical sewer plan and section
By courtesy of Messrs. John Taylor & Sons, Consulting Engineers

On the direction in which the sections should be plotted there is some difference of opinion; some engineers prefer always to plot the sections so that the flow of sewage is from right to left, and the manholes number from left to right, regardless of the direction in which the flow occurs on plan.

This method gives a very good appearance to the section drawings and can be applied consistently. Other engineers prefer to take the direction of flow on the sections as being that which is taken by the main line on the plan, and to number manholes from left to right, or right to left, on all sections alike, according to the direction so determined in the first place. A third method which is also used, but which cannot be applied consistently to all schemes, is to attempt to draw all the sections so as to lie in the same direction as on the plan. This would be possible if sewers always flowed east and west in straight lines, but, as they frequently flow north and south and deviate in different directions throughout their lengths, the system almost invariably breaks down.

Sections should always be plotted to the same horizontal scale as the Ordnance sheet on which the plan is drawn, and preferably the vertical scale should be distorted to ten times the horizontal scale, and not to an arbitrary figure of distortion. This is in order that one boxwood scale can be used for measuring both horizontal and vertical dimensions, saving time and avoiding danger of confusion. Without using this method, confusion is frequent, particularly when scales somewhat similar in appearance, such as 1/2500 horizontal and 20-feet-to-the-inch vertical, have been selected.

Unless it proves very inconvenient, all sections should be plotted to the same datum line; the designer should plot to Ordnance Datum or to a line, drawn, say, 100 feet above Ordnance Datum, and stick to that throughout. He should not draw a base line for one section 60 feet above Ordnance Datum, another 80 feet from Ordnance Datum, and so forth, for not only is this likely to confuse him and result in an incorrect drawing, but it is still more probable that it will confuse the contractor and resident engineer when the work is carried out.

Once the rough sections have been prepared, the designer should go over the work, improving the spaces of his manholes, the gradients of his sewers, and so forth, economising in materials and excavation to the utmost, but at the same time making sure that the sewers will serve all properties, that they can be laid to the lines shown on the drawings, that the gradients and sizes are adequate for the flows estimated, and that the gradients will be self-cleansing under working conditions.

Rate of flow of sewage. The quantity of foul sewage produced by a town or developed area during dry weather is very much in accordance with the water supply, both in total quantity per annum and variation of rate of flow from hour to hour (see Vol. I, Chapter X). The greater part of the water that is supplied to domestic premises is not delayed for long. During the processes of cleaning vegetables and washing dishes some water runs straight from the tap to the drain. Water run into lavatory basins or baths may be retained for some minutes or a fraction of an hour, as also happens when a water-closet or urinal is flushed, but generally not sufficiently long appreciably to affect the difference between the general curve of water demand and that of discharge of sewage. Some water is lost by soakage into the ground, evaporation, etc.

The main causes of difference between the curve of water demand and the curve of flow of sewage over the same period are the effect of time of concentration of the sewerage system, smoothing out the midday peak, and infiltration of subsoil water where the sewers leak. To some extent sewers leak outwards as, for example, in chalk districts, but leakage of subsoil water into the sewers is far more frequent, and at times the amount of subsoil water entering old sewers is very high. This infiltration is comparatively steady throughout the twenty-four hours during dry weather, but it can increase greatly in wet weather. On the average, infiltration in old sewerage systems appears to add $33\frac{1}{3}$ per cent to the dry-weather flow as based on water demand and not taking water losses or water from private wells into account.

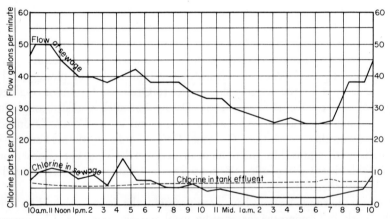

FIG. 3—Hourly variations in flow and strength of sewage

The flow of sewage due to water supply varies from a very small quantity round about the early hours of the morning to a peak round about midday, when the rate of flow is about twice* the average dry-weather flow. The time that the peak occurs depends, among other factors, on the size of the sewerage system. If the sewers are long, the time taken for the sewage to flow through them delays the peak, causing it to occur during the early afternoon, instead of at 11 a.m. or midday. It may also smooth the peak a little. The general form of the normal curve of flow plotted against time is an even rise from about 6 a.m. to 11 a.m., noon or 1 p.m. and an even fall until 6 a.m. once more, with not infrequently a minor peak round about 5 p.m.

* In those parts of Greater London that drain to the Beckton and Crossness Sewage-treatment Works, in which the times of concentration are in the region of 6 and $6\frac{1}{2}$ hours respectively, the peak daily dry-weather flow is about one and a third times the average dry-weather flow.

Besides variation of flow throughout the day, there are comparatively regular variations from day to day owing to domestic and industrial routines.

It is extremely difficult to exclude rain-water from sewers and, even where the separate system is strictly enforced, the rate of flow tends to increase in wet weather. Where systems of sewerage are not strictly separate, the flows due to rainfall can be very considerable. It has been the practice of some engineers to assume that wet-weather flows in the soil sewers of separate systems amount to six times dry-weather flow. On the other hand, many soil-sewerage schemes have been designed on the basis of the sewers being so sized that they will accommodate four times the average dry-weather flow, and this is now the most usual practice.

Figure 4 shows a hypothetical soil-sewerage scheme, and Table 3 the calculation sheet for that scheme. In this example it is assumed that there is

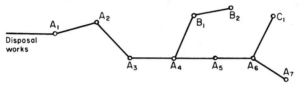

FIG. 4—Drainage diagram illustrating soil-sewerage calculation

an average of 3·7 persons per house, the dry-weather flow (or water consumption, which is often taken to be the same) is 25 gallons per head per day, and the engineer is allowing for the sewers to have a capacity equal to four times dry-weather flow.

Then, taking the flow in the length of sewer C1 to A6, there are 510 houses flanking the road in which this length is laid.* Therefore the flow is:

$$\frac{510 \times 3 \cdot 7 \times 25 \times 4}{9000} = 21 \text{ c.f.m.}$$

(*Note:* 9000 gallons per day almost exactly equals 1 cubic foot per minute.)

The sewer would be laid at least 5 feet deep to invert, so as to have the cover required by the Department of the Environment, and so as to be sufficiently deep to permit the connections from house-drains without rendering the gradients of the drains unusually slack. In the event of houses being below the road, the sewer would have to be deeper, or else an additional sewer would have to be laid behind the houses. Generally, the crown of the sewer should be at least 3 feet, plus the necessary fall of the house-drain, below the ground-level at the most distant connection to the drains.

* The minimum size of soil sewers is sufficiently large to take the flow from a quite extensive estate of many roads. In the present instance it is required to give an example calculation involving larger sizes than the minimum size, and, to give a true example, it would be necessary to make the calculation for a very extensive sewerage scheme. To simplify the calculation, the number of houses is increased to ten times the number that would normally be erected in the lengths of roads concerned.

In the present case it is assumed that it is satisfactory for the sewer to be laid 5 feet deep to invert, that the length from C1 to A6 is 360 feet, the ground level at C1 is 213·54 feet above Ordnance Datum and the ground level at A6 is 210·14 feet above Ordnance Datum. Then the invert of the sewer at C1 would be 208·54 O.D., and the invert at A6 205·14 O.D., giving a fall of 3·40 feet in a length of 360 feet, or 1 in 106.

At a gradient of 1 in 106, the smallest diameter of public sewer in usual practice—i.e. 6 inches diameter—is adequate to take the flow of 21 cubic feet per minute and according to Vol. I, Fig. 68 this will discharge 35·6 cubic feet per minute.

In the above manner the required size of sewer for the length C1 to A6 is calculated. The calculations for the remaining lengths in the system shown in Fig. 4 are then similarly calculated, commencing at the top end and working downstream until the calculation for the system is completed as shown in Table 3.

TABLE 3

CALCULATIONS FOR SOIL SEWERS

Location (manhole to manhole)	Maximum flow (c.f.m.)	Available gradient 1 in:	Diameter of sewer (inches)	Capacity (c.f.m.)
C1–A6	21	106	6	35·6
B2–B1	18	54	6	49·7
B1–A4	43	149	9	87·9
A7–A6	20	200	6	25·8
A6–A5	63	200	9	76·1
A5–A4	82	190	12	168·0
A4–A3	144	284	15	249·0
A3–A2	162	300	15	243·0
A2–A1	181	150	12	189·0
A1–Disposal works	203	310	15	239·0

Where the lengths of sewer connect at manholes, in the majority of cases the invert of the incoming sewer will be the same as that of the outgoing sewer if the two are of equal diameter and, similarly, any side connection joining the main line will have its invert level with that of the main sewer, unless circumstances permit economy of excavation by raising the invert of incoming lateral sewers or upstream lengths. In such circumstances a small difference of level is arranged for in the construction of the invert of the manhole, but where a lateral connection or an upstream length has its invert more than about 30 inches higher than that of the outgoing sewer, a backdrop or tumbling bay manhole, as will be described later, has to be constructed.

As before mentioned, when the incoming sewer has a smaller diameter than the outgoing sewer its invert is higher than that of the outgoing sewer by at least the difference of diameter.

Towards the end of Table 3 is a comparatively exceptional case, in which the outgoing sewer from manhole A2 is smaller than the incoming sewer. In this case the invert of the outgoing sewer will be level with the invert of the incoming sewer, *while its crown will be lower than that of the incoming sewer.*

Preparation of drawings. General plans are best prepared on Ordnance Survey maps and reproduced by tracing or obtaining photographic copies on tracing cloth from which further prints can be taken. In this case fees have to be paid to the Ordnance Survey for the reproduction of negatives and prints. Sewers should be shown as thick lines on negatives or as coloured lines on Ordnance Survey maps, and the manholes as small circles lettered and numbered appropriately. A north point should be put on all tracings. Sections should be drawn with the sewers indicated by a line or two lines, as needs be, according to the diameters, the lower line representing the invert. Distances between manholes should be indicated (this is preferable to chainages, which involve much adjustment during preparation and much future calculations by persons reading the drawings), and all invert-levels and ground-levels at manholes. Gradients, sizes of pipe, materials and extent of concrete protection should be given (see Fig. 2). The results of trial holes should also be shown on sections. The locations of the sewers should be indicated on the sections, whether in road (name of road given) or in private land, etc. Fences, hedges, ditches, streams, etc., should be indicated, with inverts, top-water and flood-levels of all watercourses.

CHAPTER III

SURFACE-WATER SEWERAGE: THE LLOYD-DAVIES METHOD AND THE RATIONAL METHOD

SURFACE-WATER sewers and combined sewers are in many respects designed and constructed in the same manner as soil sewers. They are laid in straight lines between manholes, the last being spaced at intervals of not more than 360 feet. The sewers are given cover and protection as required by the Department of the Environment and usually, even more so than in the case of soil sewers, they are laid in the highway. When they are associated with soil sewers in a "separate" scheme they are laid as near the centre of the carriage-way as the soil-sewer trench will permit; obviously they must be sufficiently far from the latter if they are not to be undermined, for the smaller sizes of surface-water sewers are, if anything, laid slightly shallower than soil sewers because the house connections leading to them usually begin at higher levels than the soil connections. If they are close to the soil sewers, the higher sewers must be given concrete support to prevent them from subsiding into the lower trench.

Surface-water sewers should be laid to self-cleansing gradients similar to those recommended for soil sewers running full. Some designers consider that, as surface-water sewers do not contain offensive matter, silting is not important; but it should be remembered that road grit, which not infrequently finds its way into surface-water sewers, is heavier than organic matter, and requires a good velocity to scour it away: also the organic material in soil or combined sewers tends to sweep away sand or silt. In "combined" schemes good gradients are essential, for the sewers must be large to accommodate the surface-water run-off, and therefore during periods of dry weather there is comparatively little flow in them, and consequently a tendency to silt. In these circumstances, the sewers should be laid at the gradients required to produce self-cleansing velocities when running partly full during times of dry-weather flow.

Noticeable differences between soil and surface-water schemes are that while in the former the whole of the sewage (as, of course, in a combined scheme) is usually collected together in one system, discharging to a disposal works or point of outfall, and the flow in each branch sewer is almost directly proportional to the number of properties served, in the latter the area to be drained is often divided up into a number of drainage areas, each of which can be drained to a convenient point on the natural system of watercourses, the sea, etc., and in some cases soakaways, *and the main sewers have discharge capacities less than the aggregate capacities of the branches connecting to them.* This last fact, which is based on both experience and reason, forms the starting point of all surface-water sewer design.

Preliminaries. In preparing a surface-water drainage scheme, the first part of the procedure is to take a contoured map of the whole area, indicate clearly the natural watercourses or other points of outfall available, and divide the area into a number of drainage areas, each of which can afterwards be treated as an individual scheme independent of the others. (In the final stage, when all the individual schemes have been designed and the flows estimated, the capabilities of all the streams to take these flows can be investigated, and should any one point of outfall be found unsatisfactory it may prove necessary for two or more drainage areas to be linked or else natural watercourses improved.)

Selecting any one of the drainage areas discharging to one point of outfall, the engineer can commence to design his system of sewers. In the same manner as for soil sewers, the directions of flow of each branch can be decided first by indicating the direction of fall of each road on a transparent overlay, as before described, and then all the branch sewers are linked together by main sewers discharging to the outfall. When the general outline of the system has been determined, the designer should commence at the *top end* of his system, working out the sizes and gradients. This involves measuring the area, estimating the impermeability of the surface drained and deciding the intensity of rainfall for which allowance should be made.

Surface-water sewerage theory. The design of surface-water sewers involves theory far more complicated than that of soil-sewers. The first reason for this is that rainfall lasts for short periods only and, on the average, the intenser the rainfall the shorter the duration of storm. The effect of this is that a very high rate of precipitation may cause rapid run-off from impermeable surfaces into the sewers, but the run-off from the sewers may endure for a longer time and never reach the same peak intensity. Early designers of sewers observed this fact and devised empiric rules based on experience. At the present day in Great Britain it is usual to adopt the Lloyd-Davies method or some modification thereof in which rainfall statistics are mathematically related to the characteristics of the sewerage system, and the sewers are designed so as, on the average, to run full less frequently than once a year, to be surcharged still less frequently, and rarely to an extent sufficient to cause flooding. In America a very similar system, known as the "rational method," is used. As will be described, the author now considers this method preferable.

The construction of surface-water sewers may be considered as an insurance or a gamble. Reasonable expenditure is made on works as a precaution against flooding and the damage or other hazards that could result therefrom. But it is not practicable to construct sewers large enough to accommodate the most intense or disastrous storm that could be envisaged: this would involve expenditure out of all proportion to the need. On the other hand, if works are constructed to what is considered economic proportions, *there can be no guarantee that flooding and damage will not occur*, perhaps shortly after completion of construction. The aim of the engineer is to arrive at the happy medium between inadequacy and extravagance.

It is difficult to determine the truly economic figure, but over the years opinions have been expressed by eminent engineers which have not varied too much from a consensus. And experience has shown that works which are reasonably economical very rarely prove inadequate. The author knows of only three instances in which drains or sewers newly constructed to recognised standards were involved in flooding shortly after completion and, in each case, there was a reasonable explanation. In two cases, flooding occurred on the occasion of an exceptional storm on sites where the drains in question connected to existing sewers of doubtful adequacy. In the third, a very large unexpected factory area was connected to sewers designed to take the flow from a residential area only.

Impermeability of surfaces. The amount of run-off of rainfall from the developed areas to the sewers depends on the extent and impermeability of the surface of the area drained. If two areas of land, equal in size, differ in that one has twice the amount of impermeable surface of the other—i.e. twice the amount of road surface, surface of roofs, etc.—and the garden land is of a porous nature in both, one would expect that area to give twice the run-off of the other under the same conditions of rainfall.

The method employed in determining the "impervious area," as it is called, is to take the gross area of the drainage area and multiply it by an impermeability factor, giving the impervious area, which is usually expressed in acres. The impermeability factor is found sometimes by experiment, but

TABLE 4

IMPERMEABILITY FACTOR CALCULATION

Type of surface	Area in acres A	Impermeability factor	Impervious area in acres A_p
Roofs	$2\frac{3}{4}$	0·95	2·61
Carriage-way	1	0·60	0·60
Footway	$\frac{3}{4}$	0·80	0·60
Paved yards	$\frac{1}{2}$	0·90	0·45
Garden land	4	0·10	0·40
TOTAL	9		4·66

$$\text{Then: impermeability factor} = \frac{\text{impervious area}}{\text{gross area}}$$
$$= \frac{4\cdot66}{9}$$
$$= 0\cdot518$$

most frequently, owing to the lack of means of experiment, by estimation in the following manner. The roof areas, paved areas and unpaved areas of a sample area are measured, and each component is multiplied by an impermeability factor which, at some time, has been determined more or less accurately by experiment. Then the impermeability factor for the sample area is calculated as set out in Table 4.

The impermeability of the different kinds of surfaces may vary somewhat from place to place, and even from time to time in the same place, but the figures given in Table 5, or figures similar thereto, are not infrequently used as a basis of estimation.

TABLE 5

TYPICAL IMPERMEABILITY FACTORS

Type of surface	Impermeability factor
Watertight roof surfaces	0·70 to 0·95
Asphalt pavements in good order	0·85 to 0·90
Stone, brick and wood-block pavements with tightly cemented joints	0·75 to 0·85
Same with open or uncemented joints	0·50 to 0·70
Inferior block pavements with open joints	0·40 to 0·50
Macadam roadways	0·25 to 0·60
Gravel roadways and walks	0·15 to 0·30
Parks, gardens, lawns, meadows, depending on surface, slope and character of subsoil	0·05 to 0·25
Wooded areas depending as before	0·01 to 0·20

Examination of the results of recorded rainfall and run-off in selected areas, after the exclusion of all questionable data, showed that the most consistent results were obtained when it was assumed that all run-off to sewers (during the time of concentration) was from paved and roofed surfaces, and none from garden or park land. The results for the selected sites were as follow:

Percentage "ultimate" impermeability
78·8 (Blackpool)
70·3 (Doncaster)
83·1 (Leicester)
86·2 (Oxhey Estate)
81·0 (North Woolwich)
—————
Average 79·88, say 80

This suggests that the most reasonable estimate of impermeable area is made by allowing an 80 per cent run-off from all roofed and paved areas.

In residential areas the percentage impermeability of each drainage area varies in accordance with the rule determined prior to the Second World War on the basis of Kuichling and Bryant's figures:

Percentage impermeability $= 10.\sqrt{n}$
where: $n =$ number of houses per acre.

This rule was rechecked in 1966, using the assumption of 80 per cent run-off from paved and roofed surfaces and measuring surfaces in various localities in Greater London. The figures were:

Houses per acre	Percentage impermeability
1·83	12·70
1·83	13·72
2·84	16·05
6·43	25·70
7·43	32·30
9·00	27·40
14·70	35·50
18·75	45·60
27·00	55·60

Early combined and surface-water sewerage practice. It is desirable to study early practice not merely as a matter of historical interest but as a check on theoretical suggestions that otherwise may tend to run wild. While the early engineers did not have the benefit of present-day research they based their designs on experience, and the similarity of practice in different parts where storms were of similar intensity suggests that this practice was satisfactory.

At the beginning of the century most engineers were allowing "flat-rates" of rainfall for the design of surface-water and combined sewers. In the sewerage of Berlin, carried out in the 19th century, the maximum rainfall allowed for was $\frac{7}{8}$ inch per hour, of which one-third was supposed to enter the sewers; in the sewerage of Dresden a maximum of 0·7 inch per hour was allowed.

In London practice has changed over the years. Worth and Crimp found that there was a maximum run-off of 0·4 inch per hour from a 160-acre built-up area of Bayswater when the rainfall was 2 inches per hour. D. C. Graham in 1912 found the discharge of the Effra Branch Sewer, draining 3000 acres, had seldom run at its full capacity of 0·2 inch per hour, also a sewer draining about 1000 acres in Streatham and Balham allowed for $\frac{1}{3}$ inch of rain per hour on the area, while proposed new sewers to drain the Wandle Valley Area of 1700 acres were based on a run-off of $\frac{1}{4}$ inch per hour. In a theoretical investigation which the author made of run-offs from the various parts of London County it was found that these could be expressed by the formula:

$$R = \frac{3}{\sqrt[3]{A}}$$

where: $R =$ run-off in inches per hour
$A =$ gross acreage.

This is in remarkable agreement with the above observations.

In the inter-war period, drainage of L.C.C. housing estates was based on flat-rates, in some cases as low as ½ inch per hour over the impermeable area and, although the council was now beginning to adopt the Lloyd-Davies method, in the design of the drainage of the Oxhey and Kidbrooke estates, carried out after the Second World War, flat-rates of 1 inch and ¾ inch were allowed respectively.

What is significant is that no flooding has been known to take place on L.C.C. housing estates except as a result of stoppages, or unforeseen development, showing that the old rule-of-thumb practice, while crude and unscientific, was satisfactory, particularly as it was more economical than many techniques based on the Lloyd-Davies method in the 1930s.

An improvement on the use of flat-rates of rainfall for areas of all sizes was the substitution of a varying scale of rainfall intensity according to the size of the drainage areas. For example, in Leeds[224] it was once the practice to allow for 1 inch of rainfall over 20 acres, ¾ inch over 35 acres, ½ inch over 50 acres, and so on to ¼ inch over 600 acres. These figures are somewhat irregular but average 24 per cent lower than if they had been calculated by the author's formula above.

Method of application. If rain falls at the rate of 1 inch per hour on an acre of land, the rate of run-off in cubic feet per minute is:

$$\frac{43{,}560}{12 \times 60} = 60 \cdot 5 \text{ cubic feet per minute}$$

If one-quarter of the area is paved so as to be completely impervious and the rest is completely permeable so that the rain that falls on it soaks away, the "impermeability factor" is 0·25 and the "impervious area" is 0·25 acres. Thus, the run-off of rainfall from an area of land that is partly paved or developed can be expressed by the formula:

$$Q = 60 \cdot 5 \times Ap \times R$$

where: $60 \cdot 5$ = a constant which converts inches rainfall per hour per acre to cubic feet per minute
Q = run-off in cubic feet per minute
Ap = "impervious area" or area drained in acres × impermeability factor
R = intensity of rainfall in inches per hour.

The metric equivalent of this formula is:

$$Q_1 = 0 \cdot 1666 \, Ap_1 R_1$$

where: Q_1 = run-off in cubic metres per minute
Ap_1 = impervious area in hectares
R_1 = rainfall intensity in millimetres per hour.

The Lloyd-Davies method. As will be seen from the foregoing, it was appreciated at an early date that the rainfall intensity to be allowed in sewer design could be reduced inversely as a function of the area of the catchment. The reason for this relation between applicable rainfall intensity and area is that, owing to the form of rainfall statistics, the storm that produces the greatest run-off of all storms of the same frequency of occurrence is the one the duration of which is equal to the time of concentration of the area, that is, the time taken for water to flow from the top end of the sewer to the point at which flow is desired to be known. And as, on the average, time of concentration must increase with area, rainfall, on the average, is similarly related to area.

It was Lloyd-Davies who first set out in acceptable form the theory that the storm, duration of which is equal to the time of concentration, is the one which causes the greatest run-off from the area concerned. The above formula is usually known as the Lloyd-Davies formula for, although it is applicable to flat-rate calculations, it is essentially his method and was given a slightly different form in his paper.[163]

This formula is also used with the somewhat similar American "rational" method.

Rainfall intensities. Essential to the Lloyd-Davies method is a means of estimating the intensities of storms of various durations and suitable for use in design. On the basis of limited data, Lloyd-Davies prepared a curve of "maximum rainfall intensity." This, he pointed out, would give too high results for actual work, and for design purposes he suggested a *working curve*.

This curve was expressed by the "Birmingham" formula:

$$R = \frac{40}{t + 20}$$

where: R = intensity of rainfall in inches per hour
t = time of concentration (or duration of storm) in minutes.

In 1929 the Ministry of Health Departmental Committee on Rainfall and Run-off[184] observed that local authority surveyors were using curves similar in type to the Birmingham curve but varying considerably as a result of having been based on inadequate local data and questionable methods of interpretation. The committee, therefore, in an endeavour to reach uniformity, obtained from the Meteorological Office records that showed that one standard curve could be adopted for all localities in Great Britain subject to the proviso that, where adequate records existed, exceptional cases should be considered on their own merits. The committee accepted the Birmingham formula as satisfactory to cover storms of duration from 20 to 100 minutes, but for storms lasting from 5 to 20 minutes they substituted the formula:

$$R = \frac{30}{t + 10}$$

This, together with the Birmingham formula, made up what became known as the "Ministry of Health Standard Curve." This was not stated as being a curve giving a storm of any specific occurrence, but an enveloping curve which included "all except the two or three highest storms during the seven years' period for the average of all the stations."

It would appear that the committee had overlooked the fact that Lloyd-Davies' working curve was not a curve of rainfall statistics but a curve for the design of sewers. Lloyd-Davies considered that the maximum rainfall intensity curve gave "too high a result for actual work" and suggested an "economical-intensity curve" which he compared with the frequency curves for storms per year. This allowed for storms liable to occur about twice a year for durations up to 5 minutes, once a year for durations up to 13 minutes and once every 15 months for durations up to 55 minutes. Lloyd-Davies had noticed that impermeability of surfaces increased with duration of rainfall and, while he did not actually say so, his working curve made reasonable allowance for this.

Thus, the effect of the modification by the Ministry was not to improve accuracy but to introduce a new error which, augmented by further modifications of practice in the early 1930s (as will be described in Chapter IV) led to a period of several years during which sewers were generally oversized by about 50 per cent.

It is, however, logical to begin surface-water sewer design with a study of accurate rainfall statistics and afterwards make as accurate an allowance as practicable for change of impermeability during rainfall.

In 1935 E. G. Bilham[23] published in "British Rainfall" an examination of rainfall statistics entitled "The Classification of Heavy Falls in Short Periods." This gave the results of a very extensive investigation he had made in view of the unsatisfactory statistics of short, intense storms available at that time. He expressed the results in the formula now usually referred to as Bilham's formula:

$$N = 1.25 \, T \, (r + 0.1)^{-3.55}$$

where: N = number of occurrences of storms of this intensity in 10 years
T = time of duration in hours
r = total rainfall in inches during T.

Bilham's formula did not immediately come to the notice of sewerage engineers because the Ministry of Health formula was good enough for the design of the majority of moderate-size surface-water sewerage schemes in which time of concentration seldom exceeded 100 minutes. But in 1938 the writer, while investigating more complex problems than were involved in the general run of design, found the need for a formula more mathematically convenient than the Ministry of Health formula and, unlike the latter, capable of extrapolation: and while studying "British Rainfall" in search of data he encountered Bilham's work.

The accuracy of Bilham's formula, between the limits of storms liable to

TABLE 6
INTENSITIES OF RAINFALL
Inches per hour of rainfall according to Bilham's formula

Time in mins	Once a year storm	Once in 1·33 years storm	Once in 2 years storm	Once in 3 years storm	Once in 4 years storm	Once in 5 years storm
5·0	2·117	2·397	2·833	3·321	3·702	4·020
5·5	2·007	2·268	2·675	3·131	3·487	3·784
6·0	1·910	2·156	2·538	2·966	3·300	3·579
6·5	1·824	2·056	2·417	2·821	3·137	3·400
7·0	1·748	1·968	2·310	2·693	2·993	3·242
7·5	1·679	1·888	2·214	2·578	2·864	3·101
8·0	1·617	1·817	2·127	2·475	2·748	2·974
8·5	1·560	1·751	2·049	2·382	2·643	2·860
9·0	1·508	1·692	1·977	2·297	2·547	2·756
9·5	1·460	1·637	1·911	2·219	2·460	2·660
10·0	1·416	1·587	1·851	2·148	2·380	2·573
11·0	1·337	1·496	1·743	2·020	2·237	2·417
12·0	1·269	1·418	1·650	1·910	2·114	2·283
13·0	1·208	1·349	1·569	1·814	2·006	2·166
14·0	1·155	1·288	1·496	1·729	1·911	1·063
15·0	1·107	1·234	1·432	1·653	1·827	1·971
16·0	1·064	1·185	1·374	1·585	1·751	1·889
17·0	1·024	1·141	1·321	1·524	1·682	1·814
18·0	0·989	1·100	1·274	1·468	1·620	1·747
19·0	0·956	1·063	1·230	1·417	1·563	1·685
20·0	0·926	1·029	1·190	1·370	1·511	1·629
21·0	0·898	0·998	1·153	1·327	1·463	1·576
22·0	0·872	0·968	1·118	1·287	1·418	1·528
23·0	0·848	0·941	1·087	1·250	1·377	1·483
24·0	0·825	0·916	1·057	1·215	1·339	1·442
25·0	0·804	0·892	1·029	1·183	1·303	1·403
26·0	0·784	0·870	1·003	1·152	1·269	1·367
27·0	0·766	0·849	0·979	1·124	1·238	1·332
28·0	0·748	0·829	0·956	1·097	1·208	1·300
29·0	0·732	0·811	0·934	1·072	1·180	1·270
30·0	0·716	0·793	0·913	1·048	1·153	1·241
32·0	0·687	0·761	0·875	1·004	1·105	1·188
34·0	0·661	0·731	0·841	0·964	1·061	1·141
36·0	0·637	0·705	0·810	0·928	1·021	1·098
38·0	0·615	0·680	0·782	0·895	0·984	1·058

TABLE 6

INTENSITIES OF RAINFALL
(continued)

Time in mins	Once a year storm	Once in 1·33 years storm	Once in 2 years storm	Once in 3 years storm	Once in 4 years storm	Once in 5 years storm
40	0·595	0·658	0·756	0·865	0·951	1·022
42	0·576	0·637	0·731	0·837	0·920	0·989
44	0·559	0·618	0·709	0·812	0·892	0·958
46	0·543	0·600	0·689	0·788	0·865	0·930
48	0·528	0·584	0·669	0·765	0·841	0·903
50	0·515	0·568	0·651	0·745	0·818	0·879
55	0·483	0·534	0·611	0·698	0·767	0·823
60	0·457	0·504	0·577	0·659	0·723	0·776
65	0·433	0·478	0·547	0·624	0·684	0·753
70	0·413	0·455	0·520	0·593	0·651	0·698
75	0·394	0·434	0·496	0·566	0·621	0·666
80	0·378	0·416	0·475	0·542	0·594	0·637
85	0·363	0·399	0·456	0·520	0·570	0·612
90	0·349	0·384	0·439	0·500	0·548	0·588
95	0·337	0·371	0·423	0·482	0·528	0·567
100	0·326	0·358	0·409	0·466	0·510	0·547
110	0·306	0·336	0·383	0·436	0·478	0·512
120	0·288	0·317	0·361	0·411	0·450	0·482
130	0·273	0·300	0·342	0·389	0·426	0·457
140	0·260	0·286	0·325	0·370	0·405	0·434
150	0·248	0·273	0·310	0·353	0·386	0·414
160	0·238	0·261	0·297	0·338	0·369	0·396
170	0·228	0·250	0·285	0·324	0·354	0·379
180	0·220	0·241	0·274	0·311	0·340	0·365
190	0·212	0·232	0·264	0·300	0·328	0·351
200	0·204	0·224	0·255	0·289	0·316	0·339
210	0·198	0·217	0·247	0·280	0·306	0·328
220	0·192	0·210	0·239	0·271	0·296	0·317
230	0·186	0·204	0·232	0·263	0·287	0·308
240	0·181	0·198	0·255	0·255	0·279	0·299
250	0·176	0·193	0·219	0·248	0·271	0·290
260	0·171	0·187	0·213	0·242	0·264	0·282
270	0·167	0·183	0·208	0·235	0·257	0·275
280	0·163	0·178	0·202	0·229	0·251	0·268
290	0·159	0·174	0·198	0·224	0·245	0·262

occur once a year and once in ten years, has been confirmed and the formula is now accepted and used by engineers whose work includes problems more complex than the design of small estate sewers.

From the foregoing formula can be derived:

$$R = \frac{10 \cdot 54 \, N^{0 \cdot 2817}}{t^{0 \cdot 7183}} - \frac{6}{t}$$

where: R = inches of rainfall per hour
N = number of years between storms of this occurrence
t = duration of storm in minutes.

Table 6 is based on this formula.

It is of importance that engineers and research workers should know the means by which the rainfall data were obtained and interpreted, for there have been many misunderstandings on this matter, and these have led to controversy and the formulation of unsound theories. One erroneous, but by no means uncommon, belief is that only one storm corresponding to a single point on the once-a-year curve occurs during a year. Another fallacy is that any storm complying with the curve is an uninterrupted period of rainfall during which intensity rises to a peak occurring before half-time and then falls off to nil.

With regard to the latter belief, it is true that autographic rainfall records show that individual storms do have a more or less characteristic "shape;" but such uninterrupted storms must not be confused with statistical periods of rainfall and cannot be substituted therefore in rainfall calculations—at any rate at our present state of knowledge.

Bilham arranged for the collection of data giving the shortest intervals of time in which specified amounts of rain fell. This did not mean that there was continual rainfall during the period: rain might commence, fall off and recommence—in other words, there might be several ("storms" in the popular sense) included in the observed rain. In addition, the method of investigation could not make any allowance for differences in, or average amounts of, rainfall before or after the specified period. The curves prepared from these data showed the amount of rainfall likely to occur at all specified times, and the rainfalls given by *all* points on, for example, the once-a-year curve would be expected to occur with equal frequency—once a year, at any one rainfall station, although not necessarily at the same time.

On the basis of this class of statistics, the logic of the Lloyd-Davies method becomes obvious. Whatever the length of the time of concentration of a catchment area, there will occur on the average once a year a rainfall which, during that time, has an average precipitation complying with the formula. Thus, by working to the Lloyd-Davies method and the Bilham rainfall curve, the engineers can design sewerage systems in which all the sewers run full once a year (although not necessarily at the same time), partially full several times a year and are surcharged less frequently than once a year.

SURFACE-WATER SEWERAGE: THE LLOYD-DAVIES METHOD

Proof of the Lloyd-Davies theory. Suppose a storm of constant precipitation and 30 minutes duration were to fall on a drainage area having a time of concentration of 30 minutes and the flow in the sewer was observed at the point of outgo: at first the trickle of run-off from near-by would appear in the sewer. Then, as time passed, more distant parts of the area would contribute, together with the near part. At the end of 30 minutes the whole area would be contributing and the flow in the sewer would have reached its maximum. After 30 minutes the storm would have ceased and accordingly, there being no flow coming in from the near-by part of the area, flow would reduce although there would still be water coming from the more distant parts until, at the end of 60 minutes, run-off would have finished. From this it will be seen that the curve of run-off for a storm of constant precipitation and having duration equal to the time of concentration is a straight line sloping up to a peak, followed by a straight line sloping down to zero, and the total time of run-off is time of concentration plus duration of storm (see Curve B, Fig. 5).

Of course, this is not the exact truth. Many factors such as draw-down, surge, or wave-action, and variations of velocity due to the sewer being partially full by different amounts in different places, would distort the curve so that the peak would be rounded off and the tail-off after the peak lengthened. But these factors are too complex and indeterminate to be taken into consideration in a reasonable calculation and, as it has been found they have little effect, they can be neglected.

Should a storm of shorter duration than the time of concentration fall on

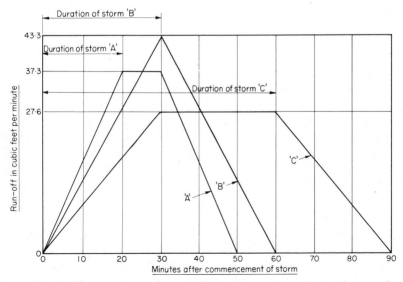

FIG. 5—Run-off from one impervious acre for once-a-year storms shorter than, equal to and longer than the time of concentration (30 minutes in the example)

the area, the rate of run-off at the point of outgo builds up for the duration of the storm. Once the storm is over, the near-by area ceases to contribute, but flow continues to be brought from the distant parts for a time with no increase or diminution until tail-off occurs. The curve in this case is a straight line sloping up to a maximum for the duration of the storm, followed by a straight line horizontal to the time axis for a period of time equal to the time of concentration minus the duration of the storm, and finally a line sloping down to zero for the duration of the storm: and the maximum rate or run-off is given by the formula:

$$Q = 60 \cdot 5 \times ApR \times \frac{t}{t_c}$$

where: t = duration of storm in minutes
t_c = time of concentration in minutes.

Because the area contributing by the time peak is reached is less than the total, the peak run-off is less than in the case when duration of storm equals time of concentration, in spite of the shorter storm having a greater intensity (see Curve A, Fig. 5).

If a storm of greater duration than the time of concentration falls on the area, the curve builds up to a maximum at the end of time of concentration, continues steady until the end of the storm—i.e. for a period of time equal to duration of storm minus time of concentration—then falls off to zero in a time equal to time of concentration. The total period of run-off is again equal to time of storm plus time of concentration. In this case the maximum rate of run-off is given by the formula:

$$Q = 60 \cdot 5 \times ApR$$

but as the value of R is less than that for the storm having duration equal to time of concentration, the run-off is less (see Curve C, Fig. 5).

Anyone who has not done so should work out for himself examples as given above, so as to satisfy himself that it is the storm which lasts as long as the time of concentration that produces the maximum run-off.

Method of procedure. The means of using the Lloyd-Davies method is as follows. Suppose the top end of the drainage system concerned is a patch of land measuring 5 acres, and the impermeability factor is 0·318. Then Ap for that area will be:

$$5 \times 0 \cdot 318 = 1 \cdot 59 \text{ acres.}$$

The value of R cannot be known until the diameter of sewer is known, because the diameter determines the velocity of flow, the velocity determines the time taken for the sewage to flow from the top end to the bottom end of length of sewer, or "time of concentration," and the time of concentration determines the intensity of rainfall for which allowance should be made. The

SURFACE-WATER SEWERAGE: THE LLOYD-DAVIES METHOD

designer must therefore first assume an intensity of rainfall, design a sewer to take that intensity, calculate the velocity of flow in the sewer at the gradient available, find the time of concentration and, if this assumed rate of rainfall is not right, start all over again with a new, more accurate assumption. Let us assume that in this particular calculation the remaining data are as follows:

> Length of sewer: 1200 feet.
> Available gradient 1 in 200.

Suppose the designer first assumes 1 inch of rain, then the run-off will be calculated as follows:

$$Q = 60 \cdot 5 \times Ap \times R$$
$$= 60 \cdot 5 \times 1 \cdot 59 \times 1$$
$$= 96 \text{ cubic feet per minute.}$$

This can be accommodated by a 12-inch diameter pipe which, according to Crimp and Bruges' formula (see Vol. I, Fig. 68), at a gradient of 1 in 200 passes 164 cubic feet per minute at a velocity of 209 feet per minute.

The time of flow through 1200 feet of sewer at 209 feet per minute is about 6 minutes. To this used to be added a "time of entry" figure, to allow for the initial time required for rain-water to flow over roofs, along gutters, down drains and over the surface of the land before reaching the sewers. If a time of entry of 3 minutes is accepted the total time of concentration will be 9 minutes for which, if the engineer is designing on the once-a-year storm in Table 6, the appropriate rate of rainfall is 1·51 inches per hour. Using this rate of rainfall, the run-off becomes:

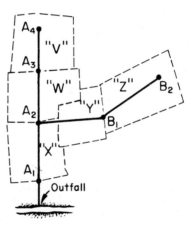

FIG. 6—Drainage area

$$60 \cdot 5 \times 1 \cdot 59 \times 1 \cdot 51$$
$$= 145 \text{ cubic feet per minute.}$$

This quantity can be accommodated by the 12-inch pipe, which can now be taken as being of the correct size for draining the area concerned.

But drainage areas are not drained by sewers of even size throughout their lengths, and therefore calculations have to be made for each length in turn. The procedure is to commence at the top end of the system, working from manhole to manhole.

Example calculation. When thus working out the sizes of the surface-water sewers for a complete scheme, the engineer prepares a calculation sheet.

Take, for example, the drainage area and sewerage scheme illustrated in Figure 6. The conditions in this case are given in Table 7.

TABLE 7

EXAMPLE CALCULATION

Drainage area	Gross area in acres	Percentage impermeability	Impervious area in acres
V	1·5	33·3	0·5
W	1·5	33·3	0·5
X	2·0	40·0	0·8
Y	1·2	33·3	0·4
Z	1·2	25·0	0·3

The lengths and gradients of the sewers are as in Table 8.

TABLE 8

EXAMPLE CALCULATION

Location	Gradient 1 in:	Length in feet
A_4–A_3	250	310
A_3–A_2	300	340
B_2–B_1	175	360
B_1–A_2	180	360
A_2–A_1	245	330
A_1–outfall	250	150

The complete calculation is set out as in Table 9, and proceeds as follows. Taking first the length of sewer from A_4 to A_3, the designer proceeds by trial and error, trying a suitable small size of sewer, 7 inch,* 9 inch or 12 inch. Suppose he commences by trying a 7-inch-diameter sewer: a 7-inch pipe at a gradient of 1 in 250 discharges 34·8 cubic feet a minute at a velocity of 130 feet a minute. As the length between manholes is 310 feet, the time of flow is about $2\frac{1}{2}$ minutes, which, added to a time of entry of 3 minutes, gives a time of concentration of $5\frac{1}{2}$ minutes. From Table 6 it will be seen that the intensity of rainfall to be allowed for a once-a-year storm of $5\frac{1}{2}$ minutes duration is 2·01 inches per hour. Then by the Lloyd-Davies formula:

$$Q = 60.5 \times 0.5 \times 2.01$$
$$= 60.8 \text{ cubic feet per minute}$$

which is more than the 7-inch pipe can pass without surcharge.

Suppose next a 12-inch pipe is tried. A 12-inch pipe at a gradient of 1 in 250 discharges 147 cubic feet a minute at a velocity of 187 feet a minute. Then the time of flow through the pipe is $1\frac{2}{3}$ minutes, which, added to 3 minutes time of entry, gives less than 5 minutes (the minimum ever allowed) as the time of concentration. Thus the intensity of rainfall is 2·12 inches per

---

* 7 inch was once frequently used as the minimum diameter.

SURFACE WATER SEWERAGE: THE LLOYD-DAVIES METHOD 39

hour, which, applied to the Lloyd-Davies formula, gives a run-off of 64 cubic feet per minute. This is less than the capacity of the 12-inch pipe, and therefore that size is excessive.

Next the designer can try what is by now obviously the correct size, 9 inches diameter. He will quickly find that a 9-inch sewer laid to the available gradient of 1 in 250 discharges 68·1 cubic feet per minute at a velocity of 154 feet per minute. This gives a time of flow through the length of sewer of about 2 minutes, which, added to the time of entry—3 minutes—gives the time of concentration of 5 minutes. For 5 minutes the intensity of rainfall is 2·12 inches per hour, giving again a run-off of 64 cubic feet per minute, which is comfortably within the capacity of the pipe. So a 9-inch-diameter pipe is chosen.

TABLE 9

EXAMPLE CALCULATION FOR SURFACE-WATER SEWERAGE SCHEME

(1) Location (manhole to manhole)	(2) Drainage areas	(3) Impervious areas (A_p) (acres)	(4) Time of concentration (t) (mins)	(5) Rainfall intensity (R) (ins per hr)	(6) Run-off (c.f.m.)
A_4–A_3	V	0·5	5 (3*+2)	2·12	64
A_3–A_2	$V+W$	1·0	7 (5+2)	1·75	106
B_2–B_1	Z	0·3	5½ (3*+2·3)	2·01	36
B_1–A_2	$Z+Y$	0·7	7½ (5½+2)	1·68	71
A_2–A_1	$V+W+X+Y+Z$	2·5	9 (7½+1½)	1·51	228
A_1–outfall	$V+W+X+Y+Z$	2·5	9	1·51	228

(1) Location (manhole to manhole)	(7) Dia. of sewer (ins)	(8) Gradient (1 in:)	(9) Discharge (c.f.m.)	(10) Velocity (ft min)	(11) Length between manholes (ft)	(12) Time of flow through length (mins)
A_4–A_3	9	250	68·1	154	310	2·0
A_3–A_2	12	300	134·0	170	340	2·0
B_2–B_1	7	175	41·6	156	360	2·3
B_1–A_2	9	180	80·3	182	360	2·0
A_2–A_1	15	245	269·0	219	330	1·5
A_1–outfall	15	250	266·0	217	150	0·7

* Equals 3 minutes time of entry

This has given the run-off from area V and the required size of sewer from A_4 to A_3. Next has to be calculated the combined run-off from areas V and W, so as to determine the required size of sewer from A_3 to A_2. To do this the total impervious area for areas V and W is taken, and the total time of concentration, which consists of the time of entry, the time of flow from A_4

to A_3 and the time of flow from A_3 to A_2. The last of these three is found by trial and error, as before.

Having found the sizes of sewer for the branch line A_4 to A_2, the sizes of sewer making up the branch line B_2 to A_2 are found in a similar manner. The next calculation is to find the diameter of the sewer A_2 to A_1. This sewer takes the flow from the whole area V, W, X, Y, Z, and therefore the total impervious area of these components is summed for the purpose of the calculation. But it will be noted that there are now two times of concentration, one of which was calculated for the A line, the other of which was calculated for the B line. For the purpose of calculating the run-off of the whole area, the *longer* time of concentration is selected, which happens to be that of the B line plus the time of flow from A_2 to A_1.

The final calculation is that for finding the size of sewer discharging from A_1 to outfall. As there is no additional area contributing throughout the length of A_1 to outfall, there is no increased time of concentration, and therefore the rate of flow is taken as being merely the quantity discharging from manhole A_1—i.e. 228 cubic feet per minute—and the diameter of pipe is found by the flow tables according to the gradient available.

Change of impermeability during rainfall. The writer admits that, having been too impressed by the Lloyd-Davies method and some of the elaborations thereof devised during the 1930s, he failed to appreciate the comparative value of the best examples of the American "rational" method and therefore continued to carry out research until he arrived at the conclusions that had already been reached in America many years ago. By careful study of data given in the Road Research Laboratory Research Notes RN/2361/LHW. B.W.9, RN/2362/LHW. B.W.11 and RN/2366/LHW. B.W.14, for the Kidbrooke, Oxhey and Blackpool housing sites respectively, he found that there was a difference between the actual run-off hydrographs as recorded by flow recorders on the sewerage system and the hydrographs as calculated from the curves drawn by autographic rain gauges. He could not accept the arbitrary adjustment that the laboratory made for this, on the ground that it was based on a mathematical fallacy, and found that this difference could be explained only by change of run-off coefficient or "impermeability" during rainfall as had already been observed by Lloyd-Davies. In the third edition of this book he expressed the opinion that:

> "change of impermeability appeared to be due in the main to storage on the impermeable surfaces. At the commencement of rainfall, drops were held apart by surface tension and did not flow at once (this was carefully observed on a paved surface falling at 1 in 7); there was storage to be filled in rainwater guttering, gullies and puddles on surfaces not laid to adequate falls."

This is in line with American opinion as given in "Handbook of Applied Hydraulics," edited by Calvin Victor Davis, second edition, 1952, which is as follows:

"Pondage in depressions, evaporation, and absorption, and other factors reduce the run-off, so that not all of the rainfall, even on impervious surfaces, reaches the sewer inlets. The longer the storm duration, the larger may be the percentage of rainfall run-off."

The data from Kidbrooke, Oxhey and Blackpool showed that heavy storms, of the kind that caused intense run-off, almost invariably fell after a dry period and that there was a time-lag of at least 2 minutes before any recordable flow arrived in the sewers. From then on impermeability factor (using the term to mean the difference between rainfall and run-off to sewers)

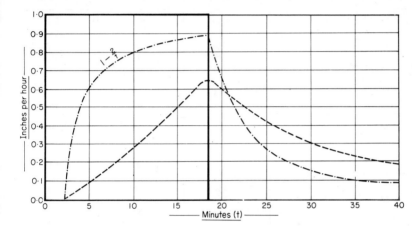

FIG. 7—Change of momentary impermeability during rainfall

The full line represents rainfall precipitation—a storm of constant precipitation for 18 minutes at the rate of 1 inch per hour. The dot-and-dash line indicates the rate of run-off from the impermeable area into the sewerage system. The broken line indicates the run-off at the point of outgo from the sewerage system. It is assumed that the rainfall duration was equal to the time of concentration.

built up according to a regular curve. Eventually, on cessation of rainfall and after peak run-off, the surface storage drained away for a considerable period.

The formula:

$$\text{Momentary impermeability} = i\left(1 - \frac{2}{t}\right)$$

where: i = ultimate impermeability factor found by dividing total run-off from a paved surface over a long period of time by rainfall on paved surface in the same period
t = time in minutes from start of rainfall.

gives a fair interpretation of the facts as far as have been ascertained (see Fig. 7).

To make practical use of the above formula it is necessary to integrate it for the whole of the time-of-concentration so as to give the average impermeability during that time. The integrated formula reads:

TABLE 10
PROBABLE RUN-OFF TO SEWERS

Duration of rainfall (minutes)	Average impermeability during storm as a percentage of ultimate impermeability	Once-in-three-years rainfall according to Bilham's formula (inches per hour)	Probable run-off from impermeable area to sewer	
			(inches per hour)	(mm per hour)
11	50·822	2·020	1·03	26·1
12	53·47	1·910	1·02	25·9
13	55·818	1·814	1·01	25·7
14	57·915	1·729	1·00	25·4
15	59·8	1·653	0·99	25·2
16	61·507	1·585	0·97	24·8
17	63·057	1·524	0·96	24·4
18	64·474	1·468	0·95	24·0
19	65·775	1·417	0·93	23·6
20	66·974	1·370	0·92	23·3
21	68·082	1·327	0·90	23·0
22	69·11	1·287	0·89	22·6
23	70·066	1·250	0·88	22·4
24	70·959	1·215	0·86	21·9
25	71·793	1·183	0·85	21·6
26	72·577	1·152	0·84	21·3
27	73·313	1·124	0·82	21·0
28	74·007	1·097	0·81	20·7
29	74·661	1·072	0·80	20·4
30	75·28	1·048	0·79	20·0
32	76·422	1·004	0·76	19·5
34	77·452	0·964	0·75	18·9
36	78·387	0·928	0·73	18·5
38	79·239	0·895	0·71	18·0
40	80·021	0·865	0·69	17·6
42	80·740	0·837	0·68	17·2
44	81·404	0·812	0·66	16·8
46	82·019	0·788	0·65	16·4
48	82·591	0·765	0·63	16·0
50	83·124	0·745	0·62	15·7

TABLE 10 (continued)
PROBABLE RUN-OFF TO SEWERS

Duration of rainfall (minutes)	Average impermeability during storm as a percentage of ultimate impermeability	Once-in-three-years rainfall according to Bilham's formula (inches per hour)	Probable run-off from impermeable area to sewer	
			(inches per hour)	(mm per hour)
55	84·312	0·698	0·59	14·9
60	85·329	0·659	0·56	14·3
65	86·212	0·624	0·54	13·7
70	86·985	0·593	0·52	13·1
75	87·669	0·566	0·50	12·6
80	88·278	0·542	0·48	12·2
85	88·825	0·520	0·46	11·7
90	89·319	0·500	0·45	11·3
95	89·767	0·482	0·43	11·0
100	90·176	0·466	0·42	10·7
110	90·896	0·436	0·40	10·1
120	91·509	0·411	0·38	9·5
130	92·04	0·389	0·36	9·1
140	92·503	0·370	0·34	8·7
150	92·910	0·353	0·33	8·3
160	93·272	0·338	0·32	8·0
170	93·597	0·324	0·30	7·7
180	93·889	0·311	0·29	7·4
190	94·154	0·300	0·28	7·2
200	94·395	0·289	0·27	6·9
210	94·616	0·280	0·26	6·7
220	94·818	0·271	0·26	6·5
230	95·004	0·263	0·25	6·3
240	95·177	0·255	0·24	6·2
250	95·338	0·248	0·24	6·0
260	95·487	0·242	0·23	5·9
270	95·626	0·235	0·22	5·7
280	95·756	0·229	0·22	5·6
290	95·878	0·224	0·21	5·5
300	95·993	0·219	0·21	5·3

$$\text{Average impermeability during rainfall} = i \left(\frac{t - 2 + 4\cdot6052 \log_{10}\frac{2}{t}}{t} \right)$$

From this Table 10 has been prepared.

The writer's recommendation is that, when the original Lloyd-Davies method be used, the appropriate *run-off* for the once-in-three-years storm as given in Table 10 should be substituted for R in the Lloyd-Davies formula.

When this is done, no time of entry should be allowed as this has already been included in the allowance for change of impermeability. It has also been found that the shape of the run-off curve is so different from any rainfall curve that the use of any of the methods involving a time–area graph (or the equivalent) generally has negligible effect on the calculation and that therefore such complications usually should be avoided.

The rational method. Since the Second World War some writers have miscalled the Lloyd-Davies method the rational method. This is misleading because, although both are time-of-concentration methods in which the rate of run-off per acre varies as a function of the time taken for the flow from the farthest part of the catchment to reach the point at which flow is desired to be known, the Lloyd-Davies and the American rational method are not identical in that the rational method, in its best forms, makes allowance not only for a varying rainfall intensity but also for a change of run-off factor with time.

There is, admittedly, some difficulty in that several varieties of time-of-concentration method have been used and described as the "rational method" in the United States and Canada. In some instances, for example, at St. Louis, Kansas City, Louisville, Grand Rapids, Peroria and Decatur in the United States, and at London, Ontario and Montreal in Canada, the method allowed for variation in run-off factor according to duration of rainfall while, at Buffalo, Cincinnati, Detroit, District of Columbia and Milwaukee, long times of entry have been used which have a practical effect similar to change of impermeability during rainfall. In other localities changes of rainfall and impermeability have been embodied in curves which give run-off in cusecs per acre varying with time of concentration.

The rational method is represented by the formula:

$$Q = ciA$$

where: Q = run-off in cusecs
  c = a run-off coefficient representing the ratio of run-off to rainfall
  i = rainfall in cusecs per acre
  A = drainage area in acres.

Table 11 gives the values c and i which the author considers are applicable to British rainfall and the average run-off from paved areas in Great Britain.

SURFACE-WATER SEWERAGE: THE LLOYD-DAVIES METHOD

It will be seen that in the foregoing formula there are two factors which vary with time. If one of these is eliminated the calculation is slightly simplified and therefore the method which the author prefers is to allow a run-off factor made by adjustment of the rainfall figure and use an unvarying ultimate impermeability factor.

Table 10 gives the probable run-off from impermeable area to sewer which, if used as R in the Lloyd-Davies formula, will virtually convert the Lloyd-Davies method to the rational method as used in various parts of the

TABLE 11
RUN-OFF AND RAINFALL COEFFICIENTS FOR RATIONAL METHOD

Duration of rainfall (minutes)	Paved area run-off coefficient "c"	Once-in-three-years rainfall intensity "i" (cusecs per acre)	Duration of rainfall (minutes)	Paved area run-off coefficient "c"	Once-in-three-years rainfall intensity "i" (cusecs per acre)
11★	0·41	2·04	55	0·67	0·70
12	0·43	1·93	60	0·68	0·66
13	0·45	1·83	65	0·69	0·63
14	0·46	1·74	70	0·70	0·60
15	0·48	1·67	75	0·70	0·57
16	0·49	1·60	80	0·71	0·55
17	0·50	1·54	85	0·71	0·52
18	0·52	1·48	90	0·71	0·50
19	0·53	1·43	95	0·72	0·49
20	0·54	1·38	100	0·72	0·47
21	0·54	1·34	110	0·73	0·44
22	0·55	1·30	120	0·73	0·41
23	0·56	1·26	130	0·74	0·39
24	0·57	1·22	140	0·74	0·37
25	0·57	1·19	150	0·74	0·36
26	0·58	1·16	160	0·75	0·34
27	0·59	1·13	170	0·75	0·33
28	0·59	1·12	180	0·75	0·31
29	0·60	1·08	190	0·75	0·30
30	0·60	1·06	200	0·76	0·29
32	0·61	1·01	210	0·76	0·28
34	0·62	0·97	220	0·76	0·27
36	0·63	0·94	230	0·76	0·27
38	0·63	0·90	240	0·76	0·26
40	0·64	0·87	250	0·76	0·25
42	0·65	0·84	260	0·76	0·24
44	0·65	0·82	270	0·77	0·24
46	0·66	0·79	280	0·77	0·23
48	0·66	0·77	290	0·77	0·23
50	0·67	0·75	300	0·77	0·22

★The product of "c" and "i" for 11 minutes is appropriate to all durations of less than 11 minutes.

United States and Canada. This is the method which the writer recommends for the general design of surface water and combined sewers for urban areas in Great Britain.

An example calculation follows, using the drainage area in Figure 6, but with new values for impervious areas, gradients and sewer lengths as given in Tables 12 and 13. The calculation is then as given in Table 14, in which the figures for the equivalent metric calculation are given in italics.

The reason it was necessary to use longer distances and appropriately larger areas than given in Tables 7 and 8 is that, had the original figures in Tables 7 and 8 been used, there would have been no change of rainfall intensity and therefore the calculation would have been a flat-rate calculation, not a time-of-concentration calculation. *This demonstrates the important point that the recommended method does away with the need for time-of-concentration calculations and permits the use of flat-rates for the vast majority of sewerage calculations,* for the rainfall intensity of 1·03 inches per hour applies for all times of

TABLE 12
EXAMPLE CALCULATION

Drainage area	Gross area in acres	Paved area in acres	Percentage impermeability of paved areas	Impermeable area in acres
V	12	3·0	80	2·4
W	20	2·5	80	2·0
X	30	5·0	80	4·0
Y	15	2·5	80	2·0
Z	14	2·0	80	1·6

TABLE 13
EXAMPLE CALCULATION

Location	Gradient 1 in:	Length in feet
A_4–A_3	450	970
A_3–A_2	600	950
B_2–B_1	350	800
B_1–A_2	360	1270
A_2–A_1	360	750
A_1–outfall	500	800

SURFACE-WATER SEWERAGE: THE LLOYD-DAVIES METHOD

concentration up to 11 minutes and could be extended to, say, 15 minutes without being extravagant. Thus, the engineer could allow the flat-rate of 62·3 cubic feet per minute per acre (4·36 cubic metres per minute per hectare) of impervious area for all cases where the time-of-concentration did not

TABLE 14
EXAMPLE CALCULATION FOR SURFACE-WATER SEWERAGE SCHEME

(1) Location (manhole to manhole)	(2) Drainage areas	(3) Impervious areas (A_p) (acres) (hectares)	(4) Time of concentration (t) (mins)	(5) Run-off intensity*(R) (ins per hr) (mm per hr)	(6) Run-off (cu ft min) (cu metres min)
A_4–A_3	V	2·4 0·97	6	1·03 26·1	150 4·2
A_3–A_2	$V+W$	4·4 1·78	6+6 = 12	1·02 25·9	272 7·7
B_2–B_1	Z	1·6 0·65	5	1·03 26·1	99 2·8
B_1–A_2	$Z+Y$	3·6 1·46	5+7 = 12	1·02 25·9	222 6·3
A_2–A_1	$V+W+X+Y+Z$	12·0 4·86	12+3 = 15	0·99 25·2	718 20·4
A_1–outfall	$V+W+X+Y+Z$	12·0 4·86	15	0·99 25·2	718 20·4

(1) Location (manhole to manhole)	(7) Dia. of sewer (ins) (mm nominal)	(8) Gradient (1 in:)	(9) Discharge (cu ft min) (cu metres min)	(10) Velocity (ft min) (metres min)	(11) Length between manholes(ft) (metres)	(12) Time of flow through length (mins)
A_4–A_3	15 375	450	198 5·6	162 49	970 296	6
A_3–A_2	18 450	600	279 7·9	158 48	950 290	6
B_2–B_1	12 300	350	124 3·5	158 48	800 244	5
B_1–A_2	15 375	360	222 6·3	181 55	1270 387	7
A_2–A_1	24 600	360	776 22·0	247 75	750 229	3
A_1–outfall	27 675	500	901 25·5	227 69	800 244	3½

*See Table 10.

exceed 15 minutes, thereby considerably reducing the amount of work involved in designing almost any drainage scheme.

As has already been mentioned, the recommended method virtually does away with the need for any check with the aid of a time–area graph or similar device. This can be illustrated with an example based on the time–area graph illustrated in Fig. 10. As is shown in this calculation, the run-off from the whole area of 31·2 acres caused by a "Ministry of Health Storm" of 0·435 inch per hour gives a run-off of 821 cubic feet per minute requiring, for the gradient of 1 in 500, a 27-inch diameter pipe, but that the effective impervious area of 19 acres under a "Ministry of Health Storm" of 1 inch per hour gives a run-off of 1150 cubic feet per minute, requiring a 30-inch diameter sewer. With the recommended method, the whole area of 31·2 acres would receive a run-off from the sewers of 0·51 inch per hour, giving a run-off of 963 cubic feet per minute, requiring a 30-inch diameter sewer. The effective area of 19 acres receiving a run-off from the sewers of 0·92 inch per hour gives a run-off of 1058 cubic feet per minute, again requiring a 30-inch diameter sewer. Thus, the recommended method achieves the same result as obtained with the time–area graph but with a fraction of the work.

Should, however, it be considered desirable that the time–area method should be used, the *ordinary* tangent method can be applied because the formulae:

$$R = \frac{62}{t + 50} \text{ and } R = \frac{76 \cdot 88}{t + 86}$$

approximate very closely to the once-in-three-years run-off as given in the right-hand column of Table 10 for times of concentration of 11 to 100 minutes and 100 to 200 minutes respectively. To do this tangent lines 50 minutes apart or 85 minutes apart in time respectively have to be applied as described in Chapter IV. If the example that has just been worked is so applied to the time–area graph shown in Fig. 10 it will be found that a 30-inch diameter sewer is still required, showing how seldom the time–area graph is needed.

CHAPTER IV

SURFACE-WATER SEWERAGE: ELABORATIONS OF THEORY

WHEN applied to compact, regularly shaped housing or factory estates or developed areas generally of moderate size, the Lloyd-Davies method often gives very fair estimates of run-off, and results in suitable sizes of sewers. This is not due to inherent accuracy but to the balancing of errors which it so happens are roughly equal and opposite in the circumstances described above. The accuracy of the Lloyd-Davies method becomes less impressive when the catchment is large and the impervious area irregularly distributed. It can then be proved *theoretically* that unless modifications are made the application of the method could result in underestimation of flows and sewer sizes.

When some method is used to adjust for irregular distribution of impervious area, however, it is often found that the flows and sewer sizes have, *in fact*, been overestimated by an appreciable amount. This, at first, led to some distrust of the time-of-concentration methods, and some engineers made arbitrary deductions in the light of their experience. There were also some not too erudite speculations as to why overestimation did occur.

The reason is now quite clear. Overestimation of flows is almost invariably due to overestimation of impermeability, first, because of a natural tendency to estimate on the "safe" side, and second, because the effective impermeability during a short intense storm is, as before described, less than the impermeability as ascertained by comparing the total rainfall with the total run-off after a storm of appreciable duration.

While the methods for allowing for the effect of irregular distribution of impervious area are much less important than was once thought and are seldom required, some knowledge of them is desirable.

Allowance for irregular distribution of impervious area. When the impervious area is equally distributed throughout the length of the sewer or sewerage system the time of concentration as estimated is practically the same as the "effective time of concentration," and the storm, the length of which is equal to the time of concentration, is the one which causes the greatest run-off. But the estimation of the time of concentration by calculating the time of flow through the sewers can be misleading, for part of the time included may be time of flow through lengths of sewer which are collecting little or no flow from laterals. Take, for example, the drainage area illustrated in Fig. 8. This shows a drainage area sewered by a system of sewers, including the main line *ABCD*. In calculating the time of concentration an engineer might calculate the time of flow from *A* to *D*. But the area through which the length *AB* flows and the area through which the length *CD* flows are small, contributing comparatively little flow to the total. If these areas were neglected a calculation could be made for the run-off of the main drainage area contributing between B and C. This would reduce, by a little, the total impervious area contributing, but would greatly reduce the time of concentration, increasing the rate of rainfall to be allowed for, and more than probably resulting in a greater calculated (and actual) run-off than if the whole area were included.

Means of estimating the effect of such irregularities of distribution of impervious area have been the subject of very much discussion. Numerous methods have been suggested; some approximate, some logically unsound, others of practical and academic interest. Nearly all the methods require in the first place the construction of a time–area graph.

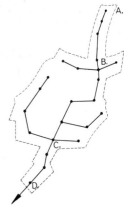

FIG. 8—Drainage area: irregular distribution

Time–area graph. A time–area graph is a graph which shows the sum-total of "impervious acres" contributing to a selected point in a sewerage system at various periods of time after the commencement of a storm, and usually up to the end of the calculated time of concentration of the area contributing to that point. To take the simplest case, suppose there is a length of sewer draining a long street, with houses and road gullies contributing throughout its length: at the commencement of the storm rain-water will be arriving at the lower end of the sewer only from the houses and gullies in the immediate vicinity of the lower end; but as time passes, water from up the street will arrive, until at the end of the time of concentration water from the top end of the sewer will be arriving at the bottom end and mingling with the water coming in at the bottom end and all other parts of the lengths of the sewer. The rate of run-off will then have reached its maximum and, for as long as the storm continues at the same rate of precipitation, the rate of flow at the bottom end will remain the same. Thus, the time–area graph of a single sewer or an area in which the distribution of impervious area is not, and does not require to be, known, is represented by two straight lines, one which slopes from *zero time, zero acres* to a point which represents completion of time of concentration and total acres impervious area. At the end of the time of concentration the total area continues to contribute for as long as rain continues to fall, and therefore with further increase of time the impervious area remains constant during the continuance of the storm. The second straight line is therefore a line drawn parallel to the time axis and representing the total impervious area. (See the individual time–area curves for areas V, W, X, Y, Z on the bottom half of Fig. 9.)

The bottom portion of Fig. 9—i.e. that portion not ruled with closely spaced lines—represents the construction of the time–area graph for the drainage area illustrated in Fig. 6, taken about the point of outfall. The time–area curve could have been calculated about the point A_1 and the same

SURFACE-WATER SEWERAGE: ELABORATIONS OF THEORY 51

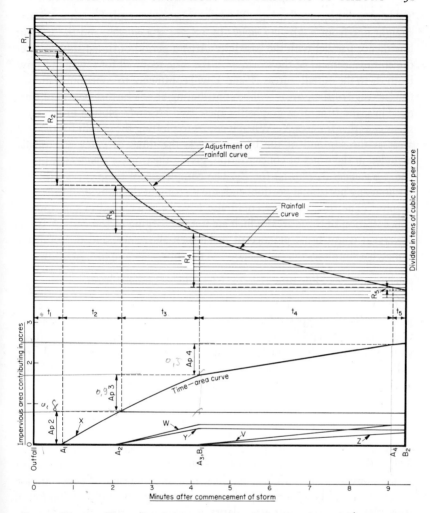

FIG. 9—Diagram illustrating time–area graph and the Ormsby and Hart method

result obtained, but for the purpose of demonstrating that the time of flow from A_1 to point of outfall through a line of sewer which has no lateral connections has no effect on the calculation, the point of outfall has been chosen for the origin of the curve. The method of construction of the time–area graph for such a composite area commences with constructing the individual time–area graphs for the component areas.

After the commencement of a storm there will be no flow from the outfall until the nearest area—area X—begins to contribute. This will be *0·7 minute* (t_1) after the storm has started. For *1·5 minutes* (t_2) area X will contribute at an increasing rate until at the end of t_2, or *2·2 minutes* (t_1 plus t_2) after the commencement of the storm, the whole of area X (0·8 impervious acre, see

Table 9) will be contributing. Thus the time–area graph for area X is the straight line from *0 acre, 0·7 minute* to *0·8 acre, 2·2 minutes*, and the straight line thence parallel with the base line. After a period of time t_1 plus t_2 after the commencement of the storm—i.e. the time required for sewage to flow from A_2 to the outfall—areas W and Y will commence to contribute, and the total acreage of these areas will be contributing after a further period of time t_3. Thus, the time–area graphs for these two areas are plotted in the same manner as that for area X. When it comes to plotting the graphs for the two areas at the top end of the branch sewers the times of entry are usually included in the times of concentration, giving the times *5 minutes* (t_4) for V; and *5·3 minutes* (t_4 plus t_5) for Z. The completion of these curves, the last without any horizontal portion, ends the plotting of the individual time–area graphs. (The time of entry could also have been included in the times of concentration of those areas which are not discharging to top ends of lateral sewers: it is not usual practice to make such allowance, and this omission somewhat simplifies the construction of the time–area graph. The allowance can be made by adding time of entry to $t + Y$ in the rainfall formula $R = \dfrac{X}{t+Y}$.)

The time–area graph for the whole area is merely a sum curve of the components. This is prepared by measuring the totals at all places of change of slope—i.e. at the ends of t_1, t_2, t_3, t_4 and t_5—with the aid of dividers. (Care should be taken when totalling that lines which coincide—for example, the V line and the W line at the end of t_4—are not counted, in error, as single lines.) When this is done the height of the time–area curve at the end of the time of concentration should equal the total impervious area, otherwise a mistake must have been made.

It will be seen from Fig. 9 that the preparation of a time–area graph shows at a glance any irregularity in the distribution of impervious area. When the area is regularly distributed the graph is merely a straight line sloping upwards to the right from the origin to "total time of concentration –total acreage," but the greater the degree of irregularity of distribution of impervious area the more irregularity is there in the time–area graph, the steep portions showing the positions in time of the areas which matter.

Time–area graphs or tabulated data of similar significance are used in various methods for finding the effect of irregular distribution of impervious area on run-off from sewerage systems.

Of these methods the "tangent methods" are of most practical use.

Tangent method. If one is using a rainfall or run-off formula of the type $R = \dfrac{X}{t+Y}$ the run-off from a catchment in cubic feet per minute is given by the formula:

$$Q = 60 \cdot 5 \times Ap \times \left(\dfrac{X}{t+Y}\right).$$

Then, taking a time–area graph for a simple "theoretically rectangular" catchment, if one measures minus *Y minutes* from the origin, and from the point so obtained draws a straight line to the point *Apt*, that is, maximum impervious area, maximum time of concentration, the tangent of the angle between this line and the time axis = $\dfrac{Ap}{t_c + Y}$ and therefore indicates the run-off from the catchment.

Now take a more complex time–area graph made up from a number of subsidiary catchments and therefore being in the form of an irregular curve. If two parallel lines are drawn *Y minutes* apart in the time scale, one touching the point of maximum convexity on the upper side of the time–area graph, and the other touching a point of convexity on the under side of the curve, the tangent of the largest angle which these lines can form with the time axis (the above conditions being maintained) represents the maximum run-off from the area concerned. Then, the "effective time of concentration" (t_e) is the time measured between the two points of contact between the "tangent lines" with the time–area graph; and the "effective impervious area (Ap_e);" similarly, the area as measured between these two points (see Fig. 10).

When the time–area graph is a straight line, time of concentration and impervious area are unaltered and the tangent method gives the same answer as the Lloyd-Davies method in its original form; but when the time–area graph is much bent or curved the reduced effective impervious area and reduced effective time of concentration substituted in the Lloyd-Davies formula gives a greater run-off than the total impervious area and total time of concentration.

This method was first used with the Ministry of Health Standard Curve, $R = \dfrac{30}{t + 10}$ and $R = \dfrac{40}{t + 20}$, being applied to those parts of the curve that represents $t = 5$–20 minutes and $t = 20$–100 minutes respectively.

In the example illustrated in Fig. 10 the total time-of-concentration is 72 minutes giving, in accordance with the Ministry of Health formula, a rainfall intensity of 0·435 inch per hour and therefore a run-off of 821 cubic feet per minute. If the available gradient of the sewer is 1 in 500, the nearest diameter of pipe is 27 inches. But if one takes the *effective* time-of-concentration of 20 minutes, giving a rainfall intensity of 1 inch per hour and applies this rainfall to the *effective* impervious area of 19 acres, the resulting run-off is 1150 cubic feet per minute (or 40 per cent more), requiring a 30-inch diameter sewer.

The tangent method gives the run-off that would be produced by a storm of constant precipitation having a duration equal to the *effective* time of concentration and therefore, when applied to the Ministry of Health storm, represents the maximum run-off liable to occur on the average once in about 15 months from the catchment concerned and in the assumed conditions. It does not allow for any rainfall occurring before or after the "design storm" or for any irregularity of precipitation during the design storm. As, however, there are no reliable data on these matters and the tangent method has been found sufficiently accurate, most engineers are content to accept it.

As the tangent method can be applied to the modified rational method for the reasons given at the end of Chapter III, it has received a new lease of life and the writer considers it the most satisfactory means of using a time–area graph.

There is an unsound form of the tangent method which was, in fact, the original form used until Wearing Riley[212] pointed out its error. In this, the "tangent line" was drawn from a point *"minus Y minutes"* on the time scale to the point of maximum convexity on the top side of the curve, and not from

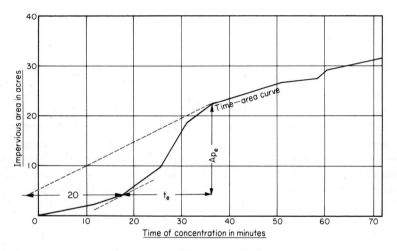

FIG. 10—Tangent method

the point of maximum convexity on the under side. Thus, it took no account of sparsely developed or straggling areas in the lower reaches of the catchment and therefore, where these existed, gave underestimates of run-off. The method needs to be mentioned, because it is still used by those engineers who have not kept themselves informed on the literature of the subject.

Revised or modified tangent method. The tangent method is based on the following proposition:

$$\text{As } R = \frac{x}{t+y}$$

1. $Q = C \tan \varphi$
2. The greater the value of φ the greater the value of Q.
3. Therefore the greatest angle that can be produced by drawing a line tangential to the time–area curve from a point at "minus twenty minutes" from any part of the time–area curve denotes the maximum run-off from the area concerned.

SURFACE-WATER SEWERAGE: ELABORATIONS OF THEORY

It follows that it is not applicable to rainfall formulae other than $R = \dfrac{x}{t+y}$. *But the tendency in Great Britain has been to abandon this type of formula and to substitute:*

$$R = \frac{x}{t^n}$$

or:

$$R = \frac{x}{t^n} - \frac{y}{t}$$

Formulae of these varieties produce curved "tangent lines" making a new method necessary. Accordingly the author devised the "revised" or "modified" tangent method which could be used with any rainfall formula and, once the necessary transparent overlay had been made, it could be used again and again. In view, however, of the fact that the "tangent curves" plotted from the figures in the right-hand column of Table 10 are almost straight lines, making it possible to revive the ordinary tangent method, this modified tangent method would appear to be redundant. *The writer does not recommend any of the other methods that follow in this chapter for present-day design (i.e. the Coleman and Johnson, Ormsby and Hart and Hydrograph methods) because they involve too much work to no advantage.*

In the modified tangent method a transparent overlay is prepared on which a number of "tangent curves" of equal rate of run-off are drawn to the same scales as the time–area curve to be tested. In Fig. 11 these curves were plotted to the formula $R = \dfrac{5 \cdot 9}{t^{0 \cdot 625}}$ (an approximation to the once-a-year storm according to Bilham's formula) as follows. Commencing first with a run-off of 1000 cubic feet per minute, this figure is first divided by 60·5. To find the point on the curve at 120 minutes, the result is divided by the appropriate rainfall figure (0·296), giving 55·8 impervious acres. Thus, the first point on the curve is *55·8 acres, 120 minutes*. The remainder of the "1000-cubic-feet-per-minute curve" is similarly plotted for all intervals of time down to 5 minutes. The remaining curves of 900 cubic feet per minute, 800 cubic feet per minute, etc., are very easily found by dividing the amplitude of the 1000-cubic-feet-per-minute curve into ten equal divisions at each interval of time, and similarly curves for quantities in excess of 1000 cubic feet per minute can be drawn by extrapolation of the same intervals above the curve. The resulting transparent overlay appears as in Fig. 11.

Once a transparency such as that in Fig. 11 has been prepared, velographs of the original can be taken, and the figure for run-off and impervious area altered by doubling, and doubling again, so as to give curves suitable for various calculations to various vertical scales.

The method of using Fig. 11 is as follows. Suppose it is required to find the effective time of concentration and resultant run-off for the time–area graph given in Fig. 10. The overlay is placed over the time–area curve with the origins of the two curves coinciding and the base lines coinciding. Then it will be seen at once that, if there is any marked irregularity in the form of the time–area curve, the convex part of the curve cuts a higher run-off than the end of the time–area curve. Next, the origin of the tangent diagram is slid up the time–area curve, the two base lines being kept parallel all the time. It will be observed that the convex part of the time–area curve cuts run-off curves of higher and higher values until a maximum is reached, when the origin of the tangent diagram has reached the point of maximum concavity of the time–area curve (see Fig. 12). If the origin is slid any farther beyond this point the run-off value reduces. Thus the maximum value is quickly

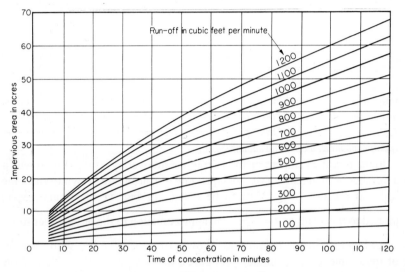

FIG. 11—"Tangent curves"

found, and can be read off with moderate accuracy directly from the tangent curve. But as the point of maximum concavity must be at one of the angles on the time–area curve, its exact position can be determined and, from the measurable values of time and area, the run-off can be accurately calculated. For example, in the case of Fig. 12, the point of maximum concavity occurs at *9·4 impervious acres, 26 minutes*, and the point that the tangent strikes is *18·2 impervious acres, 32 minutes*. This gives a difference of 18·2 − 9·4 = 8·8 impervious acres and 32 − 26 = 6 minutes, and the run-off is 60·5 × 8·8 × 1·92 = 1,022 cubic feet per minute.

Coleman and Johnson and Ormsby and Hart methods. The Coleman and Johnson method and the Ormsby and Hart method are means by which the run-off of either an actual or a hypothetical storm can be found for any catchment of which the data are available for the construction of a time–area graph. Both methods can be used to produce the same results, although the approaches are different. With Coleman and Johnson's system a time–area

SURFACE-WATER SEWERAGE: ELABORATIONS OF THEORY

graph does not have to be drawn, for each subsidiary catchment is dealt with separately. The Ormsby and Hart method consists of the application of a rainfall curve to a time–area curve (or tabulated data of such a curve). As usually applied, it gives the peak run-off only. It can, however, be used to give a run-off curve for the catchment which should agree exactly with that produced by the Coleman and Johnson method, provided, of course, the data are identical.

These methods have been used in the design of sewers with apparent satisfaction when applied to small areas. But as they necessitate the use of either "local typical" or "hypothetical" storms which are statistically unsound, they are not fit for general use in this connection.

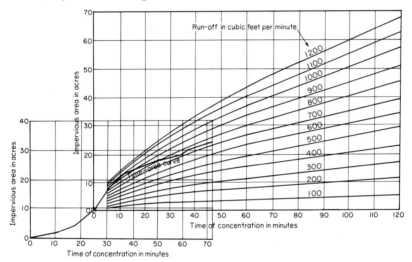

FIG. 12—Application of tangent curves to time–area graph

The Coleman and Johnson method, the Ormsby and Hart method and, more particularly, some modifications of the latter, despite their drawbacks from the design point of view, appear to be of use for research purposes, for they give with very fair accuracy the run-off curves that would result from storms falling on catchments, provided that:

1. Impermeable area and its distribution are accurately estimated.
2. Impermeability factor remains constant throughout the period of rainfall and run-off.*
3. Time of concentration and its components are accurately estimated.

Thus, if records of rainfall and run-off are taken with the aid of automatic gauges, theoretical run-off curves can be drawn, these compared with actual

---

* This never happens, but it was by calculating the theoretical run-off curve, as described at the end of this chapter, for actual storms and comparing with actual run-off curves, that the author was able to ascertain the usual rate of change of impermeability with time.

58 PUBLIC HEALTH ENGINEERING PRACTICE

run-off curves and informative facts ascertained by the agreements and disagreements of the two curves.

Coleman and Johnson method. Take, for example, a simple catchment which consists of two subsidiary catchments, the first having an impervious area of 5 acres, a time of concentration of 5 minutes and commencing to discharge at commencement of run-off for the total area; the second having an impervious area of 3 acres, a time of concentration of 6 minutes and commencing to discharge 1 minute after the start of run-off for the total area. Apply to this the heavy storm of 10 minutes' duration, the particulars of which are given in Table 15.

TABLE 15
EXAMPLE DATA

Minutes after start of storm	Rainfall intensity in inches per hour
1	2
2	4
3	7
4	5
5	3
6	3
7	2
8	1
9	1
10	1

Dealing first with subsidiary area 1: this is divided into part-areas in equal divisions of time. For simplicity in the present instance we will divide it into five parts of equal area and representing the parts which contribute one after another after each of the 5 minutes of the time of concentration.

The storm is then applied to these areas as shown in Table 16. In the first minute of run-off the first part-area discharges the rainfall due to the first minute of the storm, which has an intensity in this example of 2 inches per hour; in the second minute the second part-area discharges the rainfall due to the first minute of the storm and the first part-area discharges the rainfall due to the second minute of the storm. Thus, in the first minute the run-off is 2 inches per hour of rainfall over one-fifth of the total area: in the second minute 2 inches of rainfall over one-fifth of the area and 4 inches over another fifth, which is the equivalent of 6 inches in all over one-fifth of the total. This procedure is continued until the storm has passed over the whole area.

The figures in the "total" column multiplied by the total impervious acres, divided by the number of part-areas and multiplied by 60·5 give the run-off in cubic feet per minute. Thus it will be seen from the last column of Table 16 that the run-off builds up from zero to a maximum figure, and falls off to zero again, and the total time of run-off is equal to time of concentration for the subsidiary area plus duration of storm.

SURFACE-WATER SEWERAGE: ELABORATIONS OF THEORY

TABLE 16
RUN-OFF FROM SUBSIDIARY AREA 1

Minutes after start of run-off	Part-areas					Total $\left(\times \dfrac{5}{5} \times 60\cdot 5\right)=$	Run-off in cubic feet per minute
	1	2	3	4	5		
0						0	0·0
1	2					2	121·0
2	4	2				6	363·0
3	7	4	2			13	786·5
4	5	7	4	2		18	1089·0
5	3	5	7	4	2	21	1270·5
6	3	3	5	7	4	22	1331·0
7	2	3	3	5	7	20	1210·0
8	1	2	3	3	5	14	847·0
9	1	1	2	3	3	10	605·0
10	1	1	1	2	3	8	484·0
11		1	1	1	2	5	302·5
12			1	1	1	3	181·5
13				1	1	2	121·0
14					1	1	60·5
15						0	0·0

The same procedure is applied to subsidiary area 2, but in this instance as the run-off does not reach the point of outgo from the total area until 1 minute later than the first area, the position of the time scale is adjusted accordingly (see Table 17).

TABLE 17
RUN-OFF FROM SUBSIDIARY AREA 2

Minutes after start of run-off	Part-areas						Total $\left(\times \dfrac{3}{6} \times 60\cdot 5\right)=$	Run-off in cubic feet per minute
	1	2	3	4	5	6		
1							0	0·0
2	2						2	60·5
3	4	2					6	181·5
4	7	4	2				13	393·25
5	5	7	4	2			18	544·5
6	3	5	7	4	2		21	635·25
7	3	3	5	7	4	2	24	726·0
8	2	3	3	5	7	4	24	726·0
9	1	2	3	3	5	7	21	635·25
10	1	1	2	3	3	5	15	453·75
11	1	1	1	2	3	3	11	332·75
12		1	1	1	2	3	8	242·0
13			1	1	1	2	5	151·25
14				1	1	1	3	90·75
15					1	1	2	60·5
16						1	1	30·25
17							0	0·0

In Table 18 the run-offs from subsidiary areas 1 and 2 are added together.

TABLE 18
ADDITION OF RUN-OFF FROM SUBSIDIARY AREAS 1 AND 2

Minutes after start of run-off	Area 1, cubic feet per minute	Area 2, cubic feet per minute	Total, cubic feet per minute
0	0·0	0·0	0·0
1	121·0	0·0	121·0
2	363·0	60·5	423·5
3	786·5	181·5	968·0
4	1089·0	393·25	1482·25
5	1270·5	544·5	1815·0
6	1331·0	635·25	**1966·25**
7	1210·0	726·0	1936·0
8	847·0	726·0	1573·0
9	605·0	635·25	1240·25
10	484·0	453·75	937·75
11	302·5	332·75	635·25
12	181·5	242·0	423·5
13	121·0	151·25	272·25
14	60·5	90·75	151·25
15	0·0	60·5	60·5
16	0·0	30·25	30·25
17	0·0	0·0	0·0

The last column of this table gives the theoretical shape of the run-off curve, and the maximum figure in this column (given in bold figures) is the peak run-off which is selected when the Coleman and Johnson method is used for design purposes.

It will be seen from this simple example that the Coleman and Johnson method can be used to find the peak run-off for the most complex catchment area, but that the labour involved increases in direct proportion to the number of subsidiary catchments.

Ormsby and Hart method.* The time–area graph for the catchment is drawn. On a separate sheet of paper or a transparency, a curve of total rainfall against time is drawn, the time scale being the same as that of the time–area graph but opposite hand—i.e. it reads from right to left instead of from left to right. For convenience, rainfall can be converted from inches to cubic feet per acre by multiplying by the factor 3630.

To find the peak run-off, the steepest part of the rainfall curve is located over the steepest part (generally) of the time–area curve. The time–area curve is divided into convenient divisions of time—e.g. between the various angles in the curve resulting from the addition of the component time–area curves.

Then, during time t_1 the amount of rain which falls is R_1 cubic feet per acre, and this divided by t_1 gives cubic feet per minute per acre, which multiplied by A_1 gives cubic feet per minute. Similarly, the cubic feet per minute

* This should not be confused with Professor Ormsby's method which came before but is not described herein.

SURFACE-WATER SEWERAGE: ELABORATIONS OF THEORY 61

rates for times t_2 etc., are found, and all these rates are totalled to give the total run-off from the catchment.

Table 20 shows the working for the example illustrated in Fig. 9.

Hypothetical storm. The Ormsby and Hart method was originally proposed for use in design, and for this purpose a hypothetical storm was assumed in which the intensity rose from nil to a peak and fell off again in such a manner that during any period of time taken symmetrically about the peak of the storm the average rate of precipitation was the amount likely to fall in that time according to the selected rainfall formula. A curve was then plotted showing the total inches of rain precipitated at all times after commencement of storm and, for convenience, "inches of rain" converted to "cubic feet per acre." Figure 13 illustrates the curve of precipitation of such a

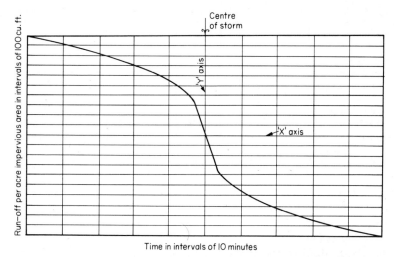

FIG. 13—Storm curve for use with Ormsby and Hart method

storm prepared from figures somewhat similar to those given in for the once-a-year storm in Table 6.* The ordinates and abscissae of this curve are given in Table 19. It will be noted that the slope of the curve is plotted in the opposite direction from the slope of the time–area graph. The reason for this is that (as will be realised on reflection after the Ormsby and Hart method has been studied) the rate of change of precipitation at the end of the storm affects the impervious area near the lowest part of the sewerage system, producing a rate of run-off which arrives at the point of outfall at the same time as the run-off due to an earlier phase of the storm falling on an upper

* This method has been used with various rainfall curves. In the example given the formula:

$$R = \frac{5 \cdot 9}{t^{0 \cdot 625}}$$

was used as an approximation to Bilham's once-a-year storm.

TABLE 19
HYPOTHETICAL STORM EMBODYING STORMS LIABLE TO OCCUR ONCE A YEAR

Minutes (X)	−60	−55	−50	−45	−40	−35	−30
Cubic feet per acre (Y)	+1074	+1042	+1004	+964	+922	+879	+829

Minutes (X)	−25	−20	−15	−10	−7·5	−5	−2·5
Cubic feet per acre (Y)	+774	+711	+639	+549	+495	+424	+327

Minutes (X)	+2·5	+5	+7·5	+10	+15	+20	+25
Cubic feet per acre (Y)	−327	−424	−495	−549	−639	−711	−774

Minutes (X)	+30	+35	+40	+45	+50	+55	+60
Cubic feet per acre (Y)	−829	−879	−922	−964	−1004	−1042	−1074

$X =$ time in minutes before and after centre of storm
$Y =$ rainfall or run-off in cubic feet per acre above and below average

portion of the sewerage system. With a symmetrical storm as illustrated in Fig. 13 this plotting in the reverse direction does not matter, but the Ormsby and Hart method is intended for use not only with hypothetical storms, but as a means of interpreting the effects of actual storms as recorded, in which case the direction of plotting does matter.

When the time–area graph has been completed, the steepest part of the rainfall curve is arranged above the steepest part of the time–area curve generally (see Fig. 9, in which the relevant portion of the rainfall curve is centred over the steepest part of the time–area curve). Here a certain amount of judgement has to be used, involving a possibility of error. For example, in Fig. 9 the steepest part of the rainfall curve is above the steepest part of the time–area curve, but it is doubtful whether one could say that it was strictly over the steepest part generally, or over the "centre of steepness."

The procedure is then as follows. Lines are projected up from every point of change of slope of the time–area curve until they meet the rainfall curve, and the calculation tabulated as in Table 20.

Then, the increment of run-off R in cubic feet per acre is divided by increment time t, giving cubic feet per minute per acre. This multiplied by increment acres impervious area Ap gives cubic feet per minute. Commencing with the first section of sewer from outfall to A_1, we have $R_1 = 65$ cubic

SURFACE-WATER SEWERAGE: ELABORATIONS OF THEORY   63

feet per acre, which, divided by 0·7 minute, gives 93 cubic feet per minute per acre, which, multiplied by 0 acre, gives 0 cubic feet per minute: thus it will be seen that the length of sewer A_1 to outfall, having no lateral contributing area, does not affect the calculation. The next section, A_2 to A_1, is treated similarly: R_2 is scaled and found to be 400 cubic feet per acre. This divided by t_2 (1·5 minutes) gives 266·6 cubic feet per minute per acre, and multiplied by Ap_2 (0·8 acre) gives 213·3 cubic feet per minute. The values of the remaining intervals of time are treated similarly and tabulated. Finally, the last column, giving cubic feet per minute, is totalled, the total being the peak run-off from the whole area.

The calculation has been made with a rainfall curve which has been plotted for all values of time between 5 and 120 minutes, and the form of the

TABLE 20

ORMSBY AND HART CALCULATION

R	t	$\dfrac{R}{t}$	Ap	$\dfrac{ApR}{t}$
65	0·7	93·0	0·0	0·0
400	1·5	266·6	0·8	213·3
140	2·0	70·0	0·9	63·0
160	5·0	32·0	Nearly 0·8	25·6
8	0·3	26·6	Negligible	Negligible
			Total c.f.m.	301·9

curve for times of duration of less than 5 minutes found by extrapolation. This extrapolation suggests that for an infinitely short period of time precipitation is at an infinite rate. This does not happen and it is usual to assume that the maximum intensity for which allowance should be made is the intensity of the 5-minute storm.* Hence, engineers may prefer to use the modified curve shown in Fig. 13 and represented by a dotted line in Fig. 9.

Theoretical run-off curve or hydrograph. If it is desired not merely to know the peak run-off but to draw the run-off curve, the rainfall curve is applied to the time–area curve in a number of positions, from the time-origin of the rainfall curve being placed over the time-origin of the time–area curve, to the end of the rainfall curve being placed over the end of the time–area curve. From the run-off obtained in this way the run-off curve is plotted.

It is, of course, not necessary to draw the rainfall curve for this latter purpose: rainfall can be given in the form of a table. Similarly, the time–area curve can be converted to a table.

Returning to the example used in the Coleman and Johnson calculation, the first and second columns of Table 21 represent the time–area graph, and

* The author now considers that the greatest run-off is that due to the 11-minute storm, see Chapter III.

TABLE 21
TIME–AREA GRAPH DATA

Minutes after start of run-off	Impervious acres contributing	Increment impervious acres
1	1·0	1·0
2	2·5	1·5
3	4·0	1·5
4	5·5	1·5
5	7·0	1·5
6	7·5	0·5
7	8·0	0·5

the third *increment* impervious acres for each minute of the time of concentration. If the increment rainfall is applied according to the Ormsby and Hart method, the time scale of the rainfall curve being reversed, the run-off each minute can be found. This is done in Table 22, except that, as time intervals are regular, *intensity* of rainfall, as given in Table 15, is used. Table 22 gives increment impervious area in first column of each and gives intensity of rainfall in second and product in third. The totals of the products multiplied by 60·5 give the run-off curve as shown in Table 23. It will be seen that this run-off curve is exactly the same as that obtained by the Coleman and Johnson method; it gives the same peak run-off and again the run-off lasts for the time of concentration plus the duration of storm.

Hydrograph methods. It has been suggested that unit hydrograph and similar methods based on rainfall and run-off records as normally used for river works design could be applied with advantage to the design of surface-water sewers. A little reflection will show that this is not the case. The majority of calculations of flow in surface-water sewers are for sewers or sewerage systems *that do not exist* and therefore obviously it is not possible to gauge the flow in those systems. The time-of-concentration methods are therefore applied to assumptions based on observations made on other sewerage systems in which circumstances were similar and on which has been built up the body of data available for design purposes. Furthermore, the fact that sewers are constructed to circular or other simple forms and laid to predetermined gradients makes possible, with the aid of techniques more accurate than the hydrograph methods, much more certain determination of flow in them than is possible in the case of rivers of varied cross-section and in which the flows may depend very largely on the extent to which flooding takes place or can be permitted.

Storage of surface water. More questionable theory has been written about storage of surface water in sewers than perhaps in connection with any other branch of public health engineering. It is therefore necessary for the

TABLE 22
HYDROGRAPH METHOD CALCULATION

1st minute	2nd minute	3rd minute	4th minute
1·0 × 2 = 2·0	1·0 × 4 = 4·0	1·0 × 7 = 7·0	1·0 × 5 = 5·0
1·5	1·5 × 2 = 3·0	1·5 × 4 = 6·0	1·5 × 7 = 10·5
1·5	1·5	1·5 × 2 = 3·0	1·5 × 4 = 6·0
1·5	1·5	1·5	1·5 × 2 = 3·0
1·5	1·5	1·5	1·5
0·5	0·5	0·5	0·5
0·5	0·5	0·5	0·5
Total 2·0	Total 7·0	Total 16·0	Total 24·5

5th minute	6th minute	7th minute	8th minute
1·0 × 3 = 3·0	1·0 × 3 = 3·0	1·0 × 2 = 2·0	1·0 × 1 = 1·0
1·5 × 5 = 7·5	1·5 × 3 = 4·5	1·5 × 3 = 4·5	1·5 × 2 = 3·0
1·5 × 7 = 10·5	1·5 × 5 = 7·5	1·5 × 3 = 4·5	1·5 × 3 = 4·5
1·5 × 4 = 6·0	1·5 × 7 = 10·5	1·5 × 5 = 7·5	1·5 × 5 = 7·5
1·5 × 2 = 3·0	1·5 × 4 = 6·0	1·5 × 7 = 10·5	1·5 × 3 = 4·5
0·5	0·5 × 2 = 1·0	0·5 × 4 = 2·0	0·5 × 7 = 3·5
0·5	0·5	0·5 × 2 = 1·0	0·5 × 4 = 2·0
Total 30·0	Total 32·5	Total 32·0	Total 26·0

9th minute	10th minute	11th minute	12th minute
1·0 × 1 = 1·0	1·0 × 1 = 1·0	1·0	1·0
1·5 × 1 = 1·5	1·5 × 1 = 1·5	1·5 × 1 = 1·5	1·5
1·5 × 2 = 3·0	1·5 × 1 = 1·5	1·5 × 1 = 1·5	1·5 × 1 = 1·5
1·5 × 3 = 4·5	1·5 × 2 = 3·0	1·5 × 1 = 1·5	1·5 × 1 = 1·5
1·5 × 3 = 4·5	1·5 × 3 = 4·5	1·5 × 2 = 3·0	1·5 × 1 = 1·5
0·5 × 5 = 2·5	0·5 × 3 = 1·5	0·5 × 3 = 1·5	0·5 × 2 = 1·0
0·5 × 7 = 3·5	0·5 × 5 = 2·5	0·5 × 3 = 1·5	0·5 × 3 = 1·5
Total 20·5	Total 15·5	Total 10·5	Total 7·0

13th minute	14th minute	15th minute	16th minute
1·0	1·0	1·0	1·0
1·5	1·5	1·5	1·5
1·5	1·5	1·5	1·5
1·5 × 1 = 1·5	1·5	1·5	1·5
1·5 × 1 = 1·5	1·5 × 1 = 1·5	1·5	1·5
0·5 × 1 = 0·5	0·5 × 1 = 0·5	0·5 × 1 = 0·5	0·5
0·5 × 2 = 1·0	0·5 × 1 = 0·5	0·5 × 1 = 0·5	0·5 × 1 = 0·5
Total 4·5	Total 2·5	Total 1·0	Total 0·5

TABLE 23
HYDROGRAPH METHOD CALCULATION

Minutes after start of run-off	Totals from Table 56 × 60·5 =	Cubic feet per minute
0	0·0 × 60·5	0·0
1	2·0 × 60·5	121·0
2	7·0 × 60·5	423·5
3	16·0 × 60·5	968·0
4	24·5 × 60·5	1482·25
5	30·0 × 60·5	1815·0
6	32·5 × 60·5	**1966·25**
7	32·0 × 60·5	1936·0
8	26·0 × 60·5	1573·0
9	20·5 × 60·5	1240·25
10	15·5 × 60·5	937·75
11	10·5 × 60·5	635·25
12	7·0 × 60·5	423·5
13	4·5 × 60·5	272·25
14	2·5 × 60·5	151·25
15	1·0 × 60·5	60·5
16	0·5 × 60·5	30·25
17	0·0 × 60·5	0·0

student to approach the matter with caution. Storage calculations are, however, of considerable value when they relate to storage of storm water in ponds, tanks, tank-sewers, the suction wells of pumping-stations, etc., and knowledge of the relevant theory can save very considerable time that might otherwise be spent on laborious calculations.

There is a not uncommon belief that the time of concentration methods—the Lloyd-Davies method, and the American "rational" methods and their elaborations—do not allow for storage in the sewerage system, and a number of theories of varying degrees of absurdity have been based on this belief. The fact is that the time-of-concentration methods are all methods of estimating the effect of storage and, in the case of the Lloyd-Davies and rational methods, relating this to rainfall or run-off statistics.

To anticipate captious criticism, it must be mentioned that none of the time-of-concentration methods is absolutely accurate. Simplifying assumptions are necessary to avoid making these methods too ponderous to be of practical use. For example, time-of-concentration is calculated on the assumption that velocities are those which would occur were the sewers running full or half-full throughout—a condition which does not apply to the normal case. Also, no allowance is made for such draw-down as may be expected to occur at places where diameter increases. Observations and example calculations show that the errors introduced by these assumptions are almost negligible and, in the majority of instances, the actual time-of-concentration (and consequently the storage in a normally designed sewerage system) is slightly less, not more, than as calculated.

There are also special cases, first, tank-sewers and other storage tanks with module, penstock or orifice outlet control to which time-of-concentration methods do not necessarily apply because the restriction of flow is not that due to the hydraulic characteristics of the sewers, and there are ancient and other improperly designed sewerage systems in which, it has been considered, adventitious storage takes place as a result of a network of oversized sewers of inadequate gradient upstream and a restriction of outflow due to an inadequate sewer leading from this network. Such a circumstance could well exist although none has been proved to produce appreciable reduction of outgoing flow. It must, however, be understood that these conditions where storage is induced by abnormal circumstances are entirely different from those in a properly designed sewerage system in which the storage can be

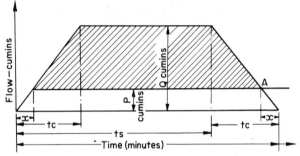

Fig. 14—Diagram illustrating basis of storm-water storage calculation
By courtesy of Mr. B. A. Copas and the Institution of Public Health Engineers

expected to be virtually the same as calculated by a suitable time-of-concentration method. What certainly is not rational is to take a sewerage system of normal design, find the effect of storage by means of a time-of-concentration method and then assume a reduction of outflow on account of a nebulous and purely arbitrary storage over and above that made possible by the hydraulic conditions or allowed for by the method used.

This is what has been done in more than one instance, where the theorists, not understanding the basic theory, have found the effect of storage by time-of-concentration methods and then made allowance for additional storage which they think must be present. As, however, in such circumstances it has not been possible for them to calculate storage, they have made some purely arbitrary allowance—for example, in one instance, that the storage is such that the outflow as calculated is always reduced to two-thirds and, in another, that the proportional depth of flow throughout the sewerage system is always the same as that at the point of outgo—an unlikely circumstance.

Storage in a surface-water sewerage system is *not* the total enclosed space or any arbitrary portion thereof, but only that part of the space that *can be occupied* by sewage in the hydraulic conditions prevailing. The storage that can take place must be below the surface of the hydraulic gradient: it depends on incoming flow and resistance to flow by friction and velocity-head losses. If there is no incoming flow there is nothing to store: if there is no resistance to flow there is no storage, because the surface-water runs out as soon as it enters. To find the storage space occupied, flow and resistance must be related together by calculation, *and that is what the Lloyd-Davies method and other time-of-concentration methods do.*

Time-of-concentration bears a similar relation to rate of flow and storage occupied in sewers as does the detention period of a sedimentation tank to

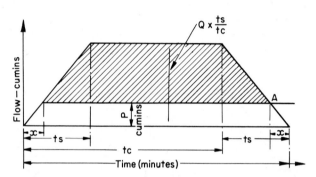

FIG. 15—Diagram illustrating basis of storm-water storage calculation
By courtesy of Mr. B. A. Copas and the Institution of Public Health Engineers

tank capacity and rate of flow. When time-of-concentration equals duration of storm, this relation is expressed:

$$tc = \frac{C}{q}$$

where: tc = time-of-concentration in minutes
C = cubic feet of storage occupied at end of time-of-concentration
q = one-half the peak rate of run-off, or $30·25 \, ApR$.

Storage of storm water in tanks. The problem of the amount of storage capacity required to store storm water from a catchment of known area and allowing for a predetermined rate of outflow from the storage was first investigated by the writer, who arrived at approximate expressions for conditions relating to the once-a-year storm by working out families of cases, plotting the results on logarithmic paper and finding the equations of the curves. Later, mathematical examination was made of these results and B. A. Copas made a full investigation of the problem, including the effect of time of concentration of the catchment. Continuing on the basis of Copas' work, the writer checked his equations and found general formulae applicable to storms of all frequencies of occurrence. The following is Copas' approach:[48]

"Consider a catchment area of Ap impermeable acres with a time-of-concentration of t_c minutes. It is desired to find the storage volume necessary when the outflow

SURFACE-WATER SEWERAGE: ELABORATIONS OF THEORY

rises sharply at the beginning of a storm to P cumins and thereafter remains at this value until the storm has passed.

Let
$$t_s = \text{duration of storm in minutes}$$
and
Q = peak flow from catchment area due to storm of duration t_s minutes.

Assuming that the area contributing to the point where storage is required increases at a uniform rate during the time of concentration, the flow cycle will be as shown in Fig. 14. It will be seen that after x minutes from the commencement of the storm the inflow exceeds the outflow and water will continue to pass into storage until point A is reached, which is x minutes before the cessation of inflow. The storage volume is represented by the hatched portion of Fig. 14. Total volume of inflow up to point A

$$= Qt_s - \frac{xP}{2}$$

Total volume of outflow up to point A

$$= P(t_s + t_c) - \frac{xP}{2} - \frac{2xP}{2}$$

Therefore storage volume
$$C = \text{inflow volume less outflow volume}$$
$$= Qt_s - P(t_s + t_c) + xP$$
But by similar triangles
$$\frac{x}{P} = \frac{t_c}{Q} \text{ and } x = \frac{Pt_c}{Q}$$

Substituting for x

$$C = Qt_s - P(t_s + t_c) + \frac{P^2 t_c}{Q} \qquad . \qquad . \qquad . \qquad . \qquad (1)$$

This is the general expression for storage which may be used for any rainfall intensity-time curve desired, and if no other better method suggests itself the maximum value of C can be found by trial and error insertion of various values of t_s.

Now consider cases where t_s is less than t_c; in other words, rainfall will have ceased before all the area of Ap acres is able to contribute (see Fig. 15).

As in the previous example, let Q = peak flow from the area of Ap acres due to storm of duration t_s, *assuming* the whole area could contribute by the end of time t_s. But the whole area *cannot* contribute by the end of time t_s and peak flow

$$= Q \times \frac{t_s}{t_c}$$

Proceeding as before, volume of inflow up to point A

$$= \frac{Qt_s}{t_c} \times t_c - \frac{xP}{2}$$

volume of outflow up to point A

$$= P(t_c + t_s) - \frac{3xP}{2}$$

Therefore
$$\text{Storage } C = Qt_s - P(t_c + t_s) + xP$$

But
$$\frac{x}{P} = \frac{t_s}{Qt_s \over t_c} \text{ or } x = \frac{Pt_c}{Q} \text{ and } C = Qt_s - P(t_c + t_s) + \frac{P^2 t_c}{Q}$$

This expression for C is identical to equation (1) and therefore the one expression is applicable to all values of t_c and t_s.

It is now proposed to consider the application of this general-storage formula to various storm curves of the type

$$R = \frac{k}{t_s^n}$$

where R is rainfall intensity in inches per hour. The form of this expression is most convenient for storage calculations and by selecting appropriate values of k and n, close approximations can be made to Bilham's formula over quite large ranges of t_s. Escritt uses a formula of this kind,

$$R = \frac{5 \cdot 9}{t_s^{0 \cdot 625}}$$

as a "once-a-year" storm curve and it is interesting to examine in some detail the general-storage formula as applied to this rainfall intensity curve.

As
$$R = \frac{5 \cdot 9}{t_s^{0 \cdot 625}}, \quad Q = \frac{5 \cdot 9 \times 60 \cdot 5 Ap}{t_s^{0 \cdot 625}} = \frac{356 \cdot 95 Ap}{t_s^{0 \cdot 625}}$$

and
$$C = 356 \cdot 95 A p t_s^{0 \cdot 375} - P(t_s + t_c) + \frac{P^2 t_c t_s^{0 \cdot 625}}{356 \cdot 95 Ap} \qquad . \qquad . \qquad (2).\text{''}$$

The values of k and n suggested by Copas were as follow:

Storm once in 1 year $k = 6 \cdot 18 \quad n = 0 \cdot 645$
,, ,, 2 ,, $k = 8 \cdot 511 \quad n = 0 \cdot 662$
,, ,, 5 ,, $k = 12 \cdot 22 \quad n = 0 \cdot 677$
,, ,, 10 ,, $k = 15 \cdot 63 \quad n = 0 \cdot 685$
,, ,, 20 ,, $k = 20 \cdot 28 \quad n = 0 \cdot 693$
,, ,, 30 ,, $k = 23 \cdot 44 \quad n = 0 \cdot 698$

Copas continued to produce a method by which storage capacity could be obtained from graphs based on his calculations.

The difficulty in the mathematical solution is that the equations must be solved laboriously by trial and error, unless the graphs mentioned above are available. The writer therefore continued from Copas' expression (2), but substituting k for $5 \cdot 9$ and n for $0 \cdot 625$ and letting $t_c = 0$, it was found that for as long as $t_c = 0$, $\frac{Q}{P} = \frac{1}{1-n}$. This equation also showed that the power of t_s must always be less than unity, for if it became unity, then P became zero which meant that a storage tank receiving flow from a sewerage system need have no outlet, for it could store all the rain that fell in geological time,

SURFACE-WATER SEWERAGE: ELABORATIONS OF THEORY

which was absurd. For similar reasons the Ministry of Health rainfall formula was entirely useless in connection with storage calculations.

Continuing,

$$\begin{aligned}C &= (Q-P)\,t_s \\ &= (Q-P)\left(\frac{60 \cdot 5\,Apk}{Q}\right)^{\frac{1}{n}} \\ &= \frac{n\,(60 \cdot 5\,Apk)^{\frac{1}{n}}}{Q^{\frac{1-n}{n}}} \\ &= \left[(60 \cdot 5\,Apk)^{\frac{1}{n}} \times n\left(\frac{1-n}{P}\right)^{\frac{1-n}{n}}\right]\end{aligned}$$

It was from here onwards that Copas and the writer diverged, for Copas found the true mathematical expression for the storm of particular duration and intensity that called for the maximum storage for any particular time-of-concentration of sewerage system, but in so doing arrived at equations that were difficult to solve and necessitated the use of graphs, while, partly by mathematics and partly by the process of "breaking down" the problem, the writer evolved formulae which, by introducing negligible error, simplified the calculation. These read:

$$C = \frac{3354 \cdot 5\,Ap^{\frac{3}{2}}\,N^{\frac{1}{2}}}{P^{\frac{1}{2}}} - \frac{2}{3}Pt_c$$

$$C = 423 \cdot 5\,Ap\,N^{\frac{1}{3}}t_c^{\frac{1}{3}} - 2Pt_c + \frac{P^2 t_c^{\frac{5}{3}}}{423 \cdot 5\,Ap\,N^{\frac{1}{3}}}$$

where: C = required storage capacity in cubic feet

Ap = impervious area, i.e. area of catchment in acres multiplied by an impermeability factor

N = number of years between occurrence of storms of appropriate intensity (e.g. should the engineer design on the storm liable to occur once every three years, $N = 3$)

P = rate of pumping or outgo from the storage well in cubic feet per minute

t_c = time-of-concentration, i.e. time for storm water to flow from the farthest part of the catchment to the pumping station, in minutes.

The required storage capacity should be calculated by each of the formulae in turn and whichever gives the higher value for storage is applicable.

In many instances time-of-concentration is negligible and then it is necessary only to use the formula:

$$C = \frac{3354 \cdot 5\,Ap^{\frac{3}{2}}\,N^{\frac{1}{2}}}{P^{\frac{1}{2}}}$$

Storage ponds. Storage in ponds is a subject of study which has been revived since the Second World War. It is understood that the interest in this matter arose out of the attempts made by certain of the New Towns to placate river boards when it was found that the run-off from the new sewerage systems would seriously overload the existing streams. In one instance the town was to be built mainly on chalk where there were no watercourses and formerly the whole of the precipitation was either evaporated or soaked into the ground, and where, in the author's opinion wrongly, it was decided to install surface-water sewers to take the run-off from all roofs and paved surfaces and to discharge to a small stream draining an insignificant area of clay in a valley. Naturally, this would have meant very serious overloading of the stream and, although such discharge could at that time have been made legally without hindrance, it was decided to provide a system of ponds capable of delaying the run-off sufficiently to prevent the stream from flooding.

This practice has been copied in somewhat similar circumstances, and it would appear that the innovation is likely to survive in view of a change in the law. For under the *Rivers (Prevention of Pollution) Act*, 1951, the consent of the river board is required for the bringing into use of any new or altered sewage outlet or discharge and this includes outlets from the surface-water sewers of a local authority. (No mention, however, is made of surface-water discharges by private parties or New Towns.)

Storage ponds were used in Denmark in the 19th century and these, which included storage tanks and tank-sewers, were first described by Heydt in 1908, who classified them into four types as follow:

1. Those placed at the upper ends of sewerage systems to balance the run-off from undeveloped areas upstream in circumstances where local watercourses were connected into sewers.

2. Those at the points of connection of new sewerage systems to existing sewerage systems.

3. Those provided in new sewerage systems to reduce the cost of sewers.

4. Those constructed at the points where sewerage systems discharged to watercourses. The last included storage tanks before sewage treatment works.

Where a storage pond is constructed so as to limit the discharge to a stream to a maximum figure that it is considered the stream can accept, it is desirable to fix a module* which will discharge at a constant rate no matter what the depth of water is in the pond. If a module is not provided the discharge would vary between the maximum and minimum according to the head over the outfall pipe and twice the storage capacity would be required. If a module is provided, the capacity of the pond can be calculated according to the foregoing theory.

* See flow-regulating devices in Chapter XXV.

The practice of constructing storage ponds for surface water is, to say the least, controversial. Storage tanks for soil- or combined-sewage are generally not desirable except when combined with pumping-stations, sewage-treatment works or sea outfalls.

CHAPTER V
STORM-RELIEF WORKS

THE sizes to which combined and partially-separate sewers have been constructed at various times have depended on the extent to which storm water could be discharged to natural watercourses without treatment, according to current practice and official opinion. At no time has practice ever been generally up to the standards considered desirable, for always compromises have had to be made between the ideal and what was at the time practicable.

Early practice. Prior to the publication of the Reports of the Royal Commission on Sewage Disposal the Local Government Board adopted what was even then considered a compromise, that three times the dry-weather flow should be fully treated at the sewage works, between three and six times the dry-weather flow partially treated through what were described as streaming storm-water filters, and the remainder discharged as storm water without treatment. It may be this early practice which was the forerunner of a present-day misconception shared by many engineers, namely, that at sewage-treatment works three times dry-weather flow should be fully treated, an additional three times passed to storm tanks and the remainder discharged at the sewage works without treatment. This is not, in fact, permissible and not good policy from any point of view.

The Royal Commission did not approve the discharge of any sewage *at treatment works* without some treatment and, in the following words in the Fifth Report, set out what became the basis of practice:

> "In our opinion, the difficulties of dealing with the increase of flow, due to storms, can best be met by the provision of stand-by tanks to receive the storm sewage in excess of that which can be passed through the ordinary tanks. The main overflow at the works should be from the stand-by tanks and not from the outfall sewer, so that no sewage may escape without some settlement."

On the matter of storm overflows on sewers, the Royal Commission made as specific recommendations the following:

> "It is probably impracticable to dispense altogether with storm overflows on branch sewers, but in our opinion these should be used sparingly, and should usually be set so as not to come into operation until the flow in the branch sewer is several times the maximum normal dry-weather flow in the sewer.
>
> No general rule can be laid down as to the increase in flow which should occur in the branch sewers before sewage is allowed to pass away by the overflow untreated. The Rivers Board, or in districts where there is no Rivers Board, the County Council, should have power to require the local authority to alter any storm over-

flows which, in their opinion, permit of an excessive amount of unpurified sewage to flow over them. The local authority should have the right to appeal to the Central Authority in any case in which they consider that the requirement of the Rivers Board is unreasonable or impracticable of fulfilment.

The general principle should be to prevent such an amount of unpurified sewage from passing over the overflow as would cause nuisance."

In spite of the Royal Commission recommendations, it is understandable that the previous practice of the Local Government Board should set some precedent when the discharge of storm overflows anywhere on sewerage systems was being considered. This much is certain: for a considerable time and to some extent to the present day, practice has been to discharge over storm weirs all flows exceeding six times dry-weather flow. At one time it was generally held that the "requirements" of the Ministry of Health relating to storm overflows were:

"Storm overflows on sewers should as far as reasonably practicable be avoided, but where they are necessary they should be placed in such positions and with the weirs so fixed that no nuisance is likely to result. In any district where there is an active river authority the Ministry will desire to be informed of the opinion of such authority in respect of any proposed overflow.

In the absence of any special circumstances overflow weirs should be fixed so as not to come into operation until the flow exceeds six times that of the average dry-weather flow.

There should be no overflow for untreated sewage or storm water at or near the disposal works." [224]

Tendency of modern practice. Since that time it has been established that any interested party such as a riparian owner can take legal action against the local authority that discharges sewage or sewage effluent to an inland watercourse so as to cause pollution. Moreover, the *Public Health Act, 1936*, contained the following sections:

30. Nothing in this Part of this Act shall authorise a local authority to construct or use any public or other sewer, or any drain or outfall, for the purpose of conveying foul water into any natural or artificial stream, watercourse, canal, pond or lake, until the water has been so treated as not to affect prejudicially the purity and quality of the water in the stream, watercourse, canal, pond or lake. *[Sewage, etc., to be purified before discharge into streams, canals, etc.]*

31. A local authority shall so discharge their functions under the foregoing provisions of this Part of this Act as not to create a nuisance. *[Local authority not to create any nuisance]*

The only exception to the above is a section in the *London Government Act*, 1963, to the effect that the Greater London Council may cause or permit storm water to be discharged into the Channelsea River or Abbey Creek in the London Borough of Newham at such times and in such manner as may be necessary to prevent flooding within the sewerage area of the Greater London Council and provided the Council shall take steps to avoid the creation of any nuisance in these rivers.

From the foregoing it would appear that it would now not be considered good practice or legal to discharge any storm water except after settlement in storm tanks. As to whether it is practicable to dispense altogether with storm overflows on branch sewers, the writer can say that, in forty-four years practice, *he has never had to consider putting an overflow for untreated storm water so as to discharge to inland waters.*

That the installation of storm overflows is more a matter of habit or bad practice than necessity is illustrated in the sewerage area of the Greater London Council. As described in Chapter I, the use of storm overflows by the London County Council originated with Bazalgette's scheme to intercept the dry-weather flows of existing sewers but to spill-over the storm water into the River Thames. Nevertheless of 220 known storm overflows in Greater London, 87 only are in the old London County Council catchment. The greater number have been installed in an area of Middlesex where it would appear to have been the practice of several local authorities to provide sewers of inadequate size and spill-over storm water wherever a stream was encountered. In addition to these known overflows there are many more that have not been located, for it was the practice of some boroughs to have manholes which jointly served both soil- and surface-water sewers and arranged in such a manner that surcharge of one sewer would cause discharge to the other.

On the south of the Thames conditions are very different. There appear to have been no more than about seven storm overflows belonging to the former Metropolitan Boroughs and nine in the outer London boroughs now in the sewerage area of the Greater London Council. Broadly, all these local authorities south of the Thames and the Metropolitan Boroughs north of the Thames provided some 2500 miles of sewers big enough to take the whole of the soil sewage and rainfall run-off in their districts, passing the flows to the London County Council for disposal. The London County Council, perhaps as a result of tradition, failed to spend that much extra to provide the 25 or so miles of large sewer that would be necessary to take these flows to the outfall works.

The policy of putting in storm overflows to relieve a sewerage system, instead of providing combined sewers of sufficient size to carry the sewage to the treatment works, would appear to be one of making the cost of sewerage schemes seem low but thereby involving someone else in expenditure. As has already been mentioned in Chapter I, the combined sewerage system involves the least cost on the community if one takes into account the cost of the drains provided by the property developer, the cost of the sewers and the cost of improvement of watercourses to which discharge of storm water is made.

Storm-tank policy. The tendency would now appear to be that, where it would be unduly expensive to carry the whole of the storm water to the sewage-treatment works or sea outfall, discharge can be made to inland waterways via properly constructed storm tanks. This procedure has been adopted in some instances with the approval of the Ministry and could well be adopted in all future cases.

As the result of changes in sewerage arrangements from time to time, it has come about that in Greater London there are storm tanks not only at nearly all the sewage-treatment works but also in other localities where sewage works no longer exist or never have existed. For example, there are storm tanks at two localities in Epsom and Ewell, one locality in Kingston, one in Brent, one in Ealing, two in Waltham Forest, one in Newham and one in Barking. It is very probable that most of these will have to be retained for several years and it may even prove desirable to consider them as permanent installations.

Wherever they are located, storm tanks serve two purposes. First, they store and return to flow for full treatment the numerous spill-overs due to lesser storms. Second, when the storm is of such magnitude as to cause them to spill over their outlet weirs, they give settlement to the sewage. Consequently storm tanks must have a very considerable effect on reduction of pollution of streams.

Where storm tanks are located away from sewage-treatment works, they cannot be manned day and night and therefore need to be designed for automatic operation. The procedure would then be as follows. Storm tanks, during rainfall days will:

(*a*) partially fill during a storm and then require emptying and finally desludging to the intercepting sewer after rainfall;

(*b*) fill completely and spill over to the watercourse and then require emptying and desludging.

Simple automation of both these procedures can be effected by having inlet to the tank by discharge over weirs and return to sewers via tide flaps. The procedure would then be:

1. On storm flow exceeding design capacity of outgoing intercepting sewer, sewage would back-up and, unable to pass tide flaps, would pass over weirs into storm tanks.

2. At end of storm, level of sewage in intercepting sewer would fall below that in tanks and contents of tanks would pass via tide flaps to sewer. The opening of flaps would operate switches starting the appropriate desludging mechanisms.

A recording mechanism, showing whether or not there had been any failure of the procedure, would be required in order that visiting staff could take any necessary action. Such an arrangement should eliminate need for local staff, because failure to sludge on any one day would not be a serious matter.

For the value and capacity of storm tanks see Chapter XXV.

Storm-overflow chambers. Storm overflows were formerly always in the form of side weirs—that is, the weirs were constructed parallel to the direction of flow. The design problem involved was so to proportion and arrange the weirs that overflow commenced at the desired rate of flow, and that at the maximum rate of flow the head over the weirs was not so great that an excessive quantity was forced down the sewer to be relieved. A number of methods were applied to effect this. In the first place, it had been pointed out that high-velocity flow parallel to the weirs had the effect of giving inaccurate gauging, and formulae were prepared to make allowance for this. In practice, however, it was satisfactory to rely on the standing-wave and turbulence produced by reduction of velocity of the flow entering

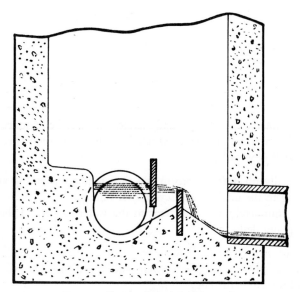

Fig. 16—Simple storm-overflow weir

the chamber from the smaller-diameter sewer, and to install weirs of calculated length, together with a baffle that would prevent too much sewage from passing down the sewer that was to be relieved.

A more recent design of storm-overflow chamber is the stilling-bay type in which a hopper-bottomed tank is constructed to reduce the velocity of flow. The flow to be passed on for full treatment is taken from the bottom of the tank via an orifice or pipe of the size required to restrict the flow. Storm water spills over peripheral weirs.

All storm weirs must be provided with scum-boards or other means of preventing large solids from being discharged.

Fig. 16 illustrates in outline a simple form of storm-overflow weir. More elaborate constructions become necessary where large sewers are concerned, and at sewage-treatment works. The latter case is dealt with under the heading of storm-water separation, in the section on sewage treatment.

Storm weirs can be simplified when the flow of sewage that is not to be discharged as storm water has to be pumped. In this case the maximum capacity of the pumps is so arranged as to be exactly equal to the rate of flow to be discharged to the treatment works. Any excess on this quantity backs up in the suction well, and can be passed either over short-length weirs to a storm-relief sewer or to pumps installed to discharge storm water.

Storm-relief sewers. Storm-relief sewers are sewers constructed so as to connect from storm-overflow chambers to selected points of discharge (with storm tanks) to natural waters. Their capacities are calculated according to a time-of-concentration method as hereinbefore described.

The outfalls of storm-relief sewers should be constructed in such a manner that there is no possibility of danger or damage. Although storm sewers or tanks discharge their maximum flows infrequently, the volume and velocity of discharge can be very great, and may damage the banks of streams to which the flow is passed. For this reason the direction of outflow should be well considered, being preferably in the direction of flow of the stream, and not towards the opposite bank. The outfall should be constructed very solidly of concrete or brickwork, and should have deep foundations, so as not to be undermined by the flows of the stream or the sewage. Generally, a tide flap should be provided, more for the purpose of preventing the access of vermin and small boys than preventing back-flow. It also deflects flow downwards. This is preferable to the provision of a screen or grid, which may collect debris and lead to chokage.

It is sometimes advisable to protect the bed of the stream, in which case the protective work is best in the form of an apron with deep foundations at the edges: aprons without deep foundations become undermined and break up.

Intercepting sewers. An intercepting sewer is a sewer which is constructed for the purpose of diverting a flow or supplementing an existing sewer at times of maximum flow. Intercepting sewers are used in connection with both combined and separate soil sewers, more rarely in connection with surface-water schemes. Overflow weirs or take-off chambers are provided at the head of an intercepting sewer and arranged to pass the quantity of sewage that the intercepting sewer is designed to accommodate. The lower end of the intercepting sewer may discharge back into the sewer which it has relieved at a point where the flow can be accommodated or, alternatively, to another soil or combined sewer.

CHAPTER VI
SOAKAWAY SYSTEMS FOR SURFACE WATER

ARTIFICIAL surface-drainage schemes are similar in broad outline to natural drainage, and in those places where, owing to the permeability of the earth, there are few streams, or no streams at all, surface-water sewers are not constructed unless there are special reasons for their use. Instead soakaways are provided, and roofs and paved surfaces are drained to them. In some areas soakaway systems may be the rule, but in some other cases it may have to be decided whether soakaways are preferable to surface-water sewers.

Broadly, the rule is that soakaways are constructed where there are no streams or ditches, as the absence of these indicates suitable conditions for soakage. Otherwise the decision depends on the desirability or otherwise of discharging surface water into the earth and on the relative costs of the two systems.

Roof and road-surface water is often soaked away in places where the underground water is used for water-supply purposes, but some surface water can be foul, and should preferably be discharged to combined sewers. The comparative costs of soakaways and surface-water sewers can be determined only when the efficiency of soakaways in the particular district has been found by experiment.

Soakaways are often dotted about individually and a number of gullies connected to them. But a system sometimes favoured is to place the soakaways at regular intervals and to connect them together by small-diameter drains into which the gully connections discharge. This latter system has the advantage that if one soakaway is inefficient it overflows to the next one down the gradient. When this method is adopted it is usual to install an extra-large soakaway at the bottom of each line, because the last soakaway cannot overflow to another. If the linking drains are constructed so as to leak, they may well save the situation should the soakaways become inadequate.

Storage capacity. The utility of a soakaway depends on the rate at which water escapes from it into the ground and on its storage capacity. Provided that the soakaway can discharge water as fast as run-off reaches it during the most intense rainfall, no storage capacity is required. If, on the other hand, the soakaway is so large that it is able to store the run-off due to some inches of rain, its efficiency in discharging to the earth need not be high. In designing soakaways one has to consider rate of percolation and storage together.

Common practice is to allow a storage capacity equal to $\frac{1}{2}$ inch of rainfall over the impermeable area draining to the soakaway. This practice works very well wherever the subsoil is highly permeable and suitable for soakaway construction. There are, however, circumstances where, although the ground appears permeable, doubt remains as to the efficiency of soakaways, and tests and theoretical considerations are advisable. The formula at the end of Chapter IV gives a rough guide to the storage capacity for soakaways of which the initial rate of percolation is known if the following notation is used:

C = storage capacity in cubic feet
Ap = impervious area served by soakaway, in acres
P = one-half the initial rate of soakage, just after soakaway is filled with water, in cubic feet per minute
N = number of years between occurrence of the design storm.

Before designing soakaways for general use, a number of samples have to be constructed, filled to the top working level, and the rate of percolation found by experiment. The capacity may then be calculated on the basis of the results obtained. As soakaways cannot be surcharged and do not profit by delayed run-off, a fairly infrequent storm—e.g. one in 20 years—should be used to give a fair factor of safety to cover both the reduction of the value of the soakaways as they become silted and the high run-off due to storms of rare occurrence.

Geological considerations. Often the surface material may be impervious to water, but a stratum of highly permeable rock may be some feet below the surface. In such a case the soakaway should be carried down into the permeable rock, and the perforated sides commence below the impervious material. Soakaways should not be carried below the highest natural level of the water-table, otherwise they will be partly filled with water even during dry periods, which condition may prove a nuisance and interfere with cleansing operations. Thus, the design of every soakaway has to be revised according to local conditions as ascertained on excavation.

Construction of soakaways. Soakaways have been constructed by the simple expedient of excavating holes in the ground and filling them with large stones. This method, although crude, is not necessarily economical, for the stones occupy space, and thereby reduce the capacity of the soakaway, and often stone filling may cost more than brick or concrete construction, for the quantity of material required to fill a large cavity is much more than that necessary to line it.

Soakaways generally are large circular manholes constructed of brick or concrete, with open bottoms and perforated sides.

The most usual design of a simple soakaway is constructed of open-steined brickwork. The bottom is a circular chamber unpaved. The inflowing water enters by the small-diameter drain at the top and falls straight to the bottom. Large soakaways are often provided with backdrop entries similar

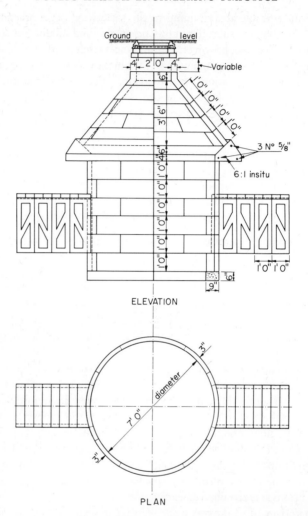

Fig. 17—Soakaway constructed of pre-cast concrete segments and suitable for use in limestone formations

Note adits on two sides of chamber

By courtesy of the late Gilbert A. Ballard

in detail to that shown in Fig. 45 or, alternatively, down-pipes and water-cushions are provided. In limestone areas circular brick walls may not be necessary, a supporting framework of reinforced concrete being sufficient to hold up the superimposed filling. This method facilitates subsequent construction of adits to increase inadequate capacity (see Fig. 17). Where there is much organic material such as leaves, etc., liable to collect in soakaways, the floors may consist of a hard, but not waterproof, pavement to make cleansing easy, or a deep layer of sand or ballast.

CHAPTER VII
SEWERS ACROSS VALLEYS

WHEN it is found impracticable for a main sewer to be laid underground at a fall towards the disposal works or point of outfall, owing to a valley, river, railway cutting or other declivity in the ground, the engineer has the choice of several courses of action. In some circumstances he may be able to make a deviation in the line and continue gravitation. In others, pumping may be the solution. But where pumping is unnecessary and a deviation is out of the question, the alternatives left are continuing the gradient of the sewer by carrying the sewer above ground-level on a bridge or aqueduct or else permitting surcharge and constructing what is usually referred to as an inverted siphon. Both these methods are last resorts, adopted only when absolutely necessary, for they involve undesirable features. The sewer above ground-level is almost invariably unsightly and, for this reason, liable to rouse objection from local residents. The inverted siphon is a possible cause of trouble, calling for careful design if it is not to require frequent attention to prevent or remedy silting. Thus, it will be seen that the design of sewers above ground-level and the design of inverted siphons are important matters, deserving study.

Sewers constructed above ground-level. When sewers are constructed above ground-level, and the gradients and cross-sections are maintained, no particular hydraulic problems are involved. The engineer has merely to evolve a sound engineering design, economical in cost, and as inoffensive to the eye as practicable. Such sewers are usually constructed of cast-iron or steel, according to local conditions, and may be bare pipes supported at intervals on piers; pipes carried on lattice girders; or pipes enclosed in brick or concrete structures.

Pipes supported on piers. When steel or cast-iron pipes are carried above ground-level on piers it must be remembered that they are exposed to the rays of the sun, and allowance must therefore be made for expansion. In the case of steel pipes, the pipes are jointed with welded or other standard joints, but one of the various proprietary designs of expansion joints is adopted and provided at calculated intervals, while intermediately between each expansion joint the length of pipe is firmly anchored to prevent creep longitudinally, one way or the other. The distance between supporting piers is calculated according to the bending moment of the steel pipes. The piers are arranged to be at or close to every joint, and intermediately if necessary.

When arrangements are made for expansion, particularly in the case of long lengths of steel pipes, the pipes should be free to move longitudinally over the tops of the piers (except where anchored), and for this purpose, roller bearings may be provided.

The protection given to steel pipes varies according to circumstances. Generally, they should be bitumen and glass- or asbestos-fibre-sheathed externally, or otherwise protected against the weather.

Cast-iron pipes with spigot and socket joints are to some extent restricted to those cases where the line of pipes is not very far above ground-level, for each pipe must be supported behind the joint, and the joints of cast-iron pipes are at closer intervals than those of steel pipes: thus, a greater number of piers is required. It is not always considered necessary to provide expansion joints for short, straight lengths of spigot and socket cast-iron pipes, which always have appreciable flow in them, for expansion stresses in such circumstances may be small. But, generally, expansion arrangements should be provided.

The piers supporting pipes do not need to be excessively heavy. They should be designed to withstand wind pressure and to give adequate haunching to the pipe. On the other hand, the foundations must be carried below the surface soil down to natural bed-rock, otherwise the piers will sink irregularly, even when the load is light. If angles on the pipe-line are unavoidable, heavy piers should be provided thereat capable of withstanding thrusts due to expansion and contraction.

Near the ends of all lines of pipes above ground-level a wrought-iron collar of radiating bars should be arranged, to prevent children from climbing on and walking along the pipes.

The strength of pipes serving as beams to carry their own weight including contents can be easily calculated and some manufacturers give the safe spans in their catalogues. But it is sometimes overlooked that the spot load introduced by the support can be more than a large-diameter pipe can tolerate and in some instances the supporting piers have been driven into the pipes. This circumstance particularly applies at the ends of pipe-lines above ground-level or in other places where pipes are laid above the ground but under an embankment which puts extra load on the pipes. The solutions in these circumstances are either to provide spreading pier heads to transmit the load to a large area of pipe or to lay the pipes on beams.

Pipes carried on lattice girders. This is a straightforward engineering problem involved where the spans are too great for the pipes themselves to take the bending moments. The lattice girders are designed according to theory, and each pipe supported at close intervals, as is convenient. The expansion joints on the pipes are arranged to coincide with the points at which expansion is permitted at the ends of the girder or girders.

Pipes enclosed in brick or concrete structures. When a large sewer is carried at a considerable height above ground-level, and appearance has to be considered, it may be found advisable to go to the expense of constructing a brick or concrete aqueduct. In this case the sewer can still be in the form of a pipe or, in the case of larger sizes where cast-iron pipes would be out of the question, the construction can be of cast-iron segments. It should, however, be noted that the type of cast-iron segments used for constructing large sewers in heading should not be adopted in this case, for such segments of necessity have their flanges on the inside, which means that the sewers have to be lined with concrete or brickwork. The segments for constructing large sewers above ground-level should have their flanges on the outside so that, when the segments are bolted together the inside of the barrel has a smooth face of cast-iron coated in the usual manner as are cast-iron pipes.

When sewers are carried on aqueducts it is simple to arrange for the whole of the cast-iron work to be protected against the weather, doing away with the necessity for special arrangements for expansion. There is, however, danger of expansion of the whole of the structure, which has on occasion been overlooked. Generally, this is not serious, and can be neglected provided the aqueduct is straight throughout its length and abuts on solid ground at each end. If, however, an aqueduct were to be constructed with angles or on a curve, there would be danger of movement towards the convex side, due to expansion. This happened in the case of the Wandsworth Aqueduct built in 1882–5 to carry the Southern High-level Sewer. For some unknown reason this immense structure consisting of brick arches supported on double piers was constructed to an arc of a circle on plan, the sewers at the ends entering tangentially and the convex side facing south.

The heat of the sun caused the aqueduct to expand longitudinally and, as the ends abutted into the earth, the middle moved outwards towards the south. Every summer the aqueduct moved farther south, returning only part of the way back in the winter until the deflection became serious, breaking the northern legs of the piers.

Supporting buttresses were provided on the south side and of such strength that when a high explosive bomb hit one during the Second World War it only produced a scab of 2- or 3-feet diameter. Nevertheless the movement went on and the aqueduct had to be demolished.

Inverted siphons. The main trouble with inverted siphons is silting, and this is because these siphons usually have to be made sufficiently large to accommodate the maximum rate of flow, with the result that at the normal rate of flow the velocity is not "self-cleansing." In addition, the rising leg of the siphon provides a gradient up which heavy particles have to be carried, and therefore the self-cleansing velocity must be made greater than that of a normal sewer.

The inverted siphon problem varies according to circumstances, the main considerations being the amount of fall available and the difference between maximum rate of flow and daily peak during dry-weather flow. Where the maximum rate of flow is not high, being, say, only twice the peak flow at midday during dry weather, and where there is plenty of fall, the problem is a simple one; for it is merely necessary to make the siphon of small diameter and steep hydraulic gradient, so as to pass the maximum rate of flow at a

Fig. 18—Steel sewer supported on piers above ground-level
By courtesy of Stewarts & Lloyds, Ltd.

velocity of, say, 6 feet per second and the daily peak at a velocity of, say, 3 feet per second. Such a siphon would be self-cleansing, and give little trouble.

Take the example illustrated in Fig. 19. The peak conditions in this case are that the dry-weather flow is 30 c.f.m., the peak daily flow 90 c.f.m. and the maximum flow to be accommodated during wet weather 270 c.f.m. The sewer entering at the top end of the siphon is of 12 inches diameter, laid at a fall of 1 in 74; its crown level at point of connection to the siphon is 20·10 O.D. The sewer connecting from the bottom end of the siphon is 18 inches diameter and has a gradient of 1 in 640. The crown level at point of connection is 16·50 O.D. It will be noted that both these sewers run full at the maximum rate of flow of 270 c.f.m. Therefore the hydraulic gradient through the siphon at that time is taken from crown to crown. The siphon is 100 feet long. Consequently the hydraulic gradient at maximum rate of flow is 1 in 27·8.

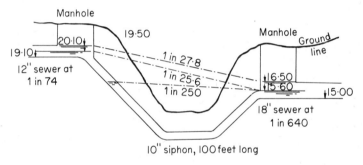

Fig. 19—Single inverted siphon

By reference to Vol. I, Fig. 68, we find that a 10-inch pipe laid at a gradient of 1 in 28 discharges 270·0 c.f.m. at a velocity of 494 feet per minute. Therefore a 10-inch-diameter siphon pipe is chosen.

The question that next arises is whether this 10-inch siphon will be self-cleansing during dry weather. (It should here be mentioned that in many cases so little fall is available when siphons have to be designed that it is necessary to calculate hydraulic gradients not merely from crown to crown but from actual top water-level to actual top water-level in the incoming and outgoing sewers respectively. In this particular example it will be seen that this is not necessary. Nevertheless, the procedure will be followed.) At the peak daily flow during dry weather (90 c.f.m.) both sewers will be running partly full, and by reference to Table 1 it will be observed that the 18-inch sewer will be flowing about 0·4 of its depth full, which is 0·6 above invert—i.e. 15·60 O.D. The 12-inch pipe will also be flowing about 0·4 of its depth full, which is 0·4 feet above invert—i.e. 19·50 O.D. This gives a fall of 3·90 feet in 100, or in 1 in 25·6. This is more than adequate to produce the required flow of 90 c.f.m. in the siphon at peak daily flow. Referring again to Vol. I,

Fig. 68, we find that a 10-inch pipe requires a hydraulic gradient of 1 in 250 to pass 90 c.f.m., and this flow is passed at a velocity of 165 feet per minute, which is self-cleansing. Therefore the 10-inch-diameter pipe can be adopted as satisfactory.

Multiple siphons. The difficulties arise when there is little fall available and, as in the case of combined-sewers, the maximum rate of flow greatly exceeds the peak dry-weather flow. Then calculations have to be carefully made, and two or more siphons provided to accommodate the fluctuating flow.

A fair example is given in Fig. 20. Here the conditions are as follows. The dry-weather flow is 23 c.f.m., the peak daily flow 69 c.f.m. and the maximum flow to be accommodated during wet weather 218·5 c.f.m. The sewer entering at the top end of the siphon is of 12 inches diameter, laid at a fall of 1 in 110; its crown level at point of connection to siphon is 209·50

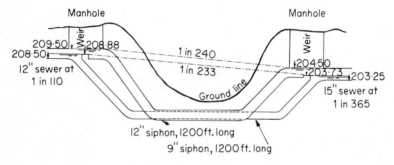

FIG. 20—Double inverted siphon

O.D. The sewer connecting from the bottom end of the siphon is 15 inches diameter and has a gradient of 1 in 365. The crown level at point of connection is 204·50 O.D. It will be noted that again both sewers run almost full at the maximum rate of flow of 218·5 c.f.m., and that therefore the hydraulic gradient through the siphon at maximum rate of flow may be taken from crown to crown. The fall is therefore 5 feet, and as the length of the siphon is 1200 feet, the hydraulic gradient at maximum rate of flow is 1 in 240.

Suppose the designer begins in this case by trying out the possibility of using a single siphon pipe; he would find that at the available gradient a 14-inch siphon would comfortably accommodate the maximum rate of flow. But at the peak daily flow of 69 c.f.m. the velocity would be slightly less than 65 feet per minute, which would mean that heavy silting would take place. The designer would therefore have to find the size of siphon required to pass the daily peak flow during dry weather at a self-cleansing velocity, and follow this by determining the size of additional siphon that would be required to supplement the dry-weather siphon at time of maximum flow.

The 15-inch sewer at a gradient of 1 in 365 passes 220 c.f.m. At 69 c.f.m. it therefore flows fractionally full, the depth of flow being 0·385 approximately of the diameter of the sewer, or 0·48 above invert, which is 203·73 O.D. Similarly, the water-level in the 12-inch sewer is 208·88 O.D. This gives a fall of 5·15 feet in 1200 feet, or 1 in 233. This is the gradient available for passing the peak dry-weather flow through the dry-weather siphon. Referring to Vol. I, Fig. 68, it will be found that at a gradient of 1 in 235 a 9-inch pipe is required, and that at a flow of 69 c.f.m. the velocity in a 9-inch pipe is slightly more than 156 feet per minute, and therefore self-cleansing.

It is now necessary to find the diameter of the additional siphon. The hydraulic gradient from crown to crown of 1 in 240 produces a flow of 69·5 c.f.m. through the 9-inch siphon. There therefore remains a quantity of 149 c.f.m. to be passed at times of storm. Vol. I, Fig. 68 shows that a 12-inch pipe at a hydraulic gradient of 1 in 240 passes 150 c.f.m. at a velocity of 191 feet per minute, and therefore a diameter of 12 inches may be chosen for the additional siphon as being of the right size.

Normal precaution against silting of storm siphons. The sizes of the siphons having been determined, it is next necessary to decide the levels of top and bottom ends of siphons and the arrangements required to prevent silting of the larger siphon during dry weather. It will be seen that the 9-inch siphon is flowing all the time, and is flushed every day by the peak flow. In order that the self-cleansing velocity in the 9-inch siphon shall be ensured during dry weather, the 12-inch siphon must not function at that time. The 12-inch siphon should also not receive *any* flow during dry weather, even by back-wash from the lower end, as otherwise it might become loaded with silt.

For the 12-inch siphon to function satisfactorily during storm flow, its crown levels at both top and bottom ends must be below the crown levels of both the incoming and outgoing sewers to which it connects. This means that its inverts at both ends will be below top water-level at peak daily flow, and unless precautions are taken it will accept flow at that time. This may be prevented by the construction of weirs set not lower than the levels of sewage in the incoming and outgoing sewers at daily peak dry-weather flow (see Fig. 20). These weirs should, if possible, be of such length that when they are submerged at time of maximum rate of flow the depth over them is so great that loss of head is negligible. If this is not practicable, the loss of head at maximum rate of flow over *both* weirs must be calculated and deducted from the available head, thereby reducing the hydraulic gradient through the larger siphon only, and the diameter of siphon must be increased accordingly.

Siphon-flushed inverted siphons. Another method which has been used to ensure self-cleansing conditions in inverted siphons is to arrange a true siphon and flushing-tank at the head of the inverted siphon to flush the siphon with crude sewage. The inverted siphon is then designed as a single pipe of the size required to take the maximum rate of flow, and it is prevented from becoming silted by the flushing arrangement, which ensures that the sewage passes through the inverted siphon in periodic doses, always at the maximum rate of flow.

This method is very attractive to the designer, for it gets over the difficulties so far discussed. But in practice it has limitations for, in those cases where flushing would be necessary, little head is available and flushing-tanks and true siphons require some feet of head for proper operation. The loss of head through a low-draught siphon may not be very much unless it is one of large size, and even then the siphon can be specially designed for the purpose, so as not to use up too much head: on the other hand, flushing-tanks, to be effective, must be of reasonable capacity—say at least one minute's discharge of siphon at the maximum rate—and if possible not less than the storage capacity of the inverted siphon. For this capacity to be accommodated, in a flushing-tank which is shallow so as not to use up head, the tank must be of large surface area; consequently the tank, unless carefully designed, will function as a settling-tank, and it will be difficult to prevent it from becoming foul.

The ideal construction is a tank with steeply sloping bottom terminating in a sump in which is fitted the dome of the siphon. But as such a tank requires some feet of depth for its construction, which depth must be deducted from the available head, it is usually found that where the depth is available for the satisfactory construction of a flushing-tank there is also sufficient depth to provide the head required for the construction of a self-cleansing siphon as illustrated in Fig. 19.

Details. Because inverted siphons are surcharged, they should be constructed of strong materials, such as cast-iron, steel tubes, reinforced concrete, etc., according to circumstances. Attention should be given to the construction of the manholes at top and bottom ends. Siphons should have bellmouth or, preferably, tapering entrances and well-designed outlets, or be otherwise arranged so as to make velocity head losses negligible. Where two or more siphons are provided all should be arranged with penstocks or sluice-valves at both ends, in order that they may be isolated to permit flushing of each siphon independently, and general maintenance. This also provides means for cleansing in the case of multiple siphons. In one instance a multiple siphon gave considerable trouble and, to overcome this, the various pipes were flushed at intervals to remove large quantities of fibrous and other solid matter. Eventually an engineer pointed out that if some of the siphon pipes were closed those remaining would be self-cleansing and, at the same time, adequate to take maximum flow. This was tried, after which there was no further trouble.

Except when absolutely impracticable, wash-outs should be arranged at the lowest point on each siphon. Manholes at both ends of siphons should be provided with ventilating columns or other adequate means of ventilation, for the siphons interrupt the natural flow of air through the sewerage system. Long siphons should have hatch-boxes at intervals not exceeding 360 feet to make rodding possible. Arrangements should be made for flushing with water.

Care should be taken in the design of the manhole at the up-stream end of the siphon to ensure that scum cannot be trapped: a good velocity of inlet is necessary at some time each day to prevent this. The crown at inlet end of the siphon should not be below water-level for, if it is, floating solids will accumulate in the upstream manhole or chamber even if there is a fair inlet velocity. This calls for special design with a change of diameter at the inlet to feed in the floating solids.

CHAPTER VIII
VENTILATION AND FLUSHING OF SEWERS

VENTILATION of sewers is essential for the purpose of preventing air-lock and permitting free flow of sewage. It is also needed to keep the air in the sewers free from gases and vapours liable to explode or poison men working in the sewers and chambers.

If a sewer is not self-cleansing, and the sewage remains stagnant in it for long periods, deposits will form and decompose, giving off gases, including methane and sulphuretted hydrogen. Methane forms an explosive mixture with air. Sulphuretted hydrogen, even in very small quantities, is a deadly poison. Moreover, this gas has a very destructive effect on cement mortar and concrete. In manufacturing towns trade effluents containing poisonous gases or vapours have been admitted to the sewers, sometimes with fatal results. Coal-gas from leaky gas-mains sometimes finds its way into the sewers. Another source of danger, of increasing frequency nowadays, is petrol from road vehicles. This is not only explosive, but also a rapid poison (see Chapter XIX). Most of these dangers can be minimised by adequate ventilation.

Sewer gases, being warm, tend to rise, and therefore run up the gradients of the sewers and discharge at the high points. To a lesser extent the flow of sewage may carry the gas down the gradients by friction. The bellows action caused by the filling and emptying of sewers as a result of the different rates of flow at various times of the day has a marked effect on ventilation. Except on the calmest days there are the differences of barometric pressure at the various parts of the sewerage system and differences of pressure at places of ventilation due to dissipation of kinetic energy of wind. (As laid down in Buys-Ballot's Law, if one stands in the northern hemisphere with one's back to the wind, the barometric pressure on one's right hand is higher than that on the left. Thus, as flow of air in a sewerage network is from high pressure to low pressure, the direction of flow in the sewerage system depends on, but is not in the same direction as, the wind.) These various factors influence ventilation in an unpredictable manner.

The ventilation of sewers is usually provided by the opening at the lowest end where it discharges to the treatment works or outfall, together with various means of ventilation intentionally provided at the top ends of the branches and elsewhere on the system. In areas where interceptors on drains are not used, practically all the ventilation that is necessary is given by the soil-and-vent pipes on the houses; but where interceptors are normally installed, the sewers need to be ventilated by ventilating columns erected at the head of each branch sewer, and sometimes at key points on the system where pressure or suction may occur as a result of flow. Alternatively, ventilating manhole covers may be used in positions where there will not be any risk of nuisance.

Sewer flushing. Where there is any doubt that a sewer will be self-cleansing, means for flushing should be provided. It was formerly customary to place penstocks or valves in some of the manholes, by means of which the sewage might be headed up in the sewer and suddenly released. This method is not good, for the retained sewage is liable to form deposits in the sewer above the penstock. Preferably, sewers should be flushed with clean water supplied from a company's main or, where possible, a spring.

In the past the capacities of flushing-tanks were determined by various rules of thumb, according to the size of sewer to be flushed and the gradient: that is, the larger the sewer the larger the flushing-tank, and the slacker the gradient the larger the flush. An approximation to the average of such practice is to allow a flushing-tank capacity equal to one minute's discharge of sewer, assuming it to be flowing at $2\frac{1}{2}$ feet per second, and to increase the capacity in inverse proportion if the actual velocity of flow at the available gradient is less.

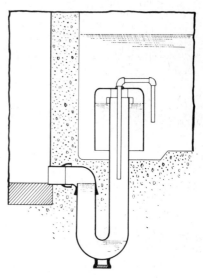

FIG. 21—Typical flushing siphon
By courtesy of Adams-Hydraulics, Ltd.

The chief difficulty, however, was deciding the size of flushing-tank necessary to give an adequate flush at the bottom end of the length of sewer to be cleansed. When a sewer is flushed the water running down the sewer is not all travelling at the same velocity: there is the "draw-down," which leads the main body of the flow, and the "tail-off," which trickles behind, and what may be an adequate flush at the top end of the sewer is smoothed out at the lower end. The rule that is therefore recommended for general application is that the capacity of the flushing-tank should not be less than one-tenth of the cubic capacity of the total length of sewer to be flushed, and generally not more than twice this amount.

Automatic flushing-tanks are discharged by means of siphons, the capacities of which are again determined in accordance with the discharge capacity of the sewer. Generally, the siphon should be capable of giving a flush which, together with the flow of sewage in the sewer, will not cause surcharge, but will at least cause the sewer to run half-full.

The usual form of siphon employed works on the following principle. Air is compressed, being confined between the water in the dome, or outer tube, and that in the deep U-trap underneath. The difference between the levels of the water inside and outside the dome is always equal to that between the levels of the water in the two legs of the trap. As the water in the tank rises over the dome, the pressure of the air inside the latter increases,

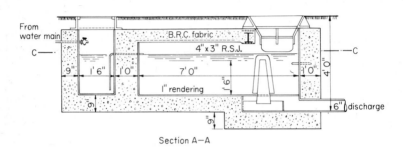

FIG. 22—Typical flushing-tank
By courtesy of John Taylor & Sons, Consulting Engineers

and the level of the water in the descending leg of the trap is forced down until it reaches the bottom of the seal. When this point is reached, air begins to pass through the trap, and the bubbles, as they rise in the ascending leg, reduce the weight of the column of water therein, and at the same time the pressure on the imprisoned air is reduced. The latter thereupon expands, and more air is forced through the trap, with a further lightening of the column of water, and a still further reduction in the pressure. As the pressure falls, the water rises under the dome until it reaches the top of the inner tube, down which it rushes, blowing out the remaining air and releasing the contents of the tank.

As water passes from the flushing-tank to the manhole or chamber into which discharge is made, air has to be admitted to the tank, otherwise the siphon may fail to function satisfactorily. One way in which this can be effected is the provision of a manhole cover of ventilating type.

When an automatic flushing-chamber is constructed, the company's water supply must be metered and discharged into a small chamber or manhole isolated from the flushing-tank by an unbreakable trap or water-seal, in order that there shall be no danger whatsoever of contamination of the water supply. The installation must be to the satisfaction of the water authority.

Other methods of sewer-flushing are more generally applicable, the automatic tank being provided only at the heads of main lines of sewer unavoidably laid to slack gradients or which will not receive much flow when first constructed. At the head of *every* sewer it is advisable to install a flushing penstock in the top manhole, so that the manhole may be filled by hose from a hydrant or by tank vehicle, after which the quick opening of the valve gives an effective flush to the sewer. Suitably designed gully-emptiers will flush the sewers where no flushing penstocks have been installed.

CHAPTER IX

DRAINAGE OF ROAD SURFACES

The proper drainage of the surfaces of roads is of great importance, for, without good drainage, puddles may form, to the inconvenience of road users and the possible deterioration of the surfacing materials.

The spacing of road gullies is a matter of practice which is very varied. Several rules have been published, some of which are very arbitrary.

Gullies and channels. The maximum distance between gullies is determined by a practical consideration which applies when roads are laid in level country and the fall of the road is not sufficient to give adequate fall to the channels so as to ensure that the run-off to the gullies is efficient. In this circumstance it is necessary for the channels to be laid to reasonable falls in spite of the kerb and the crown of the road being level. There is some difference of opinion on what is a reasonable fall. The Ministry of Transport has laid down that channels shall fall at not less than 1 in 250. Some authorities prefer gradients twice as steep. In the present instance it is suggested that a fall of 1 in 240 should be used as the basis of design wherever good workmanship is to be expected, but that this gradient should be increased at the discretion of the designer when there is any doubt as to the quality of labour and supervision.

If the kerb is level but the channel falls, the depth of the kerb must be greatest at the gully, and least half-way between gullies. The maximum variation of depth of kerb permitted is 4 inches—that is, the kerb may vary from 3 inches deep to 7 inches deep—but in common practice it is usual to limit the variation to 3 inches.

A channel laid at a gradient of 1 in 240 falls 3 inches from the point half-way between gullies to the gullies in 60 feet, and, as a consequence, the condition as to fall cannot be satisfied when the road is level unless the gullies are spaced not more than 120 feet apart. It so happens that this spacing of 120 feet between gullies is, according to one hydraulic theory of gully spacing, satisfactory for roads having a drainage width (i.e. distance from crown or high side to back of paved footway) of not more than 24 feet.

When the road slopes longitudinally at gradients greater than 1 in 240, the spacing of gullies is not so limited, and the distances can be calculated according to theory. It is not necessary to enter here into the elaborate theory that has been applied to this problem, but the figures in Table 24 give an indication of the maximum distance between gullies for roads having a cross-fall of 1 in 50. In practice, spacing is usually closer.

The most interesting condition (and the one which takes up the most time in the drawing-office) is that in which the road is not dead level, but falls at a gradient of less than the selected minimum gradient of channel. If the channels are then laid to the ruling gradient, the peak points where the kerb depth is minimum cease to be half-way between the gullies, but approach towards the upstream gullies. As the gradient of the road is steepened, so the peak points move farther towards the upstream gullies until, when the gradient of road equals the gradient of channel, the peak point is at the gully—

TABLE 24

APPROXIMATE MAXIMUM THEORETICAL GULLY SPACING IN FEET FOR ROAD WITH AVERAGE CROSS-FALL OF 1 IN 50

Drainage width in feet	Gradient of channel 1 in:						
	25	50	75	100	150	200	250
15	—	—	300	262	212	185	164
20	—	242	200	172	139	122	108
25	262	186	150	130	107	93	82

in other words, we have the same condition as on a steep road: the channel falls at the same grade as the kerb. In the foregoing conditions the position of the peak may be calculated by the following formula:

$$X = \frac{A - DR}{2}$$

where: X = distance in feet from upstream gully to peak point
A = distance between gullies
D = fall between upstream and downstream gully in feet
R = minimum gradient of channel (i.e. if gradient is 1 in 240, R = 240).

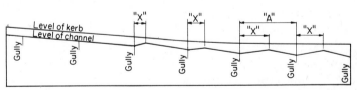

FIG. 23—Longitudinal section of road showing falls of channel and kerb line

The line of the kerb is the most visible line on the road, and it must be maintained in regular straight lines or uniformly sweeping curves for the sake of the aesthetic effect: it must not rise and fall between gullies and high points on the channel. The rise and fall of the channel below a uniformly graded kerb is, however, not obvious, even to the trained eye. Thus the above-described method ensures efficient drainage combined with good appearance.

Cross-fall or camber. We now have to deal with the drainage of the surface of the road from the crown, or highest part, to the channel. This is more complex than is generally appreciated by those who have not had much experience of urban highway design. It is simplest where roads are laid in virgin country and constructed of a material such as water-bound macadam, which permits the road being cambered with crown at centre halfway between channels, the channels being laid level one with the other. But when roads are laid in towns or between the buildings of a factory site or an estate, the channels can seldom be laid level with one another, for their levels are determined by the entrances to premises on each side of the road.

Take, for example, a residential road the houses of which are higher on one side than the other; in this case the back of the footway on the high side of the road must be level with the entrances to the front gardens, so as to avoid awkward steps or undrained pockets. Similarly, the back of the footway on the low side of the road must again be reasonably in conformity with the levels of the entrances. The footways fall from back to kerb at suitable even

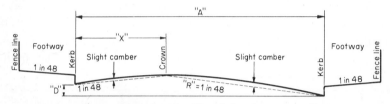

FIG. 24—Cross-section of cambered road showing crown off centre

gradients (e.g. not less than 1 in 48 and not more than 1 in 24 for paved surfaces, and not less than 1 in 24 or more than 1 in 16 for unpaved verges) and consequently the kerbs and channels on the two sides of the road will be at different levels.

When this is the case *the crown or highest part of the road ceases to be in the centre, and its position is found by the same formula as was used for finding the peak point of the channel, the notation now being:*

$$X = \frac{A - DR}{2}$$

where: X = distance from higher kerb to crown in feet
A = width of carriage-way in feet
D = difference of level between higher and lower channels in feet
R = the minimum permissible cross-fall gradient (or average camber) of the face of road from crown to channel.
(See Fig. 24 where R = 48.)

The value of R depends on the surface material of the road, figures such as the following having been recommended:

For granite sets, tar macadam, tar-sprayed macadam and grouted
    surfaces . . . . 1 in 40
For asphalt, similar smooth pavings and concrete surfaces . . 1 in 48

These minimum cross-falls or cambers are measured from *crown* of road to *peak point* of channel between gullies, and as a result the fall from the crown is steeper elsewhere and steepest opposite the gullies, where it should never be steeper than 1 in 16, and preferably not more than twice the minimum slope.

It will be seen from the foregoing that in designing a roadway the designer

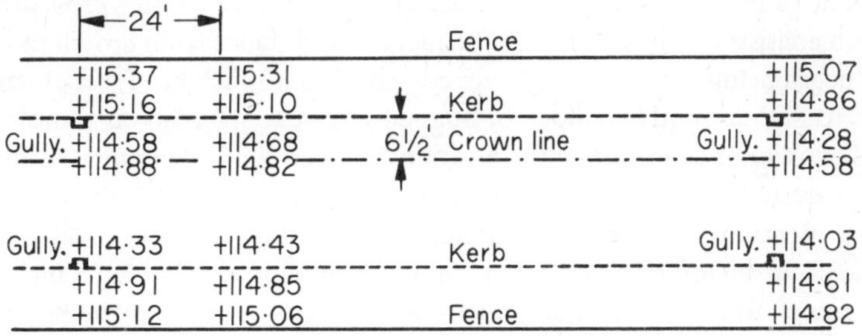

|←—24'—→|	Fence	
+115·37	+115·31		+115·07
+115·16	+115·10	Kerb	+114·86
Gully. +114·58	+114·68	6½' Crown line	Gully. +114·28
+114·88	+114·82		+114·58
Gully. +114·33	+114·43	Kerb	Gully. +114·03
+114·91	+114·85		+114·61
+115·12	+115·06	Fence	+114·82

FIG. 25—Plan showing levels on the surface of a cambered road as required to secure adequate drainage

has to calculate, and show on the plan, a number of spot levels and dimensions to which the site staff must work (see Fig. 25). These must include the following:

(*a*) At every gully or pair of gullies (for where roads are cambered gullies are preferably in pairs, one on each side of the road) a cross-section consisting of:

    1. Level on back of footway.
    2. Level at kerb.
    3. Level in channel.
    4. Level on crown.
    5. Level in channel.
    6. Level at kerb.
    7. Level on back of footway.

(*b*) Intermediately between every gully or pair of gullies at the position of peak point of channel, a similar cross-section of seven levels.

(*c*) Additional levels should be given so as to ensure regular construction of vertical longitudinal curves of roadway and elsewhere where levels are needed to ensure construction according to design.

(*d*) Dimensions should be given indicating positions of peaks of channels relative to gullies and distance of crown of road from channel.

If a road plan omits the levels given above, the design of the road falls not on the engineer, but on the clerk of works or other responsible site officer, in which case it can be said that the designer has not done his job. The practical effect of such omission is for roads to be constructed not infrequently of inferior appearance, and generally most inefficient for the drainage of the surfaces, with marked ponding in wet weather.

During construction of works the officer responsible should set pegs to the levels given on the plan and arrange longitudinal screed lines between the pegs. Should there be any longitudinal curvature of the road, additional pegs should be set between those shown on the plan, and smooth curves worked in. The camber of the road is then determined with the aid of a cross profile, which consists of a board of length equal to the distance from crown to kerb, this board being curved on its lower edge by about 1 inch in 16 feet of length, so as to give a slightly rounded or hogged surface to the road on completion. This practice of determining the position of the crown and giving a comparatively even fall from crown to channel, apart from the slight curvature above mentioned, is to be much preferred to the older practice of rolling road surfaces to the arc of a circle, for the latter method was liable to make the road much too flat at the crown, and the camber so steep near the kerb as to be objectionable to all road users and dangerous to cyclists.

Roads with an even fall from one side to the other. The rounded cambering of roads presumably originated with construction of macadam roads, for a rounded surface free from hollows or irregularity is easily secured by rolling macadams and similar materials in a longitudinal direction. When, however, modern materials, such as concrete-faced roads and mass concrete, reinforced concrete or concrete foundations faced with asphalt, etc., are used in construction, it becomes possible for the surface to be laid accurately to screed. The sharply curved camber is then no longer a necessity and, because an even fall from crown to kerb is more efficient for drainage than a varying fall, camber can be largely done away with, and instead the road given a cross-fall either from crown to kerb or from higher kerb to lower kerb with a camber of not more than 1 inch in 16 feet to prevent an appearance of concavity.

The construction of roads to a camber with more or less central crown is, in fact, limited to roads with very little difference of level between the two channels, for when the aforementioned formula and method are used, quite a slight difference of level on each side of the road throws the crown considerably off centre. If this becomes excessive, the fall from the crown to the gullies on the higher side becomes too steep, and makes the adoption of an even fall from side to side and abolition of the crown preferable. Generally, it can be said that when the designer is working to a surface slope of 1 in 48, the crown should never be permitted to approach the higher kerb by a shorter distance than 6 feet. Table 25 gives the conditions when this occurs.

When it is found impossible to reduce the difference of level of channel below the appropriate figure as given in the second column of Table 25, the next procedure is to see if it is possible to raise the higher channel 3 inches, so as to give an even cross-fall of not less than 1 in 48.

Concrete roads laid to cross-fall can be highly efficient and economical, and their use is recommended for many classes of estates and factory sites, and they reduce the number of gullies required by at least 50 per cent.

In working out the direction of cross-fall of a main motor road, super-elevation at curves is a primary consideration, but the roads of an estate, and particularly of a factory site, do not have to be designed for high-speed traffic. Super-elevation can therefore take second place, and the direction of cross-fall may be primarily determined according to the levels of the entrances of the buildings, the fall of the land (the cross-fall being in the direction of the fall of the land so as to economise in excavation), and general convenience.

TABLE 25
DIFFERENCE OF ROAD CHANNEL LEVELS

Width of road in feet	Difference of level of higher and lower channels in inches when crown is 6ft from higher kerb
25	$3\frac{1}{4}$
24	3
23	$2\frac{3}{4}$
22	$2\frac{1}{2}$
21	$2\frac{1}{4}$
20	2
19	$1\frac{3}{4}$
18	$1\frac{1}{2}$
17	$1\frac{1}{4}$
16	1
15	$\frac{3}{4}$
14	$\frac{1}{2}$
13	$\frac{1}{4}$
12 and under	0

Drainage of surfaces at intersections. In old-time highway design, cambered roads intersected in the following manner: the channels of the lateral road ran in to join the channels of the main road, while the crown of the lateral was carried through to meet the crown of the main road. This gave a mitred effect, admittedly of good appearance and easy to construct. But high-speed traffic has led to its obsolescence as regards highway design, for the crowns of laterals coming in in this manner resulted in severe inequalities in the level of the main roads, leading to discomfort, and even danger to drivers. Modern practice is for lateral roads to meet main roads as follows: the channel line of the main road is continued to a true grade across each lateral, and the crowns of the laterals are swept down to the channels of the main road to the same camber as is used elsewhere on the road. As far as roads inside factory premises are concerned, the designer may make his own choice as to which method he prefers, but modern practice should always be applied to public highways of any traffic importance.

The intersection of roads with cross-fall is somewhat different, for here we are dealing with the intersection of plain surfaces falling at varying grades in different directions, and there is therefore a number of cases to be considered. The simplest is that in which the surfaces of both roads fall in the same direction and at the same grade (the fall in this case being the resultant of the longitudinal fall and the cross-fall of the road) (see Fig. 26). This is the ideal condition to be aimed at, for it does away with complications, and the only precaution to be taken is the proper siting of a gully to prevent the water from the channel of one road running across the face of the other. In all other cases, however, the surfaces, if continued without distortion, would meet at a valley or channel line, as in Fig. 27, or a ridge or crown, as in Fig. 28, the ridge occurring when the two surfaces fall away from each other, and the channel when they fall towards each other. There is no *functional* objection to this where speed of traffic is slow, but it can look bad. Consequently, the crown or ridge should be carefully worked out to an even curve. The channel line or valley should be avoided as far as possible, but if it cannot be avoided it is best defined by a clear-cut line of granite sets to show that it is purposive and not accidental.

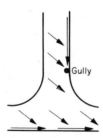

FIG. 26—Intersection of two roads with cross-falls in the same direction

Treatment of roads at entrances and gates to premises. A number of important points have to be considered when the drives of factory sites connect to public or private highways. One of the first to be kept in mind is that, whether or not the factory site is above or below the highway, the level of the access road at the line at the back of the footway of the main road should be above the channel of the main road, in order that no drainage shall take place from the main road to the factory site. The pavement crossing of a minor entrance to premises therefore rises from the channel of the main road to the back of the pavement, commencing somewhat steeply, so as to meet the level of the pavement in 2 or 3 feet, and then continuing at the gradient of the footway to the boundary of the premises, after which it may fall into the premises, so as to avoid water running out of the premises into the main road, which would be objectionable to pedestrians. Major entrances to large premises are best treated in the same manner as lateral streets, with large radius kerbs, footways and adequate sight lines.

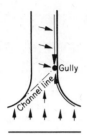

Fig. 27—Intersection of two roads with cross-falls resulting in a channel line

Where the road passes under the gates there should be no cross-fall. If the gates are at the back of the footway of the public road a level line will suffice for the surfaces drain towards the channel of the street. But if the gates are some distance back, the access road should be cambered, and at the point

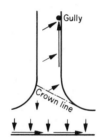

Fig. 28—Intersection of two roads with cross-falls resulting in a ridge or crown

where it passes under them should have its crown at the centre, for otherwise the space under the gate will be higher one side than the other, and this gives a bad appearance. Once inside the premises, the crown may be carried to one side or the other to secure a cross-fall. On steep hills, entrances with cross-falls are unavoidable, and special gates have to be made.

CHAPTER X

THE CONSTRUCTION OF SEWERS OF SMALL OR MEDIUM DIAMETER

As soon as the lines, levels, gradients and diameters of sewers, and the position of manholes have been determined, the designer has to decide on matters of detail. The materials of construction of the sewers are the first consideration. Pipes or tubes may have to be concreted, or cast-iron pipes used, according to the Ministry of Housing and Local Government Requirements (see Chapter I), or such requirements as may be issued by the Department of the Environment from time to time.

The official requirements are necessary precautions, the need for which has been borne out by experience. A clay pipe is too fragile to be laid near the surface without protection, and is too near its breaking load when laid in a deep trench without concrete support. Consequently, some local authorities prefer to use concrete to a greater extent as a matter of general practice. Such decisions, however, depend on circumstances. Clay-pipe sewers properly laid in good ground in rural areas and at moderate depths should not need concrete benching, and concrete surround is inconvenient when future connections have to be made by grouting a saddle on to the pipe.

Stoneware pipe-sewers. For very many years the most frequently used material for drains and small-diameter sewers were salt-glazed ware pipes and vitreous-enamelled fireclay pipes, manufactured in various grades including British Standard tested, first quality tested, British Standard, first

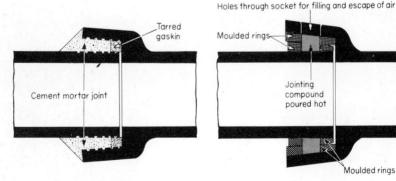

Fig. 29—Cement mortar joint on vitrified clay pipes

Fig. 30—Hot-pour joint on vitrified clay pipes

By courtesy of The Clay Pipe Development Association, Ltd.

THE CONSTRUCTION OF SEWERS OF SMALL DIAMETER   105

quality and second quality, the last being virtually throw-outs. These materials are giving place to vitrified clay pipes, which are manufactured to longer lengths and closer tolerances and, being of denser material, do not need to be glazed. While these new pipes are available for jointing in cement mortar, which was usual in the case of the previous materials, various types of special flexible joints are now available and it is expected that these will eventually replace cement joints almost entirely. The following is a list of flexible joints which have received Ministry approval at the time of writing.

Manufacturer	*Available sizes*	*Type*	*Trade name*
Church Gresley Fire Brick and Fire Clay Co., Ltd.	4″ 6″ 9″	Polyester	C. G. Swadflex
John Crankshaw Co., Ltd.	15″ 18″	Polyester	Hepseal
Donington Sanitary Pipe and Fire Brick Co., Ltd.	4″ 6″ 9″	Polyester	Donflex
Dorset Clay Products, Ltd.	4″ 6″ 9″ 12″	Polyester	D.C. Polyester Joint
Doulton Vitrified Pipes, Ltd.	4″ 6″ 7″ 9″ 12″	Polyurethane	Draw Flex
Ellistown Pipes, Ltd.	4″ 6″ 7″ 9″ 12″	Polyester	Ellflex
Eltringham Pipe Co., Ltd.	4″ 6″ 9″ 12″	Polyester	Polyester
Hawfields Brick & Pipe Works, Ltd.	4″ 6″ 9″	Polyester	Hawseal
Hepworth Iron Co., Ltd.	4″ 6″ 7″ 9″ 12″ 15″	Polyester	Hepseal
Hepworth Iron Co., Ltd.	4″ 6″	Hepsleve	Hepsleve
Kinson Pottery, Ltd.	4″ 6″ 9″	Polyester	Kinflex
John Knowles & Co. (Wooden Box), Ltd.	4″ 6″ 9″	Polyester	Vitriflex
W. T. Knowles & Sons, Ltd.	4″ 6″ 9″ 12″	Polyester	Knolflex
Lochside Coal & Fire Clay Co., Ltd.	4″ 6″ 7″ 8″ 9″ 12″ 15″ 18″	Polyester	Hepseal
Naylor Bros. (Clayware), Ltd.	4″ 6″ 7″ 8″ 9″ 12″ 15″	Polyester	Naylor Polyester Joint
James Oakes & Co. (Riddings), Ltd.	4″ 6″ 9″ 12″ 15″ 18″	Polyurethane	Fast-test
James Oakes & Co. (Riddings), Ltd.	4″ 6″ 9″ 12″ 15″ 18″	Hot pour	Oanco
Sneyd Pipeworks, Ltd.	4″ 6″ 12″	Polyester	Hepseal
Stella Tileries, Ltd.	4″	Polyester	Stella Tileries Hepseal
Sutton & Co. (Overseal), Ltd.	4″ 6″ 7″ 9″ 12″	Polyester	Ellflex
Western Pipes, Ltd.	4″ 6″ 9″	Plasticised P.V.C.	Draw Flex
Woodville Sanitary Pipe Co., Ltd.	4″ 6″ 7″ 9″ 12″	Polyester	Ellflex
Thos. Wragg & Sons, Ltd.	4″ 6″ 9″ 12″	Polyester	Easilay

On the question of how clay or concrete pipes with cement joints should be laid, there is some difference of opinion. This holds true when the pipes are to be laid on excavated surface, on concrete benching, or when they are to be benched and haunched or to be surrounded with concrete. There is a good deal to be said for more than one method of pipe-laying, and it is not easy to refute many arguments with which one is not in agreement, because most methods of pipe-laying have some drawbacks, and some, although not the best when carried out under the highest-quality supervision, have advantages over the methods preferred when supervision and workmanship are only moderately good and the best finish is not required. There may be some difference of opinion as to what may be considered the classic method of pipe-laying, but the methods to be described will probably coincide with those preferred by a good many specialists in drainage work. It must, however, be admitted that present-day standards of workmanship render the classic method of pipe-laying more and more difficult to obtain.

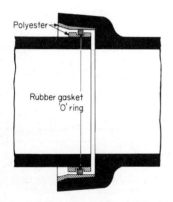

FIG. 31—Polyester mouldings and rubber "O"-ring on vitrified clay pipes

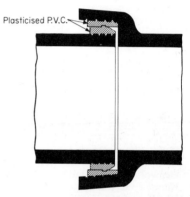

FIG. 32—Plasticised P.V.C. mouldings on vitrified clay pipes, with integral semi-circular bead on spigot providing an area of increased pressure

By courtesy of The Clay Pipe Development Association, Ltd.

Laying pipes on excavated surface. Before any pipes can be laid to true lines and levels, it is necessary for the trench to be excavated to true bottom. For this to be done, sight-rails should be erected no fewer than three at a time, truly set to the levels or falls of the proposed line of pipes, and these should be painted in two halves, black and white, so as to indicate the intended centre line of the pipe. Boning-rods should then be made of such a length that when sighted-in with the rails the feet of the rods are just at invert level of pipe. The trench should be excavated carefully to invert level of pipes, and pegs should then be driven truly in the excavation and on the dead centre-line of the pipe-line to the exact invert level, the pegs being not more than 9 feet apart. When this has been done, the excavation should be trimmed by hand to very slightly less than a pipe thickness below the level pegs, a straight-edge being rested on the pegs, so that the trimming may be

THE CONSTRUCTION OF SEWERS OF SMALL DIAMETER

executed accurately. On no account should excavation be made too deep and afterwards refilled with earth, hardcore, granular fill or any material other than cement mortar or concrete.

After the excavation has been completed, a line should be stretched parallel to the proposed line of pipes, and at such a distance from the centre-line as to be just clear of the collars of the pipes.

The next step in the pipe-laying is to cut "joint holes" in the excavation sufficiently large to receive the collars of the pipes and to permit jointing, but these joint holes should be no larger than absolutely necessary.

The pipes may then be laid and jointed, one at a time. The first pipe to be laid is set at the bottom of the gradient with the socket pointing up the gradient, and the centre-line is carefully measured from the stretched line at the spigot end, the outside of the socket being near to, but not touching, the line. The straight-edge is then placed in the invert of the pipe and resting on two of the level pegs. In this condition it should rest truly in the invert of the pipe, touching it at both ends. It should not rock up and down, but at the same time should touch both level pegs. When resting on the level pegs it should not be possible for one to spring it from side to side in the pipe; if this could be done it would indicate that the pipe had been bedded too low. Generally, the excavation should be slightly high by a matter of a small fraction of an inch, to permit trimming down as necessary or working the pipe down to a true invert.

After the first pipe has been laid, each following pipe should be laid in a somewhat similar manner. Each pipe is lowered into the trench and struck at both ends with a mallet to make sure that it is sound. The spigot, surrounded by a ring of tarred yarn, is then thrust home into the socket of the pipe previously laid and the tarred yarn driven into position with a mallet and wooden caulking tool in such quantity as to centre-up the spigot in the socket. The straight-edge is then laid in the invert of the pipe previously laid, on the nearest level peg up the gradient, and in the invert of the pipe that is being laid, and the pipe similarly tested for line and level and securely bedded as before. The first pipe-joint is then completed. The remainder of the socket is filled with cement mortar, consisting of one part of cement to two parts of sand, of such a consistency that when the joint is completed there is no tendency for the fillet to fall away from the underside of the pipe. After the socket has been completely filled with cement mortar, a fillet of cement mortar is formed round the spigot of the second pipe at an angle of approximately 45° to the barrel and extending not less than 2 inches from the face of the socket. This fillet should be flat surfaced with the aid of a trowel rather than rounded.

When the joint has been completed, any cement which may have entered the pipe through the joint should be removed with a semi-circular wooden rake. The remainder of the pipes are laid in similar manner until the length concerned is completed.

When a length of pipe has been laid and jointed complete, a test is made with air or water, and if the length of pipe is sound the trench may be refilled as described hereinafter.

Concreting. There is perhaps more difference of opinion as to how pipes should be laid when they are to be benched and haunched or surrounded with concrete than on how they should be laid on excavated earth. This may be because proper laying of pipes on concrete requires first-class labour and high-quality supervision if it is to be well done.

The method preferred for first-quality work for all forms of concreting is to commence by laying to true falls a bench of concrete, the surface of which is just the thickness of the pipe below the level pegs, the surface being truly formed a minute fraction of an inch low, and joint holes being cut in the concrete while it is green. Each pipe is then laid as before, the full length of the barrel from the back of the socket to near the end of the spigot resting on

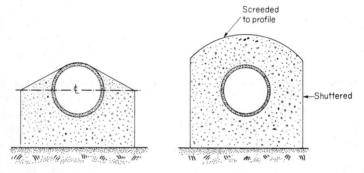

FIG. 33—Benching, haunching and concrete surround to stoneware or concrete pipe

the concrete. In order that the pipes shall be properly beeded, a thin layer of cement mortar should be spread between the underside of the pipe and the surface of the concrete: one might say that the pipe is laid like a brick on cement mortar.

After the pipes have been jointed and tested, the joint holes below the sockets are carefully filled with comparatively wet fine concrete. If the pipes are to be haunched as well as benched, shutters are placed vertically 6 inches from the outside of the pipe and up to the centre-line of the pipe, and concrete carefully filled to this level, after which it is chamfered off tangentially to the barrel of the pipe. If the pipes are to be surrounded with concrete, shutters are similarly formed, but to a higher level, and the concrete is brought over the crown of the pipe to a depth of 6 inches, the work being performed to a true curve with the aid of properly cut wooden templates and a screed. This last may be considered by some to be a refinement, but it is a means by which the engineer can make sure not only that the full depth of concrete is put in everywhere, but also that concrete is not wasted by being deposited excessively thick. Moreover, a concreted line of pipes so com-

THE CONSTRUCTION OF SEWERS OF SMALL DIAMETER   109

pleted appears during construction a far more workmanlike job than concreting finished to a spade-face. It satisfies both the workman and the engineer.

Questionable methods. The following method is described only as a warning to young engineers in charge of works what not to permit, for it is often advocated by building contractors because, in the absence of good workmanship and proper supervision, it is rapid and easy.

The procedure is as follows. First, a concrete bench is formed in the bottom of the trench, its surface being slightly lower than the underside of the sockets of the pipes. The pipes are then laid in the trench, a line being stretched in the usual manner to give a straight line on plan, while for level the pipes are boned in the usual manner, or else by methods which do not deserve description. Gross inaccuracies of gradient are more or less adjusted by propping up the sockets off the surface of the concrete with bits of wood or tile. Frequently, the whole length of pipe is laid in this manner, the yarned spigots being placed in the sockets before the joints are made in cement mortar. When the joints have been made and have set, concrete is poured on one side of the pipe and pushed down until it comes up at the other, after which the haunch or surround is completed.

An objection that has been raised to laying pipes with their sockets resting on the concrete is that the barrel of the pipe does not receive full support underneath. This objection, however, does not hold good if the concrete is pushed under in the manner above described. The real objection to this method of construction is that the pipes are not well and truly laid, although it is possible in this way to lay pipes to an *apparently* even gradient and line, to obtain very often a satisfactory test after completion and, if the concrete is good, to be sure that the work will in the end be as structurally sound as any other. When pipes are laid resting on their collars supported by the makeshift wedges above described, they are almost certainly out of level in the first instance. Moreover, when resting in this manner in the trench they are in an unstable condition, easily shifted during jointing, concreting or refilling. Laying pipes in this manner also totally prevents raking out any cement that may be in the pipe as a result of jointing.

A method of concreting that has been advocated as an alternative to that of laying a bench of concrete, forming joint holes and then laying the pipes after the concrete has set, is to lay the pipes on wet concrete. After each pipe has been laid the next bench of wet concrete is formed ahead of it, and the next pipe laid and jointed in position, and so the work continued until the whole line is completed. The concrete may be put under the barrel of the pipe only, leaving the sockets clear for jointing, and when the line has been tested, the joint holes filled and the bench-and-haunch or surround completed. The advantages claimed for this method are that the difficulty of accurately gauging the positions of joint holes is avoided, and that, the concrete being wet and plastic, it is possible for the pipes to be pushed down to the exact invert level with the knowledge that they rest securely on concrete which will shortly set.

The disadvantage of the method is most obvious to anyone inspecting the work during construction. Almost invariably the pipe-layers are treading in concrete in the bottom of the trench as the pipes are laid in position and, when caulking, are sitting on a pipe that has just been laid on wet concrete. As a result of these two operations the bottom of the trench is liable to become a miry mixture of concrete, earth and infiltration water, and there is a danger of the pipe being displaced after laying. To the engineer and the pipe-layer there is the further objection of a thoroughly unworkmanlike appearance of the line of pipes during construction.

The real test of a pipe-laying method is whether it is one which would be adopted by a craftsman attempting to do the best possible job. When viewed from this angle the best method in use is fairly obviously that in which the pipes are carefully bedded one by one to a true line and gradient, each joint completed and inspected before the next is made, and the last pipe laid always securely resting on sound earth or a set concrete bed.

Another method of official origin and backing is to lay pipes on what is known as "granular fill," a small grade aggregate with no fines. This type of bedding should never be used under pipes with cement joints for it is *certain* to subside to some extent in the course of time, causing fracture. The poorest of concrete would be superior. Neither is granular bedding to be recommended for use under pipes with flexible joints, for it could permit sufficient movement to interfere with the regularity of the invert.

The result of experience of existing sewers and drains is that where workmanship is of the highest quality comparatively low velocities of flow and comparatively slack gradients are possible, and will maintain self-cleansing conditions. Where, however, the standard of workmanship is poor, irregularities occur in the invert, and these reduce flow velocities, tend to hold up solids and make much steeper gradients necessary to ensure self-cleansing conditions. This is not theoretical supposition: it is a fact that has often been observed, and it is of great importance to the drainage authority, because when it is possible to lay pipes to comparatively slack gradients, considerable economies in excavation can be made. Conversely, if the experience of the authority is (nearly always as a result of bad pipe-laying) that steep gradients are necessary to maintain self-cleansing conditions, the designer is handicapped with considerable difficulties, and the cost of the works is increased.

Cast-iron pipes. Some engineers prefer to use cast-iron or spun-iron in place of stoneware concreted. These are, for sewerage purposes, cast-iron pipes B.S. 78 and spun-iron pipes B.S. 1211, of the types used for water supply (and not B.S. 437 as used for drainage of premises), and they are of the class referred to as "Class B" in the British Standard Specifications (see Vol. I, Chapter XI).

Spun-iron pipes B.S. 1211 laid on the ground do not differ much in cost from stoneware pipes surrounded with concrete, and they have the advantages of being more easy to lay, more permanently watertight, also more flexible and therefore less liable to injury as a result of settlement. But they are not altogether better in waterlogged ground containing sulphates, where they are liable to graphitisation, and in such circumstances stoneware pipes with special joints and surrounded with concrete made with sulphate-resisting, or in extreme cases, high-alumina cement would be preferable.

One of the chief objections to using cast-iron pipes for sewers is that the making of future lateral connections is difficult and expensive. While a saddle is easily put on a clay-pipe, the recognised method of making a lateral connection to a cast-iron pipe is to cut out a length and insert a cast-iron junction. (Incidentally, where cast-iron sewers are laid and the laterals constructed at time of laying, it is consistent to construct cast-iron laterals.)

For cast-iron pipes poured molten lead joints were once most commonly used. Now bolted-gland joints are largely used to economise in the use of lead.

Steel tubes. Steel tubes are useful, particularly for rising mains. These may have welded-sleeve joints, caulked joints or one of the many varieties of flexible joints, such as the Victaulic Joint, Johnson Coupling, etc. They are seldom, if ever, used for gravity sewers underground.

Concrete pipes. For sizes of sewer more than 9 inches diameter concrete pipes (B.S. 556) are frequently preferred to stoneware, because they are cheaper. For soil sewers of moderately large size, socketed concrete pipes are preferred. But for larger-diameter sewers—i.e. those that can be entered comfortably during construction, and particularly large surface-water sewers—ogee-jointed concrete pipes are often used, because they are cheaper than socketed pipes and can be laid in a narrower trench or heading. Pre-stressed reinforced-concrete pipes are also obtainable. For very large sewers of considerable length concrete tubes can be considered as an expensive form of permanent shuttering: it is cheaper to use mass concrete and special steel shuttering.

Pre-stressed concrete pressure pipes. Pre-stressed concrete pressure pipes are useful for large sewers and particularly rising mains, because they can stand considerable internal pressures and in this respect serve as a substitute for iron or steel. They have been made in sizes from 12 inches to 48 inches internal diameter.

Joints for concrete pipes. The first pre-cast concrete sewers were constructed of concrete pipes with ogee joints, or other joints not involving a socket. This to a great extent restricted their use to the larger sizes. The obvious advantage of using sockets similar to those of stoneware pipes brought into use the concrete pipes which had a socket of somewhat similar proportions to the socket of a stoneware pipe, which could stand up to equally strict tests for watertightness as applied to clay pipes, and which could be produced at a lower cost in the larger sizes. To make possible accurate centering of spigots a number of self-centering designs were evolved, in which either a tapered spigot was used to fit a tapered portion of the socket, or in which centering fillets were provided in the socket.

The difficulty of jointing pipes in wet trenches has resulted in the invention of a number of patent joints which may be either grouted or which, alternatively, do not require the use of cement mortar. The Marston "Ellispun" joint is a joint of the self-centering type. The joint is made by pouring grout into a hole at the top of the socket. Leakage of grout into the pipe is prevented by a ring of bitumen on the spigot, and leakage between the pipe socket and the trench is prevented by a canvas ring which is permanently fixed to the socket and tied round the spigot by means of a wire. This joint can be made in a very wet trench.

The "Stanton–Cornelius" joint serves the double purpose of providing a joint that may be made in a wet trench and of giving a degree of flexibility which will prevent breaking of joints during movement owing to unsound foundations. A rubber ring is placed round the spigot of the pipe. The spigot is driven home into the socket by means of a jack or, in the larger sizes, pulled in with a wire rope, and the sealing ring is thereby compressed between the spigot and the socket, making a tight but flexible joint. In the modified Stanton–Cornelius joint the socket may also be grouted after the joint has been made with the rubber ring.

For the construction of very large sewers or for work in heading it is an advantage to be able to dispense with sockets. This does not rule out grouting. The "Trocoll Patent Grouted Joint" is applicable to ogee joints. In the ogee joint there is a grout space, and grouting holes are provided. An outer permanent sealing band is passed round the pipe to prevent leakage of grout outwards, while inside the pipe is placed a temporary steel expanding band with rubber packing. Most grouted joints, either of pipes or tubes, may be arranged for grouting from inside the pipe instead of from the outside, in the larger sizes.

Asbestos-cement pipes for sewerage and drainage. In addition to the pressure pipes used for water-mains and sewage rising mains there are asbestos-cement pipes for drainage purposes. These are made in three grades but, unlike the pressure pipes, the internal diameters are not affected by the grade. They are listed in 4-inch to 10-inch internal diameter in 1-inch stages and 12-inch to 36-inch in 3-inch stages. Metric sizes are given but these are nominal equivalents, not additional sizes. Limited ranges of bends and angle-junctions are available.

The pipes are jointed with asbestos-cement sleeves and rubber rings. The procedure is to mount two rubber rings on each pipe end so that one ring is in close proximity to the end of the pipe and the other ring one complete roll distant from the position occupied by the first. With pipes in sizes 4-inch to 9-inch diameter this operation is facilitated by use of a taper plug. With pipes in sizes upwards of 9-inch diameter the rings are pulled over the pipe

THE CONSTRUCTION OF SEWERS OF SMALL DIAMETER 113

end by hand. Twist should then be removed by inserting the ring locator under the ring and describing one or two revolutions. The ring will then also be square to the axis of the pipe and in approximately correct location. The location should be checked by rolling the ring towards the pipe end to ensure that it comes to rest in close proximity to it.

By means of simple leverage the sleeve is pushed over the rubber rings on the end of the pipe about to be laid until the pipe end is protruding slightly. Then the sleeve is pushed over the rubber rings on the adjacent pipe to bring it into the central and final position in which the ends of the sleeves should coincide with the location marks on the pipes.

Headings. Sewers of moderate depth are generally laid in open trenches, but when depth from ground surface to invert exceeds 14 feet, excavation in heading usually reduces the cost. If this method is permitted, skilled men should be employed both to drive the headings and to refill them after the pipes have been laid. The filling must be very carefully done and closely supervised, or cavities will be left.

Timbering. Unless the ground is unusually firm, the sides in the trenches or headings and the roofs of the latter have to be supported by timbering and, if the ground is very unstable, the trenches may have to be close-timbered for their full depth. In running sand it is necessary to close-timber or to drive steel sheet piling to some feet below the bottom of the trench. If there are any buildings near-by, special care must be taken to support the sides of the trenches, and it may be necessary to leave the timber in.

Buildings near the line of a sewer should always be carefully inspected before work is begun, any existing cracks being noted and photographed; for damage done, or alleged to be done, to such buildings is a fruitful source of claims. Such damage may be caused either by the falling in of the sides of a trench or by subsidence due to the withdrawal of water or sand, by pumping.

Pumping. Where sewers have to be laid in wet ground, the water has to be kept down while the joints are being made, and until they have had time to set. This is sometimes assisted by open-jointed pipes laid beside the sewer just low enough to drain the joint holes.

In sandy ground care must be taken not to undermine the sewer by pumping the sand away from under it; and if the water which is pumped from the trench is allowed to run down the sewer, any mud or grit which it may contain should first be allowed to settle.

Filling trenches and reinstatement of surfaces. In filling trenches the greatest care should be taken not to disturb the pipes, which should first be covered to a depth of at least 9 inches with fine selected material, well trodden in. The trench should then be filled in layers not more than 6 inches deep, each of which should be thoroughly consolidated before any more earth is put in. Two men should be employed in ramming for each man filling the trench. Mechanical punners are sometimes used, but the first few feet of the filling should always be consolidated by hand. If the ramming is properly done, the whole of the excavated material, with the exception of the small quantity displaced by the pipe, should be returned to the trench.

The reinstatement of road surfaces after the laying of sewers is often an expensive matter, particularly in the case of asphalt paving laid on reinforced concrete. It is rarely possible to reinstate such a surface immediately after the trench has been filled, and it is usual to put in a temporary surfacing and to defer the permanent reinstatement until after the filling has had time to consolidate. The restoration of road surfaces is frequently the subject of disputes, and it is desirable, if possible, to arrange for the highway authority to do the work. Some county councils, indeed, insist on the reinstatement of their roads being left to them. Whether or not this is done, it should always be specified that the surfaces of all roads must be reinstated to the satisfaction of the highway authority and certificates to this effect should be called for before the final certificate is issued.

Testing lines of pipes. Sewers constructed of pipes are always tested for soundness before refilling of excavation and, in special circumstances, afterwards. Methods of testing and recommended standards are given in Vol. I, Chapter II.

CHAPTER XI

THE CONSTRUCTION OF SEWERS OF LARGE DIAMETER

Sewers of large diameter are constructed of various materials and to various forms, circular or otherwise, according to the circumstances in which they have to be built.

Modern sewers are similar to older constructions, in that the circular sewer is still the most commonly used, and next to this various forms of culvert with arched soffit. The differences between modern and previous practice are in the disuse of the egg-shaped form, the innovation of forms for the sake of structural economy, the design of the barrel to as near as possible the exact size required for maximum flow, and the utilisation of many new materials of construction.

Cross-sections of sewers. The standard sewer cross-section for all diameters is a circle. Besides being used in early large sewers and in large sewers of the present day, this form is the most common for all pipes. Other than circular forms are almost unknown for small pipe sewers, and of the sewers of intermediate size that are constructed of pre-cast concrete and similar modern materials only a very small percentage—probably less than 1 per cent—is other than circular.

The circular cross-section is that which almost invariably occurs in nature for the transmission of all kinds of fluids in animals and plants. Its advantages are the maximum hydraulic mean depth for the area contained, the optimum shape for withstanding internal pressure and, in most instances, the optimum shape for withstanding external pressure.

The reasons for adopting other shapes have been:

1. Economy in cost at a time when circular work was more expensive than straight work: this led to the construction of U-shaped sewers and sewers with vertical sides and flat inverts.

2. Shaping of the sewer to give extra headroom and safety for men walking through it: this led to the use of sewers with vertical sides and considerable height compared with width, also the egg-shaped sewer which is usually 50 per cent higher than it is wide.

3. Hydraulic advantages (often fallacious). In this class have been several special shapes of sewer, in particular the egg-shaped sewer, which at one time was very largely used because it was believed to be more self-cleansing than the circular sewer at low rates of flow. It has now been shown that this belief was erroneous and, as the egg-shaped sewer is expensive to

construct and not as strong as a circular sewer of the same capacity, the design is obsolescent.

4. Special circumstances such as low headroom under a stream, railway or structure that cannot be moved. Such cases call for designs according to circumstances and need not be discussed here.

New methods of construction are making the circular shape still more frequently selected for sewers. Wherever there is a considerable length of sewer to be laid special steel shuttering can be made and used over and over again making the cost of shuttering very small and completely doing away with the advantages of flat walls and inverts. In a recent contract there was a large-diameter reinforced-concrete culvert shown on the drawings as circular in cross-section. The contractor requested permission to put in a flat invert to facilitate construction, but after building the first length he reverted to circular section for he found that the invert could easily be cast to the true curve, vibrated concrete running in under the shuttering.

Where timber shuttering is used mostly for short lengths of sewer or culvert which do not justify special steel shuttering, an octagonal shape has very nearly the same hydraulic and structural properties as the truly circular cross-section.

Other forms approximating to a circle, in that their height is equal to their width, and others built with greater height than width, are so constructed that the section of the sewer may approximate to the dimensions of the excavation or heading; thereby expensive refilling or packing with concrete is avoided, or economy effected by the minimisation of radial work.

An additional class that appears to have been popular in America includes all designs based on the assumption of earth pressures and involving the formation of arch and invert in such a manner that the sewer takes the form of the thrust-line of the arch.

Special shapes of sewer will be further discussed under the description of mass concrete and brickwork culverts in this chapter.

Cast-iron segments. Cast-iron *segments* are used when it is advisable to use cast-iron, but when the diameter of the sewer is too large, or when other conditions are adverse for the use of cast-iron *pipes*. The particular case to which they may be applied is the construction of large-diameter sewers in heading in a soil where excavation by means of a Greathead shield is possible. The segments are cast to radius and bolted together by means of flanges on the inside of the circle. The circumferential joints usually have machined faces, which are bolted together and "rust" jointed with sal-ammoniac and iron filings, jointed with rubber preparation or provided with a groove for caulking. The longitudinal joints are usually bolted together and caulked with cement mortar, rubber preparations, lead-wool or wood fillets. Most of the faces of the longitudinal joints are radial to the circle, but in order that the work may be built up from the inside of the sewer, it is necessary to have

TABLE 26
PROPORTIONS OF CAST-IRON SEGMENTAL SEWERS

Internal diameter of sewer	5' 0"	5' 3"	5' 6"	6' 0"	6' 6"	7' 0"	7' 6"	8' 0"	8' 6"	9' 0"	9' 6"	10' 0"
Internal diameter of iron	5' 5"	5' 8"	5' 11"	6' 5"	6' 11"	7' 5"	7' 11"	8' 5"	8' 11"	9' 5"	9' 11"	10' 5"
External diameter of iron	6' 1"	6' 4"	6' 7"	7' 1"	7' 7$\frac{1}{2}$"	8' 1$\frac{1}{2}$"	8' 7$\frac{1}{2}$"	9' 2"	9' 8"	10' 2$\frac{1}{2}$"	10' 8$\frac{1}{2}$"	11' 2$\frac{1}{2}$"
Thickness of iron	$\frac{3}{4}$"	$\frac{3}{4}$"	$\frac{3}{4}$"	$\frac{3}{4}$"	$\frac{3}{4}$"	$\frac{3}{4}$"	$\frac{3}{4}$"	$\frac{3}{4}$"	$\frac{3}{4}$"	$\frac{3}{4}$"	$\frac{3}{4}$"	$\frac{3}{4}$"
Depth of flanges	3$\frac{1}{4}$"	3$\frac{1}{4}$"	3$\frac{1}{4}$"	3$\frac{1}{4}$"	3$\frac{1}{2}$"	3$\frac{1}{2}$"	3$\frac{1}{2}$"	3$\frac{3}{4}$"	3$\frac{3}{4}$"	4"	4"	4"
Thickness of flanges at base	1"	1"	1"	1$\frac{1}{8}$"	1$\frac{1}{8}$"	1$\frac{1}{8}$"	1$\frac{1}{8}$"	1$\frac{1}{8}$"	1$\frac{1}{8}$"	1$\frac{1}{8}$"	1$\frac{1}{8}$"	1$\frac{1}{8}$"
Thickness of flanges at edge	$\frac{7}{8}$"	$\frac{7}{8}$"	$\frac{7}{8}$"	1"	1"	1"	1"	1"	1"	1"	1"	1"
Diameter of bolt circle	5' 8$\frac{1}{2}$"	5' 11$\frac{1}{4}$"	6' 2$\frac{1}{4}$"	6' 8$\frac{1}{2}$"	7' 2$\frac{1}{2}$"	7' 8$\frac{1}{2}$"	8' 2$\frac{1}{2}$"	8' 8$\frac{3}{4}$"	9' 2$\frac{3}{4}$"	9' 8"	10' 3"	10' 9"
No. of bolts	19	23	23	23	23	27	29	29	34	36	39	39
Size of bolts, diameter	$\frac{3}{4}$"	$\frac{3}{4}$"	$\frac{3}{4}$"	$\frac{7}{8}$"	$\frac{7}{8}$"	$\frac{7}{8}$"	$\frac{7}{8}$"	$\frac{7}{8}$"	1"	1"	1"	1"
Size of bolts, length	4"	4"	4"	4"	4$\frac{1}{4}$"	4$\frac{1}{4}$"	4$\frac{1}{4}$"	4$\frac{1}{4}$"	4$\frac{1}{4}$"	4$\frac{1}{4}$"	4$\frac{1}{4}$"	4$\frac{1}{4}$"
No. of ordinary segments (excluding 1 key and 2 taper segments), per ring	3	4	4	4	4	4	4	4	5	5	5	5

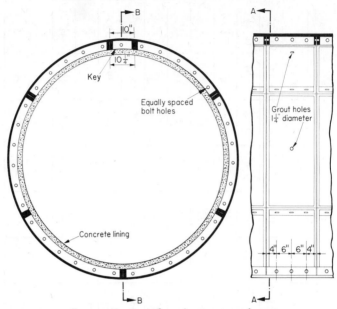

Fig. 34—Section of cast-iron segmental sewer

a "key" segment which can be inserted at the crown, and this must be of wedge form (see Figs. 34 and 35). Thus, three types of segments are required: the normal segments with radial joints, the "taper" segments next to the key segment with one radial and one non-radial joint, and the key segment. Where the sewers are laid in straight lines the key segments are staggered to break the joint.

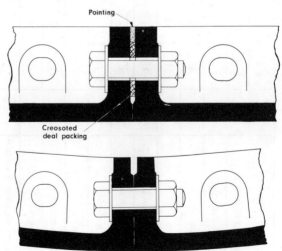

Fig. 35—Details of cast-iron tubbing
(above) transverse joint   (below) longitudinal joint

THE CONSTRUCTION OF SEWERS OF LARGE DIAMETER

As cast-iron segment sewers constructed in heading must have the flanges on the inside, it is necessary to line them with materials which will give a clean internal surface and smooth invert. Concrete is often used, and filled either to the faces of the internal flanges or else filled to cover the flanges, and so protect the iron. The centre of the invert should be protected against scour that is caused by sliding stones, etc., with vitrified brickwork. Before connecting, the space between the earth and the cast-iron is grouted under pressure.

It is possible to work in large-radius curves on plan using straight segments, the minimum radius being about twenty times the external diameter of the cast-iron ring. Sharper curves require special segments.

Junctions of small pipes to cast-iron segment sewers are made by means of special cast-iron socket pieces with curved flanges to bolt to the faces of the segments.

Pre-cast concrete segments. Desire for economy in the construction of large sewers has produced designs of pre-cast concrete segments. The McAlpine system involves the use of pre-cast segments with tongued-and-grooved edges, and which are erected to break joint longitudinally. A key segment is inserted in the crown to complete the ring, as in the method of cast-iron segment construction. Steel reinforcement bent to true radius is inserted into each circumferential groove of the joint during erection. The internal joints are pointed with cement mortar and, after this has set, the joint grooves and the cavities between the segments and the external earth are grouted under pressure, transforming the whole structure into a solid reinforced-concrete barrel of considerable strength. This method, which was introduced in 1903 in a slightly different form, claims the advantages of low cost and added strength, together with reduced gross excavation, rapid construction and minimum timbering. Excavation is carried out ring by ring. The completed barrel may be lined with $4\frac{1}{2}$ inches of brickwork, "Gunite" or otherwise as required.

The Kinnear Moodie pre-cast concrete segments are in two designs, reinforced concrete ribbed segments, arranged to bolt together by means of flanges and steel bolts, in a similar manner to cast-iron segments, and "solid" segments with recessed bolt holes. They are very largely used as an alternative to cast-iron segments where external water pressure does not exceed 30 feet.

Cost of large sewers in heading. In 1966 one engineer was using the following formula for estimating the cost of large-diameter sewers constructed in heading:

$$\text{Cost per foot run} = £(D^2 + 20)$$

where: D = diameter of sewer in feet.

For present-day work allowance should be made for variation in the value of the pound (see Vol. I, Table 3). This estimate was for pre-cast concrete segment construction lined with concrete and having a brick invert. The cost of manholes and work in compressed air would be extra, as would also be cast-iron construction.

Reinforced concrete. Reinforced concrete is useful for sewers larger than those that can be constructed of concrete tubes, or for work of special section (see Fig. 36), when high internal pressures are expected, or when it is feared that there may be irregular external loads. Reinforced concrete may also serve as a material for strengthening existing large sewers that are heavily surcharged, and particularly it may be applied to large-diameter low-pressure rising mains. Circular barrel section is usual, but a flat invert connecting to sides sloping at 45°, all tangential to the circle, can be advantageous in sewers of very large sizes. When used for soil-sewers, the work is with advantage inverted with vitrified brickwork. This need not be for more than one-sixth of the circumference—that part which is scoured by sliding stones.

Bricks for sewerage. Ordinary engineering bricks do not necessarily have the qualities required for sewerage purposes. While sound enough for building work they may be too soft to resist scour or too absorbent of water. The inverts of some of London's sewers built by the Metropolitan Board of Works of apparently hard bricks are recorded to have been scoured to a depth of several inches or as much as a course of brickwork.

A drainage brick should be smooth, vitrified, hard and heavy and a piece broken from the body of it should not absorb more than 4 per cent of its weight in water when soaked for twenty-four hours.

Many engineers consider the Accrington "Nori" as the ideal brick for use in inverts and particularly for chemically resisting structures. It is very smooth, hard and heavy, vitrified and of regular shape. A largely similar brick much used in the south of England is the Southwater. Staffordshire blue bricks are hard and heavy but lack the smoothness of the Accrington. There are several other bricks of various colours, including one that is very like salt-glazed ware, which are suitable for drainage purposes.

Mass concrete and brickwork culverts. Brick barrels constructed in either heading or open cut may be perfectly circular externally if the excavation is made so as to accommodate this form, but as a large proportion of works is constructed in either flat-bottomed trench or trapezoidal heading, the construction of a circular barrel may require considerable back-filling of concrete, particularly if, for the sake of soundness, the work is filled to excavation.

To economise in construction, it may be found advisable to use a section which, instead of being circular, approximates to a circle. Not only will this give a larger bore for the same quantity of excavation or a reduced quantity of mass concrete for the same flow capacity, but it introduces further economy in the use of plane surfaces in place of expensive radial work. Figure 37 shows one form of this kind.

In approximations to the circle in which there are vertical side walls, it is

THE CONSTRUCTION OF SEWERS OF LARGE DIAMETER 121

necessary to calculate bending moments in the walls and the floor when they are not sufficiently thick to "arch themselves." Figure 47 can be applied with discretion to such cases.

A form which was often once used, which is an approximation to the circle but does not involve plane surfaces, is the horseshoe shape, one variety of which is illustrated in Fig. 38. This section is convenient for large sewers in open cut. It is sound structurally and, owing to the fairly flat invert, is a useful form of culvert when it is intended that men should enter from time to time.

A form used in America is the inverted egg-shape or catenary arch sewer. This section can be made to the theoretically perfect form as regards structural soundness, and at the same time accommodates itself well in a normal trapezoidal heading.

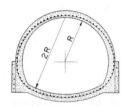
Fig. 36—Reinforced-concrete sewer

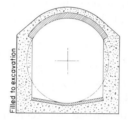
Fig. 37—Mass concrete culvert

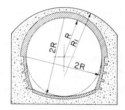
Fig. 38—Horseshoe-shaped sewer

The "normal" egg-shaped sewer,* at one time very popular in this country, is structurally poor and hydraulically has little of the advantage that has been claimed for it. It is obsolescent, although a very small percentage of precast concrete egg-shaped sewers are perhaps still made and many old egg-shaped sewers exist in large urban areas. When the normal rate of flow in an egg-shaped sewer is one-twentieth of the maximum capacity of the sewer (and it is in this condition that the egg-shape has its maximum advantage), the velocity in the egg-shaped sewer is somewhere in the region of 7 per cent higher than the velocity of a circular sewer of equal capacity laid at the same gradient. Such an advantage is negligible, but, on the other hand, the difference in cost of construction is far from negligible. In most sewerage designs it is possible to lay circular sewers at steeper gradients than egg-shaped sewers of the same capacity because of the lower crown to invert depth, and this means that it is often possible to obtain a higher velocity in a circular sewer than would be obtained if an egg-shaped sewer were used. The advantages of egg-shaped sewers are that the increased height gives more walking room for men and that, at minimum flow, solids do not tend to

* The name given to an egg-shaped sewer which is 50 per cent higher than it is wide and has an invert radius of one-half that of the crown.

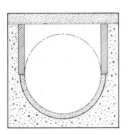
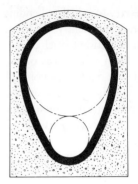

Fig. 39—Cross-section of U-shaped sewer   Fig. 40—"Normal" egg-shaped sewer

strand at the sides. They are cleaner than circular sewers only when gradients are adequate (i.e. the gradients are sufficient to be self-cleansing for a pipe of radius equal to the radius of the invert of the egg-shaped sewer) and flow periodically small.

Economy of construction in the region of about 12 per cent may be obtained when medium-depth large sewers in open cut are constructed to the U-shape illustrated in Fig. 39. The properties of design of this kind are described by Thomas Donkin.[60]

Rectangular culverts are particularly economical for construction in reinforced concrete in open cut. Pre-cast concrete segments for this section have been made. The form is not ideal for sewers because of the opportunity for silting which it provides.

The section illustrated in Fig. 41 is one that is used where there is little headroom. It has no other advantage, and usually requires detailed design of the arch. The invert, for economy, is constructed either to a large radius or sloping to a central channel or grip.

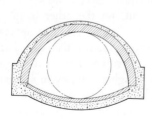
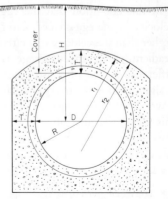

Fig. 41—Culvert with low headroom   Fig. 42—Large concrete pipe sewer

External pressures on buried pipes and culverts. The pressure on a buried pipe or culvert can be greater or less than the weight of the superimposed earth and changes with time. When a pipe is laid in the trench and the earth is newly refilled, the pressure on the pipe may be less than the column of earth vertically above it because the fill is held by friction between the sides of the trench. With time the fill will subside and eventually the pressure on the pipe should become about equal to the weight of the superimposed earth, but whether this happens in the life of the structure is a matter of conjecture.

Some practice is to assume that the load on the pipe is that which occurs at an early stage after the refilling of earth. On this assumption, the maximum load per foot run of pipe or culvert laid in trench not more than about 20 feet deep and refilled with heavy earth such as saturated clay is not likely to exceed that given by the following formulae, which was obtained by a simplification of the Marston formula, for rigid sewers on ordinary bedding occupying not less than two-thirds of the trench width:

$$W = 120 \, H^{\frac{2}{3}} \, Bt^{1\frac{1}{3}}$$

where: W = vertical pressure per foot run of sewer in lbs
H = depth of fill above top of sewer in feet
Bt = width of trench at top of sewer in feet
(maximum value of Bt cannot exceed H).

Although the theory that the load on top of a sewer that is laid in the trench is less than the weight of the column of earth vertically above the sewer is supported by experiment, it is not necessarily safe to design to this theory, for a number of reasons. First, the designer is not to know what method of excavation and refilling may be adopted by the contractor and if his design is based on the assumption of narrow trenches but the contractor excavates wide ditches with sides at the angle of repose so as to use modern machinery and avoid timbering, the actual loads will be different from those calculated. Second, the sides of trenches sometimes subside, compressing the fill after the timber has been withdrawn, again producing a condition not envisaged in some theoretical calculation.

Theory is of more importance in those instances where the load on the top of the sewer may be considerably greater than that of the column of earth vertically above. Such a condition arises when a sewer is so laid that its top is above the general level of the natural ground, so that the fill in, consolidating and subsiding as the months or years pass, gravitates past the sewer which resists its downward movement.

A very marked condition of this kind arises when a pipe is supported on piers or piles and beams at some distance above natural ground-level and earth is filled over it. When this is done, the fill subsides past the pipe on each side but the weight of a wedge of earth is carried by the pipe. This wedge has a width equal to that of the pipe or beam at its bottom but slopes out evenly and it is very wide at the surface with the result that a very great weight can bear on the pipe and beam.

In one instance a reinforced-concrete beam and the cast-iron pipe which it supported were both broken. Before further design was put in hand a scale-model test was carried out to ascertain the weight and the proportions of the superimposed wedge of earth. When this test was repeated a number of times it was found that, while in any one test the slopes of the sides of the wedges were truly straight lines, the slope varied from one test to another even when conditions were apparently identical and it was not practicable to relate this slope to the angle of repose of the material.

Owing to the very great weight of earth that would have to be allowed for in design, had it been assumed that the entire weight of the wedge bore on the pipe, a compromise was adopted which was that reinforced-concrete beams carrying pipes in fill should take the weight of wedges of earth with sides sloping out at an angle of 15° to the vertical, with the proviso that the stress in the steel should not exceed 30,000 lb per square inch under the load from a wedge of earth having sides sloping at 30° to the vertical. So far, no further failures have been reported.

In all those cases where the vertical load on the top of the sewer may equal or exceed the weight of the superimposed column of earth or where there may be irregular or heavy superimposed loads from above ground-level, theoretical calculation of earth pressure is desirable. The danger of theoretical calculations, however, is that too much faith may be put in them and, if the factor of safety that is allowed is too small, failure may result. Also earth-pressure theories often omit reference to subsoil water, the pressure of which may exceed earth pressure so much that the latter can be neglected. The pressure of earth on a sewer laid in a deep trench may be less than the weight of the earth or, if the sewer is laid in heading, particularly in rock, the earth pressure may be nil. On the other hand the water pressure will be that of the head of water from the highest level of the natural water-table or, where a sewer is laid in trench in clay, the water pressure may occasionally be that due to the entire filling of the trench with water to ground-level.

Thus, when sewers are laid at considerable depth, it is necessary to consider the crushing effect of the water pressure and, if an adequate factor of safety is allowed, it may not be necessary to make any further allowance for earth weight in many instances.

The difference between water pressure and earth pressure is that water pressure is virtually equal all round the pipe, having a crushing effect, while earth pressure bears on the crown, producing a reaction at the invert. In cases where the writer has observed failure of vitrified clay and pre-cast concrete pipes as a result of earth pressure cracks have appeared at the crown and invert and half-way down at the sides. The crown has come down, the invert up and the sides moved out.

Ancient brick culverts were often constructed of one ring of brickwork

THE CONSTRUCTION OF SEWERS OF LARGE DIAMETER 125

FIG. 43—Steel sewer being laid in trench
By courtesy of Stewarts & Lloyds, Ltd.

laid on ground carefully shaped to radius up to half diameter. After many years approaching a century some of these have failed by distortion of the arch often laterally at the top, one side of the crown tending to come down and the other moving upwards, showing some irregularity of earth pressure.

In the formulation of a rule of thumb for the thickness of circular barrels, the writer has assumed that the sewers are loaded with a head of water from ground-level to the centre of the sewer, and that when the sewers are deep no other pressures are produced (see Fig. 42).

It is next necessary to adjust the formula based on the above assumption so that it gives a reasonable result for shallow work, for which purpose it has been assumed that the minimum thickness of the barrel is one-twelfth internal diameter. This rule is good only if the work is filled to excavation to springing level, and it does not make allowance for the irregular loads due to traffic, etc., but only for loads due to earth pressure.

The formula reads:

$$480\,T - HT = HR + 80R$$

where: T = thickness in feet
H = depth in feet
R = internal radius in feet

and gives the figures in Table 27.

TABLE 27

THICKNESS OF CIRCULAR SEWERS

Depth in feet	Diameter ÷ thickness
0	12
10	10
20	9
30	8
40	7
60	6
80	5
100	4

The above formula treats the sewers as if they were thin pipes. In the case of the deeper sewers with thick barrels, it is therefore not perfectly true, because the theory for thick pipes should apply; but it gives a result near enough for practical purposes, particularly when the sewer consists, on the inner surface, of some substance of high compressive strength, such as good-quality brickwork or pre-cast concrete.

CHAPTER XII

THE CONSTRUCTION OF MANHOLES AND CHAMBERS

MANHOLES may be considered of two main classes: simple manholes for general use on small sewers; and special manholes, including manholes on large sewers, side-entrance manholes, storm-overflow chambers, backdrop manholes, cascades or tumbling-bays, pressure manholes on surcharged sewers and so forth. Manholes may also be classed according to the materials used in construction. In early days brickwork was invariably used. Now mass concrete, reinforced concrete and pre-cast concrete pipes are materials which are used extensively, while cast-iron is employed for special purposes.

The brick manhole is still one of the most common constructions for all types of work. For better-class drainage purposes impervious bricks, such as Southwater, Accrington "Nori" or Staffordshire blue bricks, are preferable. The brick manhole is a useful construction on small contracts, where local labour only is available and the quality of mass- or reinforced-concrete finish would be doubtful.

Mass concrete has come into use for manholes of all sizes and types. It has been found satisfactory and has a good appearance *if the specification and supervision are strict*. In underground work it is not as expensive as best-quality brickwork, and is a simple material for design purposes. Reinforced-concrete construction is generally unnecessary and too expensive for manholes of normal proportions. At a cost it gives added strength to a structure which is inherently strong. The placing of reinforcement in confined spaces is liable to be difficult, and the strength gained, even if it were required, could be more easily obtained and at a lower cost by merely increasing the thickness of concrete. Reinforcement is, however, useful in roof slabs for manholes constructed of concrete or other materials. All concrete should be vibrated.

The pre-cast concrete pipe manhole is very extensively used. From most points of view it is an admirable structure. It is cheap to build, requires little skilled labour and, with a reasonable degree of supervision, appears well on finish. It is particularly useful where rapid construction is of importance. Although generally illustrated in manufacturers' catalogues in the form of simple manholes, the pre-cast concrete construction is applicable to the most complicated designs, and is useful in conjunction with large sewers as well as small pipe sewers (see Figs. 44 and 45).

In many modern designs the advantages of several materials are combined in one structure, each material being employed in the situation in which it is most satisfactory. For example, foundations may be of mass concrete; inverts of suitable rendering, stoneware or blue brickwork; walls of brickwork, mass concrete brickfaced, or mass concrete with a worked face, or pre-cast concrete pipes jointed vertically; roofs of reinforced concrete, filler-joist mass concrete construction, semi-reinforced concrete, arched concrete, arched brickwork, or arched work formed of pre-cast concrete segments or cut pipes with mass concrete backing.

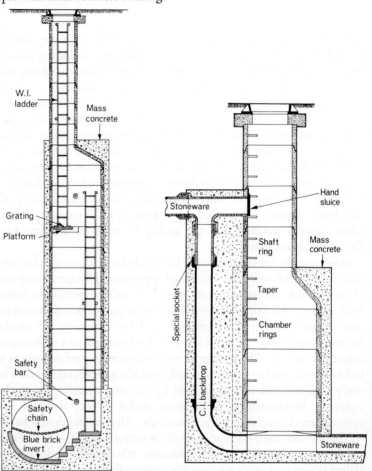

FIG. 44—Deep side-entrance manhole constructed of pre-cast concrete pipes

FIG. 45—Pre-cast concrete manhole with backdrop

Brick and mass-concrete manholes on small sewers. These manholes consist of a chamber of suitable size, often 4 feet 6 inches long by 3 feet wide, or larger or smaller, according to the opinion of the engineer or the requirements of local conditions. The floor consists of the foundation slab on which rest the walls and on which are formed the invert and benchings. Varying thicknesses have been used for this slab, but generally it may be said that the thickness of concrete under the invert of the channel in the manhole seldom needs to exceed 6 inches, because the benchings add greatly to the thickness and give adequate spread to the walls.

Inverts and benchings. Many varieties of benchings may be seen. They may be good or faulty either in design or execution of the work. The following is considered by many to be the best design (see Fig. 46).

The invert is shaped to a half-round constructed to the radius of the outgoing pipe and laid to the same fall. The sides of the channel are brought up vertically to the level of, or slightly higher than, the crown of the outgoing

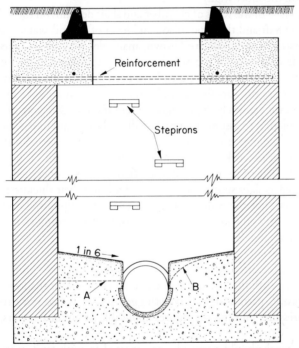

FIG. 46—Shallow manhole showing good and bad designs of benching

pipe. Here the toe of the benching is formed by a slightly rounded arris between the vertical side of the channel and the slope of the benching. The benching is sloped evenly upwards at a gradient of about 1 in 6, which gradient makes the benching reasonably dry, and at the same time not slippery: a slope of 1 in $4\frac{1}{2}$ or a steeper is dangerous. The broken line "A" shows one bad form of benching, which is too low and too flat, while the broken line "B" shows a cushion-like benching very often seen on building sites. This, while it may be clean and dry, does not give a safe foothold and has a very unpleasing appearance.

Benchings are formed of concrete rendered with cement mortar or, in some cases, of brickwork.

Chambers and shafts. The chamber of a manhole should, as far as possible, give adequate headroom—i.e. the roof should not be lower than 6 feet above the highest part of the benching unless the depth from ground-level to the benching necessarily restricts the headroom. In deep manholes and manholes of average depth an access shaft is carried up above the roof. If of brickwork, this shaft should be generally not less than 2 feet 3 inches by 2 feet $7\frac{1}{2}$ inches on plan, or 2 feet 3 inches in diameter if the construction is in concrete tubes. Larger shafts should not be required except for special-purpose manholes, and much smaller shafts are inadequate. (The ladder should be on the short wall.)

The thickness of the walls of chamber and shaft depends on the depth of the chamber or shaft and the type of soil and level of subsoil water. Where the local geological conditions are known, manholes may be designed accordingly, but in other cases it is best to assume a waterlogged soil. A number of rules for the strength of shafts, etc., have been given from time to time, but most of these are rule-of-thumb applications which do not take into account all considerations and which give inconsistent results. One way of approaching the problem is to assume that each wall is a beam, and to design the work according to the pressure set up at each depth by the external soil or water, allowing the brickwork or concrete a safe tensile strength of, say, 60 lb per square inch. Where roofed and floored chambers are concerned, walls may be considered as slabs supported on four edges, and their thicknesses reduced in proportion. For the walls of shafts, or chambers supported on two edges only, Fig. 47 gives the necessary thickness of wall according to length of the interior surface and depth.

Circular shafts are completely arched, and therefore have to resist external pressure only (or internal pressure in case of surcharge). A minimum thickness of one-twelfth the internal diameter could be used for shallow work, but for deep work, particularly in waterlogged soil, it is necessary to consider external pressure, and use the figures applicable to circular sewers, given in Table 27.

Shallow work should always be constructed to withstand more than the external pressure due to weight of water or soil because of the loads due to impact or unknown future constructions. For good work, no brickwork should be less than 9 inches thick, no mass concrete less than 6 inches thick and no reinforced concrete less than 4 inches thick.

Roof slabs. Roof slabs over chambers may be designed in accordance with normal practice for reinforced concrete, if well under the ground. But reinforced slabs near the surface should be thick and *not top reinforced* if they are liable to receive impact from heavy road loads. Nine inches might be taken as a minimum thickness.

Pre-cast concrete manholes. Pre-cast concrete manholes (B.S. 556) usually consist of rings of two sizes—i.e. chamber rings at the bottom and shaft rings at the top—together with intermediate taper pieces. The inverts

THE CONSTRUCTION OF MANHOLES AND CHAMBERS

may be formed out of mass concrete or other material, or may be purchased pre-cast to the required design. The rings are jointed with either ogee or socket joints, as used for concrete-pipe sewers.

The top of the manhole terminates with a making-up ring, made or cut to size, and a heavy circular slab to take the manhole cover. It is usual to surround part of the work with mass concrete: opinion varies as to the degree

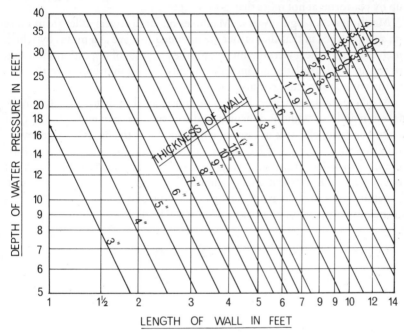

FIG. 47—Thickness of manhole walls

This diagram can be used with confidence for the determination of the thickness of grouted brick or mass-concrete walls for manholes, chambers, etc., constructed below ground-level or intended to retain water. The thicknesses are based on water pressures, which generally are greater than earth pressures. Where chambers are roofed and the height is less than the length the thickness of wall can be based on the height in place of the length.

to which this should be done. In one design which has been used extensively, and which has never been known to fail structurally, the manhole is surrounded with concrete either of a minimum thickness or filled to the excavation to 6 inches above the top joint of the taper piece, while the shaft is left unconcreted. Some engineers prefer always to concrete round the shaft as well as the chamber. The amount of concrete used should depend on particular local conditions. In main streets with heavy traffic, concreting of the shaft would add to the strength, but experience has proved that this additional concrete is quite unnecessary in rural sewerage schemes.

The manhole cover is best bedded on one or two rings of brickwork to permit future alteration of level.

The pre-cast concrete manhole may be applied to large sewers as well as to small, and to great depths as well as to medium and shallow depths. Figure 44 shows a deep manhole on a large sewer with an intermediate chamber breaking up the climb into two lengths of 20 feet. With this particular arrangement the bucket shaft as illustrated, or other means of raising buckets or lowering safety lamps from the surface, is advisable when the manhole is constructed on a large sewer. Very deep manholes should always be broken up by platforms at not more than 20-foot intervals.

Manholes on large sewers. Manholes on large sewers differ from those on small sewers, in that the former are intended to serve as a means by which men may enter the sewers themselves for the purposes of inspection, cleansing or repair, while the latter are intended merely as a means of access to the ends of the sewers for rodding or inspection.

Manholes on large sewers have to be constructed so that materials may be taken in and out without difficulty, and so that men may escape quickly, or rescue be easily effected, in cases of emergency. As the flows in large sewers may consist of considerable quantities of water travelling at high velocities, the shaft or ladder from the surface should not lead directly to the invert of the sewer. The side-entrance manhole (see Fig. 48) is the construction generally preferred. It is ideal from the point of view of safety, but it is comparatively costly, and in some designs can be dirty.

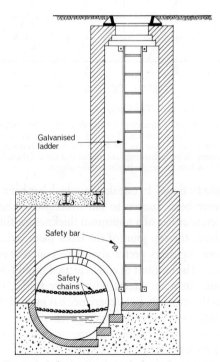

FIG. 48—Side-entrance manhole on large sewer

Where large sewers are constructed in towns, access during the day, when there is much traffic, has to be considered. This necessitates the construction of access shafts placed under the footways and provided with light covers and safety gratings and galleries leading from these shafts to the shafts or the side entrances on the sewer.

While in many cases it is useful to have bucket shafts or lamp-hole shafts, in complicated manholes it is also useful to make these shafts sufficiently large for them to be used for rescue purposes. All such manholes should generally be designed in accordance with the recommendations embodied in the Ministry of Health publication: "Accidents in Sewers. Report on the Precautions Necessary for the Safety of Persons Entering Sewers and Sewage Tanks."[183]

It is advisable to make use of safety bars and chains on all platforms in manhole shafts, and to provide safety chains on the outgoing sides of all manholes on large sewers, to prevent men from being swept down the sewer. "Warning chains" hanging from the roof are useful indications of hidden dangers.

Backdrops, cascades and water-cushions. A backdrop, cascade or tumbling bay is used when it is necessary to make connection from a high-level sewer to another sewer at a lower level. Water falling from a high level has a destructive effect on sewerage works, owing to the velocity and energy of impact producing scour, particularly when the sewage contains heavy detritus. The velocities caused by rapid change of level can greatly exceed those normally desirable in sewers. At one time it was common practice to construct a ramp (or sewer at a gradient of about 45° to the horizontal) connecting from the high-level sewer to the low-level sewer (see Fig. 49). One disadvantage of this design is that it is applicable only to drops of a few feet, because when a ramp is constructed to give a deep fall, a very large amount of concrete is required to support the length of sewer which connects into the shaft of the manhole, for this is above the ramping pipe, and therefore needs to be supported, as shown on the drawing. (Here junior assistants are warned against the very common error of showing the length of sewer which should be supported on concrete merely surrounded with 6 inches of concrete and resting on refilled earth. Such a construction would not remain in position for long.) Another disadvantage of the design illustrated is that the down-pipe is considerably smaller than the sewer from which it connects. This would mean that when the sewer was running full the down-pipe would be inadequate to take the flow owing to velocity-head loss at entrance and sewage would pour down into the manhole from the end of the sewer.

It is now common practice to construct backdrops of vertical pipes as shown in Fig. 45 with cast-iron bends at the bottom of the vertical pipes where the scour is most likely to be severe. The material of the vertical pipes can also be cast-iron.

Sewage falling down a vertical pipe tends to cling to the side or fall as free

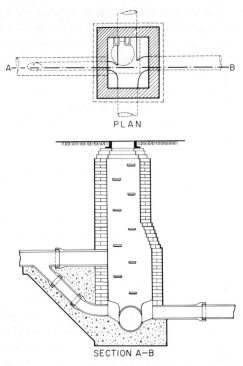

FIG. 49—Old-fashioned brick manhole with ramp or inclined backdrop

drops. This property of the vertical pipe has an advantage because it reduces the velocity of fall, and, provided the vertical pipe is not filled by the flow, there is no need for the construction of a water-cushion at the bottom of small-diameter backdrop pipes or to break up the fall into a number of smaller falls. This is, in fact, the practice adopted in America, where the soil-pipes of skyscrapers extend from roof to ground-level without a break, it having been found that no precautions have to be taken. The bottoms of backdrops should be so arranged that the high-velocity flow entering the manholes does not splash over the benching and where backdrops enter the side of a large sewer the bends should incline towards the direction of flow in the sewer.

Where large flows have to be dealt with, the vertical backdrop is not suitable unless an adequately proportioned water-cushion is formed. The water-cushion should consist of a tank-like structure into which the vertical column of water falls. The depth should be sufficient to destroy the velocity. In this connection the following formula may be used:

$$D = 2\sqrt{Q}$$

where: D = depth in feet
Q = discharge in cusecs.

THE CONSTRUCTION OF MANHOLES AND CHAMBERS 135

The form of the tank should be such that it is self-cleansing. Arrangements for the tank to empty at times of small flow should be made and precautions taken against silting. The whole of the works where the velocities are likely to be high should be constructed of vitrified bricks or other durable materials. The arrangement used should not prevent ventilation.

For large flows generally, it is best to form a cascade by breaking up the fall into a series of steps. The steps should be sufficiently deep to accommodate the flow—that is, they should never be submerged by more than half the fall—but they should not be so deep as to permit high velocities. A good form, 9 inches tread by $7\frac{1}{2}$ inches rise, makes the steps useful both as a stairway for sewer-men and as a cascade. Where very large flows have to be dealt with, the cascade must be made wide to keep a shallow flow, or the size of the steps increased to a maximum of 2 feet fall: larger falls should be avoided.

Pressure manholes. Pressure manholes are used only in exceptional circumstances. It may be found, in an existing system, that flooding occurs at one manhole only, without flooding of properties through connections. When large rising mains or siphons are constructed, these may with advantage be supplied with manholes at low points for inspection or for the removal of deposit. In both of the above cases pressure manholes have to be constructed.

A point that has often been overlooked is that when a manhole is designed to be airtight so as to resist surcharge, the pressure exerted by surcharge is greater than that which would be exerted if the manhole cover leaked slightly. When the manhole can leak air, the sewage rises to the undersurface of the cover, and the pressure-head is that measured from the undersurface of the cover to the head of the sewage. When the cover is airtight, the sewage rises in the manhole, compressing the air, but is soon brought to rest. The pressure head is then that measured between the level of the water in the manhole and the hydraulic head of the sewage.

A common occurrence at times of surcharge is for manhole covers to be blown off. When a flood occurs, a number of manhole covers may be blown high into the air, after which water gushes out of the openings. This violent lifting of the cover is not directly due to the sewage, but to the air which is compressed in the manhole by the sewage when it rises in the manhole. If the covers are of the ventilating type, or have not become airtight, they will lift gently as the sewage rises.

One type of pressure manhole is illustrated in Fig. 50. Below the manhole cover an airtight pressure door is constructed. This door is provided with clamping screws. If the door is to open upwards, the type of clamping-screws should be such that the door cannot be suddenly opened, thereby releasing compressed air. If the door opens downwards, a lifting-chain should be provided. The frame of the door should be tied down by bolts and anchor plates for a sufficient distance for the weight of concrete or brickwork to be more than that necessary to balance any upward pressure of sewage. The upper chamber should be drained.

For manholes on rising mains, where these are comparatively shallow, it may be sufficient to provide airtight manhole covers of the type that bolt down, with frames securely anchored to the rising mains themselves. On cast-iron mains hatch-boxes should be used.

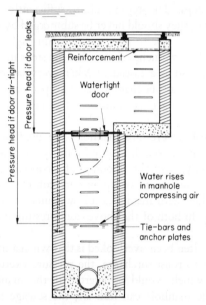

Fig. 50—Manhole designed to withstand internal pressure due to surcharge

Bellmouth chambers and bends. When two very large sewers join one another, the junction is a chamber known as a bellmouth. In this chamber the invert is formed so that the two barrels meet with a correct development that will not retard the flow in any way. The side connection joins at a radius to the main sewer, so that the side of the main sewer forms a tangent to the side of the incoming sewer. Although the inverts are usually elaborate, bellmouth chambers not uncommonly have simple flat roofs consisting of reinforced concrete or arched work formed to the radius of the crown of the outgoing sewer, or tapered arched roofs. Side-entrance access is usually provided.

Manhole iron-work. Manhole iron-work comprises manhole covers, access ladders, step-irons, gratings, safety bars, chains, etc.

Manhole covers may vary considerably in type and design, and only the required qualities can here be discussed. A cover should be sufficiently strong for its purpose. The plug should not rock or jam, and should be easily lifted. Circular covers are the strongest in proportion to their weight. Of late, triangular covers have been much advertised on the grounds that they do not rock. Nevertheless, some do rock, and the triangular shape is particularly weak. Triangular manhole covers, unless of special design, are certainly not to be recommended for large sewers that have to be entered by men, because the openings in them are too small for men to be easily able to enter wearing safety harness and they make the removal of casualties extremely slow and difficult.

Covers may be ventilating or airtight, as required, but if there is any possibility of surcharge on the sewer a *completely* airtight cover should be avoided, for the reasons mentioned under the heading of "Pressure manholes."

For good-quality work galvanised steel ladders and platforms at intervals are preferable to step-irons for all except shallow manholes or pre-cast-concrete manholes. They are more expensive initially, but they are easier to renew. Ladders should be bolted to the walls, and not built in, non-corroding bolts being preferable. Rungs should be 1 foot apart vertically or 9 inches for the sake of men wearing sewer boots.

Step-irons are usually malleable castings either large enough to take one foot at a time and built in staggered or at least 12 inches wide and not staggered. Each should be tested by being given a good blow with a hammer. Staggered step-irons should be placed at a distance apart that is normal for a person going down—i.e. $7\frac{1}{2}$ inches centre to centre as in Fig. 46—and not too wide apart as in Fig. 49—a common fault. They should not deflect without warning from the vertical and not overhang in a tapered shaft. Pre-cast concrete tube manholes are usually provided with the step-irons in position. Care should be taken that the workmen engaged on construction do not use the step-irons for hoisting purposes or otherwise loosen them, as not only does this make them dangerous but it leads to leakage of the walls.

All ladders, gratings and other galvanised iron-work should have the parts properly tennoned together: too many welded gratings have failed.

Sewage has a protective effect on iron-work as can be observed in very old large sewers. Galvanised iron ladders are found to be in perfect condition below the level to which the sewage rises at maximum rates of flow but are severely corroded above this level.

Aluminium has been used for ladders, gratings, etc. but the author does not favour the material for these purposes. First, it is necessary to select the right alloy to resist corrosion for some grades of aluminium are very quickly attacked by alkaline fluids, particularly activated sludge; also precautions have to be taken against electrolytic action. Second, it is necessary to have a much larger section of metal to give equal strength and rigidity as compared with mild steel. Third, ladders and gratings of suitable alloy and rigidity are very expensive.

CHAPTER XIII

SEWAGE PUMPING*

As far as possible, the use of pumping stations is avoided in sewerage schemes, because of the addition to running costs and supervision that they entail and because pumping stations are comparatively weak points in a sewerage system. Sometimes, however, pumping is unavoidable, and occasionally, while it might be possible to use a gravity scheme, pumping is resorted to for the sake of considerable reduction in capital cost. In those cases where it is just possible to use gravity it becomes necessary to estimate the costs of separate schemes for gravity and pumping as described in Vol. I, Chapter II. The drainage of low-lying flat areas may necessitate a multiplicity of pumping stations. Such cases require special study.

Apart from the cost, there are technical disadvantages to pumping stations. They are not so elastic as sewers, in that they cannot be overloaded. If they are designed with larger capacities than the calculated capacity of the sewer, there is an increased capital cost for pumping machinery, and, what is often more important, there may be, in the case of electrically operated plant, a heavy kVA charge on motors which seldom run. The presence of a pumping station can be objectionable should the suction well of the station become a receptacle for the settlement of solids or floating grease which has to be carted from the site. Another disadvantage which has often been overlooked is that a rising main is unlike a sewer in that it does not empty itself when there is no flow. All except the shortest rising mains have considerable storage capacity, and retain the sewage for a sufficient time for septicity to occur.

The practice of sewage pumping probably originated a little over one hundred years ago when large sewers were constructed for the conveyance of combined sewage. The pumps were very similar to those used for waterworks at that time, being actuated by immense steam engines which consisted of a rocking beam with a steam cylinder at one end and a reciprocating pump at the other. Very few of these pumps are still in existence, but it is not so very many years ago that some were still in operation.

Sewage pumping developed more or less parallel with other pumping, e.g. for water supply. Reciprocating pumps largely gave place to volute centrifugal pumps, mixed-flow and axial-flow pumps, the types used depending on the duties for which they were required. And in the course of development the importance of the difference between the pumping of clean water and of sewage heavily laden with solids became more appreciated.

Automatic operation. A marked development was in the change from manual to automatic operation, doing away with the necessity for manning

* This chapter should be read in conjunction with Vol. I, Chapter XII.

pumping stations day and night at very high labour costs, and making it possible for several pumping stations to be maintained by a few operatives travelling from one to the next, to inspect, clean up and, where necessary, make repairs or replacements.

Automatic operation was made possible by the introduction of electric power; and still today it is mostly confined to electrically-driven stations. But automatic pumping can be arranged where power is provided by internal-combustion engines; also, the arrangement of several pneumatic ejectors linked to a system of compressed-air mains makes possible the use of any prime mover.

At one time, unattended automatic electric-pumping stations were all, or nearly all, of small size, in particular those of the minimum size having 3- or 4-inch internal diameter rising mains; and there was some hesitation on the part of engineers to allow for larger stations being left unattended. However, experience has shown that large pumps are even less liable to stoppages than small pumps. Moreover, the kind of failure to be expected in pumping stations, other than main current failure due to extraneous causes, is not calamitous, and in most instances could not have been prevented even if an operative had been present and watching the machinery.

The author has designed several large, fully-automatic pumping stations, including one which had no suction-well other than the sewers themselves and which could deliver up to 340,200,000 gallons per day.

Thus, it is practicable to have an entirely automatic electric-sewage pumping station incorporating the largest pumps without making arrangements for day and night attendance. More particularly, such a station, if of great importance, can be safeguarded by the provision of electric supply from two separate sources, with automatic throw-in of one on the failure of the other or, better, the connection of one-half the number of pumps normally to one supply and the rest to the other; also the inclusion in the equipment of remote indicators and warnings that bring failure that has taken place to the attention of some person in authority.

It will be appreciated that in many instances this kind of installation can mean a very great saving in wages, apart from the number of troubles and maintenance costs that are avoided by using electric plant in place of prime movers.

Need for care in design. The automatic electric sewage pumping station was not always as perfect as some are today. Also, it was not appreciated that unsatisfactory working or even failure could result from imperfections in design or lack of knowledge of the safeguards to be used when designing such a station. Broadly, it can be said that not a few engineers failed to see that a pumping station is not merely an assembly of pumps drawing from a suction-well that connects to a sewer and delivering to a rising main; but that the sewer, suction-well, pumps and rising main must be considered as one machine, virtually designed as a whole. Certainly, pumps are more or less stock articles designed and listed by manufacturers. But the real design of a pumping station lies with the civil engineer, who should know how to relate hydraulic conditions in sewer, surface area and depth of suction-well, characteristic curve of pump, capabilities of electric-starting gear and hydraulic properties of rising main in such a way that automatic operation can continue efficiently with little risk of breakdown of any one pump and no danger of the whole pumping station closing down altogether.

The following are some of the mistakes made by various authorities which the writer has observed:

1. The engineer buying a pumping plant was advised that, to suit the small suction-well capacity, the electric starters must be of "frequent duty" type. He installed "ordinary duty" type with the result that all the starters burnt out when the pumping station first came into action.

2. The suction-well was designed to such a capacity that sedimentation took place. Having steeply sloping floors, the silt piled up until a "landslide" occurred choking all the suctions and putting the pumping station out of action.

3. The suction-well was too large, causing sedimentation, and had to be dug out manually at intervals.

4. The suction-well was too large and as the pumps were not arranged to give complete emptying, monthly scum removal was necessary.

5. No anti-surge provision was made in the case of a long rising main. On slamming of a reflux valve, a hatch-box lid was blown off and the rising main split by hydraulic shock a long distance from the pumping station.

6. No anti-surge provision was made in the case of a long rising main. The rubber seal was blown out of a hatch box and the water, leaking into the dry-well pumping station, flooded the electric motor which had been installed at too low a level.

7. No anti-surge provision was made in the case of a long rising main. Reflux valves were damaged, pump spindles broken and the rising main split by hydraulic shock a long distance from the pumping station.

8. Severe shock (however, cured before damage had occurred) was caused by a reflux valve which failed to close owing to the external arm not being properly loaded.

9. In a large pumping plant automatic sluice-valves were used instead of reflux valves. They were supposed to close by gravity but tended to stick open. Also, the electric valve-operated relays failed to work. The plant was put out of action for some considerable time until these faults were rectified.

10. In five instances pumps failed to work owing to having the wrong kind of impeller for dealing with sewage.

11. In three instances rising mains became choked with detritus in the

invert and grease in the crown, owing to the velocity being too low for self-cleansing conditions.

12. Owing to mistake in the design of the pump, thirty-two pumps delivered part of the flow round in a circle.

13. In three instances where stand-pipes had been provided to let air out of the rising mains, overflow from the top occurred because allowance had not been made for starting pressures, etc.

14. An electrical engineer interfered with the adjustment of the float arrangements of a pumping station with over-large suction-well, with the result that twelve hours' flow was pushed through the sewage-treatment works in a few minutes, rendering all the treatment plant ineffective.

15. Two digested-sludge rising mains burst owing to having valves closed at both ends when containing gassing sludge.

16. In two cases the pipework in a pumping station was blown open through having unsecured expansion joints placed on bends in the pipework.

17. A standard design of pumping station was constructed without modification in unsuitable circumstances. It was placed in a hollow with the motor floor below ground-level, with the result that surface water periodically flooded the electric plant.

18. A dry-well pumping station had a float-wire connection from the wet-well below ground-level. An electrical engineer, thinking the plant was out of action because the premises were not occupied, switched off the current. The next time there was a rain-storm the dry-well was flooded and the stand-by diesel plant ruined.

19. Some very large centrifugal pumps were protected by coarse screens. This did not prevent them from being choked by waste corks from a bottling store. (The provision of $\frac{5}{8}$-inch screens cured the trouble.)

20. Owing to manufacturer's mistake, driving shafts ran at critical speed.

21. In several instances partial failure of plant resulted from failure of electrode equipment.

22. In two cases, owing to improper design, surge occurred in the incoming sewer. In one of these the effect was to cause several pumps to cut in almost simultaneously then, after rapidly emptying the well, they cut out simultaneously, presumably because they all had the same cut-out level. (This trouble cannot develop in a pumping-station having a suction-well of adequate size and proper arrangement of cut-in and cut-out levels for the automatic starting gear.)

In the second case, a large sewer was being used as suction-well but the designer made the error of allowing this to run in a surcharged condition. On the flow reaching the crown of the sewer there was a sudden build-up of high pressure. This would have been obviated if the design had been such that the highest cut-in level had been below the crown of the sewer and the pumping rate sufficient to prevent the sewage level from rising any higher.

In addition to certainty of operation, there have to be considered efficiency and economy. Here again, earlier pumping stations were often at fault, for unduly large suction-wells led to the settlement of sludge and the accumulation of scum, calling in some instances for manual cleansing as frequently as once a month. The types of pumps used necessitated the provision of screens, which similarly had to be cleansed, and were by no means in all circumstances effective in giving protection to the pumps against water-borne solids. The screenings themselves created a problem until means were found for returning them to the flow of sewage in the rising main, either disintegrated or in their original form.

The way to avoid a dirty suction-well is to make it small, ensure high velocities of flow therein and arrange for all the contents to be pumped out before cessation of pumping. But if the well is small, pumps must start and stop much more frequently than if it is large. Again, there is the possibility of the water-level rising in a small well so rapidly for a second or even third pump to cut in before the first has started to work. Thus, there is a danger in some circumstances of causing too rapid starting and stopping of a pump.

On the matter of economy, not always was attention given to the choice of correct size of rising main to give the minimum overall annual cost, including power consumption, kVA charges, pumping-station maintenance and labour, and repayment of capital costs over the appropriate loan period.

Recently, further advantages in the use of sewage pumping stations have come to light. A pump delivers at a steady rate of flow as compared with a gravitating sewer or channel, in which the flow can vary between wide limits. Thus, a pumping station can have a regulating influence: with the provision of a suitable amount of storage capacity in the suction-well or in a tank-sewer on the suction side, it can have an evening-out effect on storm flows. Furthermore, it can effect accurate separation of storm water.

From the foregoing facts it will be seen that in the design of modern sewage pumping stations there are many factors to be considered and a number of calculations to be made, some of which, simple as they are when once known, may at first be elusive.

Types of pump. The several types of pump that can be, and are being, used for pumping sewage may be classed broadly under the following heads:

1. Reciprocating pumps.
2. Centrifugal and similar pumps.
3. Air-actuated devices, i.e. ejectors and air-lifts (see Chapter XIV).

Of these various classes, reciprocating pumps were once very largely used and still have a part to play.

The volute or truly centrifugal pump, the mixed-flow pump and the anxial-flow pump are all useful for sewage pumping: multi-stage turbine pumps are not. Sewage ejectors were specifically designed for dealing with sewage and have very wide application to small flows. Air-lifts are very useful at sewage-treatment works.

In addition to the foregoing, there are various odd kinds of pump, difficult of classification and limited in the purposes to which they can be applied— e.g. the chain pump used for emptying septic tanks and cesspools.

Reciprocating pumps. Nearly all reciprocating pumps of interest for sewage pumping are power-operated pumps of the ram type (see Vol. I, Chapter XII). Large reciprocating pumps are usually on the three-throw principle; very small pumps are often single-throw.

Medium-size and large reciprocating pumps were at one time much used for pumping crude sewage. It was usual to protect the pumps with medium bar screens with about $\frac{1}{2}$ to $\frac{3}{4}$ inch clear spaces, partly to prevent inefficiency by holding up the reflux valve with solid matter but mainly as a precaution against damage by the introduction of large solids. The use of reciprocating pumps for this purpose is now very much reduced for the reasons that centrifugal pumps of modern design and manufacture are so very much cheaper to buy and are of equal or, in some cases, even higher efficiency than the best reciprocating pumps.

In addition to crude sewage, sludge has often been pumped by reciprocating pumps, and in some instances surprisingly small pumps have been found efficient for this purpose. For use at small sewage works there are special designs of pump capable of pumping extremely low rates of flow.

Centrifugal pumps. An important consideration in the choice of pump for sewage pumping is non-liability to choking, and in some instances a compromise between this and efficiency. Volute centrifugal pumps designed for pumping clean water are not too good for sewage containing solids, for even pumps of this type of the largest size, if designed to give high efficiency, have close spaces that tend towards chokage of one kind or another. The simplest way of preventing choking of a volute pump intended to pass liquids suspending large solids is to have a very crude single- or double-bladed impeller with a large eye at the inlet. But the secret of pumping fluids containing fibrous matter—including, of course, sewage—is to have a pump that has no leading edge to the impeller over which strings, rags, etc., can fold and build up to cause a stoppage. If a pump with an impeller of the type illustrated in Vol. I, Fig. 78 is used for sewage, bits of rag, string, etc., fold over the leading edge of the blades near the eye of the impeller and are held in position by the flow. On top of these, other bits fold over in the same manner until eventually the eye of the impeller is completely blocked with a ball of rag and string. If, on the other hand, an impeller of the type illustrated in Fig. 51 is used, there is no leading edge for fibrous matter to fold over, and consequently stoppage is very rare indeed. Pumps so designed will pump almost anything, including lengths of rope, bricks, balls of rag, heavy suspensions of sand and thick slurries; but they do this at the expense of efficiency. Such types are very popular for automatic electric sewage pumping stations of the

Fig. 51—"Amphistoma" pump showing impeller for use with unscreened sewage and sludge
By courtesy of Adams-Hydraulics, Ltd.

smallest size—i.e. having rising mains of 4 inches diameter, where the annual cost of additional current required as a result of the low efficiency of the pump is easily outweighed by saving of men's time in clearing stoppages. Silicon-iron pumps are used for grit.

At very large stations it is possible to have both high efficiency and freedom from choking, but again, only if the right type of pump is selected. "Wire to water" efficiencies of clean water centrifugal pumps are about 60 per cent for 25 c.f.m., 70 per cent for 100 c.f.m., 80 per cent for 400 c.f.m. and 85 per cent for 1000 c.f.m.

Mention should be made of the Solids Diverter, in which the solids are screened out from the flow to the pump and then, by an automatic rearrangement of flow, transferred to the delivery side without passing through the pump. This has been largely used at small stations.

Economic size of rising main. It is generally accepted that a rising main should be so designed that the velocities of sewage are between limits. Apart from this, the diameter of rising main is that which will ensure the lowest annual cost in terms of pumping power and repayment of loan.

If the velocity is less than 2 feet per second, silting will take place. Many engineers, influenced by water-supply practice, do not permit velocities in excess of 6 feet per second. These limits make possible the delivery of both average and daily peak rates of flow of soil-sewage through the same rising main. But it is desirable to set the minimum velocity at about $2\frac{1}{2}$ feet per second, this velocity being both economical and definitely self-cleansing:

it is also advantageous to stretch the maximum limit to 8 feet per second, for the objection to high velocities that applies to water supply—i.e. they cause incrustation—does not apply to sewage rising mains.

The economic diameter of a rising main depends on the velocity at the *normal* rate of pumping, not the peak rate, and although on the surface this may not appear to be the case, *the economic velocity is not dependent more than in a very minor degree on the length of the rising main.* This latter fact has often been overlooked owing to an illusory belief that when a rising main is long the diameter should be large, to avoid excessive pumping costs. It is only necessary to work out the capital and running costs in a few normal cases to dispel the belief.

As in water-supply schemes, in most *soil*-sewage rising mains the economic velocity is in the region of $2\frac{1}{2}$ and 3 feet per second, and therefore, to begin with, the capital cost for a scheme involving a main with a velocity of 2·75 feet per second should be estimated. The total capital cost of stations and mains should then be converted to annual repayments of loan, and to this annual figure added the running costs. This gives a total annual cost which may be plotted on a graph. Higher and lower velocities of sewage in rising main may then be taken, the total annual costs estimated and a curve plotted. The lowest point on this curve gives the economic velocity. For economic diameters of rising mains, see Table 2.

In the case of storm-water pumping, the least costly schemes are often those which involve high velocities in the rising mains.

Arrangement of machinery. A sewage pumping station usually has to deal with various rates of flow from a minimum to a maximum, and generally it is desirable, if not essential, that changes of pumping rate from low to high shall be in even and moderate stages. This calls for a number of pumps. There are several ways in which a station can be equipped, and some have advantages over others.

Suppose a total quantity of 50 cusecs has to be pumped and that each change of rate shall be by not more than 10 cusecs at a time—i.e. the station shall be capable of pumping at 10, 20, 30, 40 or 50 cusecs—then any one of the following possible arrangements could be adopted, it being assumed that sufficient stand-by pumps are provided so that if any one pump breaks down another pump or pumps can take its place.

First case. Five duty pumps each of 10 cusecs capacity plus one standby.

Second case. One duty pump of 10 cusecs capacity, two duty pumps of 20 cusecs capacity and two stand-by pumps of 10 cusecs capacity.

Third case. One duty pump of 10 cusecs capacity, one duty pump of 20 cusecs capacity, one duty pump of 30 cusecs capacity, one stand-by pump of 10 cusecs capacity and one stand-by pump of 20 cusecs capacity.

Fourth case. Two duty pumps of 10 cusecs capacity, one duty pump of 30 cusecs capacity, one stand-by pump of 10 cusecs capacity and one stand-by pump of 30 cusecs capacity.

Out of all these examples, the first is different from the others in the following respects:

1. A total of six pumps is required in place of five.
2. The total installed horse-power, however, is less than in any of the other cases.
3. Only one size and type of pump is required, and therefore a much smaller stock of spares has to be kept.
4. In this case no electrical intercommunication is necessary between the pumping units; each pump can be arranged to start and stop according to its own float or electrode control, whereas in all the other cases the cutting-in of one pump means the cutting-out of another by means of an electric intercommunication. Thus, the first case has a marked advantage over all the others in that it is much more reliable and cannot have a complete breakdown other than failure of current supply to the station as a whole.

Arrangements of types 2, 3 and 4 are very common, but study of the foregoing points should show that in the majority of instances there are marked advantages in having a number of pumps all of the same size, similar in every respect, and completely independent in automatic operation (including such details as independent supplies of low-tension electricity for electrode contact). This independence of operation makes it possible for a fully automatic station to be left unattended, for the complete breakdown of any one pump has no effect on the rest, which cut in, in turn, as the water-level rises in the suction-well.

An exception to this rule is when a pumping station deals with both dry-weather flow and storm flow and it is intended that storm water shall be separated by the operation of the pumps. Then, the station should be considered as two separate stations—dry-weather flow and storm-water pumping stations—combined in one building. There can be two sets of pumps, so many small pumps all of one size to deal with dry-weather flow, and so many larger pumps all of another size to deal with storm water.

The number and arrangement of pumps in small sewage pumping stations having only one rising main are limited by velocities of flow in the rising main. As previously mentioned, the velocity in a sewage rising main should not drop below $2\frac{1}{2}$ feet per second, and never below 2 feet per second and, to minimise head losses and scour, the maximum velocity should not exceed 8 feet per second. Thus, the velocity is limited to a range in which the maximum rate is 3 to 4 times the minimum rate.

If it is necessary to have larger ratio of minimum to maximum velocity, difficulties can be overcome by duplication of rising mains, connecting some pumps to one main and the remainder to another. This has the additional advantage of making the station capable of partial operation in the unlikely event of a rising main being wrecked by some accident or having to be laid off for some maintenance purpose.

Pipework. Having decided the number and sizes of pump to be installed, the designer has next to determine the arrangement of pipes and pipework, valves and electric gear, and to find the size of pump chamber, motor house, etc., required to accommodate the machinery conveniently but not extravagantly. The simplest lay-out is usually the best, and in most instances this consists in having all the pumps and motors in one straight line in a narrow pumping station. At one side lies the suction-well from which the pumps draw, and on the other the common rising main or rising mains. The suctions and deliveries should be all parallel, connecting straight across from suction-well to rising main. If there are two rising mains, these are best connected from the opposite ends of the length of pipe to which the delivery branches connect.

In a dry-well pumping station with vertical-spindle pumps, each centrifugal pump has suction and delivery pipework and valves as follows. Starting at the suction-well is a downward-turned right-angle bend, preferably a "medium" bend, which in the case of large pumps for storm water can have a bellmouth entry, but which should have a plain inlet without bellmouth in the case of small sewage pumps or pumps for sludge. The mouth of this should be submerged below the lowest sewage level by an amount equal to at least the velocity-head of entry. Connected to this and passing through the wall between the wet-well and the dry-well is a straight pipe with body-flange or puddle-flange to reduce possibility of leakage of water through the wall where the pipe passes through. For drawing purposes, the flanges by which the various flanged pipes are jointed together should be shown 6 inches from the faces of walls, although in fact less space is required to permit convenient bolting of flanges together. Next comes the sluice-valve which isolates the pump from the wet-well, then any necessary making-up length of straight pipe. Finally, an upward-turning medium bend connects into the suction connection of the pump. The last is not required if the pump is of the split-suction or double-entry type.

To the delivery connection of the pump is bolted a reflux valve or non-return valve. For sewage-pumping installations reflux valves should always be placed in horizontal pipework, never in a vertical pipe, for the latter arrangement encourages delayed closing of valve should sticky matter get behind the flap, with the possibility of shock and damage to pipework, etc., should the valve close after back-flow has commenced. After the reflux valve comes the sluice-valve that isolates the pump from the rising main. Following this sluice-valve, the branch of pipework from the individual pump goes on to connect to the common rising main, to which all the pumps connect. In common arrangements this connection can be horizontal and direct into the rising main or can be turned vertically (using a medium duckfoot bend) but it is best for it to connect into the side, not the underside of the rising main at high level, for this reduces the chance of solids falling into it when not in use.

All the pipework in a pumping station should have flanged joints, because the pipes are generally not supported, and spigot and socket joints might blow open under water pressure. But, for it to be possible to take down and reassemble flanged pipework, there must be either a bend in the line or at least one flanged joint with oblique flanges or one joint of a flexible type— e.g. a Johnson Coupling. But such flexible joints should not be so arranged that they themselves could blow open.

Prevention of water-hammer in rising mains. It has been mentioned that an air vessel is a normal provision where reciprocating pumps are used so as to prevent shock on closing reflux valves. A similar provision or an alternative is advisable even with centrifugal pumps if the rising main is long, for on cessation of pumping, should a reflux valve become stuck open and then close after back-flow had commenced, a very severe shock to the pipework could be caused, with possibility of fracture.

When air-vessels are used with sewage the air is quickly dissolved and has to be replaced. For this reason an alternative for use at automatic sewage pumping stations is a stand-pipe connected to the rising main and carried to a height greater than the total starting manometric head. As in waterworks practice, the top of the pipe should be vented to prevent siphonage. Any spill-over should discharge to the suction-well.

Lay-out in preliminary design. The internal appearance of a pumping-station depends largely on the lay-out of the machinery. An ordered lay-out is also advantageous, in that it prevents errors in manual operation. Suppose that in a station containing two sizes of pumps, the pumps are arranged in one or two straight ranks of evenly spaced units, in this case the head-stocks or surface-boxes for the sluice-valves should be so placed that there is no question as to which pump each valve relates. Similarly, the starting panels should be placed so that it is obvious to which motor any starter relates; it should be possible for anyone not familiar with the station to be able to stop or start a pump at once without experimenting with several switches.

For the valves to be arranged in the above manner requires that the suction and delivery piping shall be laid out regularly. Suctions and deliveries generally should have as few bends as possible and should contain no air-locks. In the preliminary design it is best to assume standard fittings throughout. The pump manufacturer will insert all the special castings he requires, but if given a design consisting of a number of special castings he may be tempted to re-design the lay-out entirely. In an important pumping station it is an advantage if special castings can be avoided completely, because should any breakages occur these cannot be quickly renewed. The sluice-valves incorporated in the piping should be of the kind which makes it possible for heavy solids to be removed, as otherwise it may be difficult to close them.

Sizes for drawing purposes. The sizes of suction and delivery pipes other than the rising main itself depend on limiting velocities and the size of the pump installed. There are some differences in makers' sizes of pump that can have the same duty, so that if pumps are to be purchased by competitive tender the designer in the first place does not know what size of pump will be used. But he can make a fair estimate by using the formula given in Vol. I, Chapter XII.

The suction and delivery pipework can be taken as having the diameter required to give the economic velocity given in Vol. I, Table 15 except in the case of storm-water pumping stations, where higher velocities would be economic. But in no case should the velocity be less than $2\frac{1}{2}$ or more than 8 feet per second, the recognised minimum and maximum velocities of flow in sewage mains.

Automatic pumping. The use of automatic pumping units has necessitated study of the capacity and proportions of suction-wells. In many of the older pumping stations it was usual to determine the size and form of the suction-well by crude rule of thumb. The wells were generally very large, an advantage where hand operation of motors was practised, but particularly liable to cause sedimentation and septicity. The modern pumping station has a well which is as small as possible, so as to avoid the lengthy detention of sewage resulting in collection of solids or scum, and yet of sufficient size for satisfactory operation of the automatic plant. In an ejector station there is no suction well, the ejector itself giving all the necessary storage, and this fact should be sufficient to prove that large storage wells are unnecessary. Nevertheless, many existing automatic stations have large wells, the designers having adhered to practice that was usual before electric pumping had been adopted.

Pumps can be automatically controlled by floats, electrodes, pneumatic gear and other methods seldom encountered. In one type of mechanical float gear a float controlling any particular pump is suspended on a stranded non-corrodable wire which passes round a drum and has a counter-balance. The float has to be inside a vertical cast-iron tube or surrounded by a cage of guides, or otherwise it might be carried away by the turbulence of the water.

When flexible wires are used they are often brought up vertically from the float, passed over a pulley and carried horizontally through the wall of the wet-well and into the pumping station. This can be a cause of trouble. To some extent, foul odours can pass through the hole where the wire goes, but a greater danger is the possibility of water flowing from the wet-well to the dry-well should, for any reason, the pumps fail.

Another method, generally more useful, is to have floats with holes through the centre through which a rod passes. The float moves up and down the rod until it comes against adjustable stops, making the rod move upwards or downwards to move the switch to "off" or "on" respectively. The floats do not necessarily require protection against eddies in the suction-well, but they should not be so arranged as to come into contact with settled sludge at the bottom of the well or they may become lodged there.

To avoid complications, all rods are carried straight up to the float-operated switch, for angular motion by quadrants is an undesirable complication. The float switches usually require to be moved 3 inches to turn them on or off, a fact that should be kept in mind when determining tolerances of cut-in and cut-out water-levels.

Floats are made of non-ferrous metal or suitable ceramic material. They are of flattish circular shape and of such a weight that they float half-submerged so as to be equally effective in pulling down and pushing up.

In the Mobrey Magnetic Level Switch sold by Donald Trist & Co., Ltd. a botuliform float is pivoted at one end so that it can move a few inches. It contains a permanent magnet which operates a switch sealed in the unit to which the float is pivoted.

The control of pumps by electrodes is now very common. The electrodes are rods which are suspended vertically so as to dip into the suction-well at various levels. On contact of the electrode with the water a relay comes into action to operate the starting gear of the pump. The cut-out electrode stops the pump when the water loses contact with its lower end. The electrodes are supplied with low-voltage, alternating current.

Electrodes require regular maintenance. If they become coated with sewage grease they may fail to operate. Therefore they should be installed in such a manner that they can be easily cleaned in position.

In America, pneumatic devices for pump control are considered the most reliable devices for this purpose and electrodes most troublesome. Pneumatic controls are rarely used in Great Britain and the writer knows of only one type, the Brownson, made by Tuke & Bell and used by that firm when they install automatic pumping installations.

Size of suction-well. Accepting the principle that a small well is all that is needed, it is next necessary to consider how large the storage capacity should be and what the depth should be in proportion to surface area.*

In most pumping stations suction-wells are provided for one purpose only to give capacity so that pumps shall not start and stop too frequently. With the modern automatic station it is possible greatly to reduce the size of the suction-well by allowing very frequent starts, and this, together with careful design of shape of sump, makes it possible for the suction-well to be kept clean automatically. But there is a limit to the smallness of the well, for if electrical starting gear is put into operation more times per hour than it is designed for it will overheat, and this can also apply to the motors themselves. It therefore becomes necessary to specify the quality of electrical gear required and to design the suction-well to such a size that the electric gear will work well within its capacity.

* The practice described herein was first proposed by the author in "The Surveyor" for 1 July 1938.

SEWAGE PUMPING

Rheostatic starters for electric motors are made in three standard ratings as described in British Standard Specification No. 587:

    a. Ordinary Duty.
    b. Intermittent Duty.
    c. Frequent Duty.

The ordinary-duty rheostatic starters are suitable for use in service which does not ordinarily involve starting more than twice per hour, and therefore they are not suitable for use in automatic sewage pumping stations. Intermittent-duty rheostatic starters are suitable for service which may involve motors starting not more than 15 times per hour. These starters are suitable for automatic pumping-stations with fairly large suction-wells. Frequent-duty rheostatic starters are suitable for service which may involve starting not more than 40 times per hour.

In designing or specifying in connection with starting gear for automatic stations, it is best to allow a good factor of safety. Thus, the writer's practice for about thirty years has been to specify that starters and motors shall be of the "frequent" type capable of making 40 starts per hour but to make the suction-well of such a capacity that in the most adverse circumstances starting of any one pump or motor is never more frequent than 15 times per hour.*

Capacity of well between cut-in and cut-out levels of any one pump.
If a single and isolated pump is delivering from its own suction-well, it starts and stops most frequently when the inflow to the well is exactly one-half the pump delivery. This is easily proved by trial calculations as follows. Suppose a pump has a delivery of 100 cubic feet per minute and the suction-well a capacity of 100 cubic feet; then when there is no inflow the pump will be able to empty the well in one minute, after which it will remain at rest indefinitely: at the other extreme, if the inflow is 100 cubic feet a minute, the pump will run continuously without stop or restart. If, however, the inflow is 50 cubic feet per minute the rate of emptying the well will be the pumping rate less the inflow, that is:

$$100 - 50 = 50 \text{ cubic feet per minute.}$$

Then the pump will empty the well in:

$$\frac{100}{50} = 2 \text{ minutes.}$$

After this the well will fill at the rate of 50 cubic feet per minute, taking another 2 minutes to do so. Thus, the pump will start once every 4 minutes, or 15 times per hour.

* In less than 1 per cent of the pumping stations that he has designed in that period was a lower standard adopted, and this was for very special cases. A well-known firm of manufacturers of booster pumps has adopted the practice of allowing for 20 starts per hour, using frequent-duty starters (see Booster Pumps in Vol. I, Chapter XII).

If the rate of inflow is slightly greater than one-half the pumping rate, then the well will be pumped out at say:

$$100 - 55 = 45 \text{ cubic feet per minute.}$$

Then the pump will empty the well in:

$$\frac{100}{45} = 2 \cdot 2 \text{ minutes.}$$

After this the well will refill at 55 cubic feet a minute, taking 1·82 minutes to do so, giving starts at 4·04-minute intervals. Similarly, if the rate of inflow is 45 cubic feet a minute, the pump will start at 4·04-minute intervals. Greater differences between rate of inflow and half-pumping rate will give longer intervals between starting times.

Because the rate of inflow to any sewage pumping station varies from time to time, it is necessary to allow for the fact that on frequent occasions the rate of inflow will continue for some time at one-half the pumping rate of any individual pump. From the calculation given above it will be seen that in this circumstance the time between start and re-start is 4 minutes for every minute's pumping capacity in the well between the cut-in and cut-out levels of the pump. For example, if between cut-in and cut-out levels the pump could empty the well in $2\frac{1}{2}$ minutes were there no inflow, the starting to re-starting time would be 10 minutes when rate of flow into the well was one-half the pumping rate, and in no normal circumstances could starting be more frequent.

Several pumps discharging from one well. If there are several pumps discharging from the same well the circumstances relating to any one pump would remain the same, provided that the cut-in and cut-out levels of the various pumps are properly arranged so that *the first to cut in is the last to cut out*. In this arrangement, reading from the lowest level in the well, the lowest cut-out level is that of pump No. 1, the next lowest pump No. 2, the next lowest pump No. 3; and the lowest cut-in level is that of pump No. 1, the next lowest pump No. 2 and the next lowest pump No. 3, etc. (see Table 30).

Suppose that there are three duty pumps each of 100 cubic feet a minute capacity, and that the rate of flow into the well is 150 cubic feet a minute. This will have the effect of keeping one pump running all the time, so that the rate of flow into the well when No. 2 is not running will be 50 cubic feet a minute, and the rate of discharge out when No. 2 pump is running again will be 50 cubic feet a minute. Thus it will be seen that the time between start and re-start will be the same for pump No. 2 as if pump No. 1 had never existed, and therefore the *capacity of the suction-well between the cut-in and cut-out levels of that particular pump can be calculated without regard to any other pumps in the station*. The conditions will be the same for the third pump when the incoming flow is 250 cubic feet a minute. If the pumps are of various sizes the minimum capacities between their individual cut-in and cut-out levels will vary according to the pump in question.

Relative positions of cut-in levels. Having determined the storage capacity required between the cut-in and cut-out levels of each pump, it is necessary to determine also the difference between individual cut-in and cut-out levels. This is fixed by two factors: the normal performance of float switches or their equivalents in electrodes and the times required for electric pumps to get under way and deliver their full quantity. If two pumps are arranged so that one cuts in very shortly after another it can happen that the second pump will cut in before the first pump has got properly going and when the rate of inflow to the well does not justify the running of two pumps. If this occurs, the well will be emptied very rapidly by the two pumps running together, and the pumps will start too frequently.

To overcome this, cut-in levels are spaced reasonably well apart. Also, many float switches will not operate until the float has moved upwards about 3 inches; similarly, electrodes require a minimum amount of submergence. For these reasons the rule of thumb in average practice is never to have cut-in levels less than 6 inches apart vertically, but this is not always adequate.

The starting time of pumping sets is by no means an easy matter to calculate, for it depends on the construction of the starting gear, the characteristic curve of the pump and the proportion of inertia of the contents of the rising main to friction during starting. For this reason it is usual to assume a nominal starting time, as given in Table 28, in which starting times are given in seconds.

TABLE 28
STARTING TIMES OF PUMP SETS (IN SECONDS)

Type of pump	50-h.p.	150-h.p.	300-hp.	500-h.p.	1000-h.p.
Centrifugal. . .	20	30	40	60	90
Reciprocating . .	20	40	60	60	90

Following these lines of reasoning, the capacity in the suction-well between the cut-in levels of pump No. 1 and pump No. 2 should be such that when the rate of flow into the well is equal to the delivery of pump No. 1 the water should not rise so as to cut-in pump No. 2 in the time required for pump No. 1 to get going; and it should be assumed that pump No. 1 delivers nothing at all until it gets going. Further, the capacity so calculated should occupy a depth in the well of not less than 6 inches. Of course, the capacity can be made greater if it is convenient, the minimum depth of 6 inches still being maintained.

Cut-out levels. Some pumping-station designers arrange for all pumps to cut-out at the same level. This is not good practice, for it can have the effect of pumps cutting-out in the wrong order, thereby causing mutual interference. It is best to have the cut-out levels arranged at 6-inch intervals as in Table 30.

Example calculation. Suppose that a pumping station has to deal with a dry-weather flow of 150 cubic feet a minute and a maximum rate of three times dry-weather flow, or 450 cubic feet a minute, all flow in excess of this being considered as storm water and dealt with by other means: suppose that there are 1000 feet of rising main and a dead lift of 28·5 feet: also that each pump has suction and delivery pipework consisting of 7 feet of straight pipe, two sluice-valves, one reflux valve, three medium bends and one T-connection.

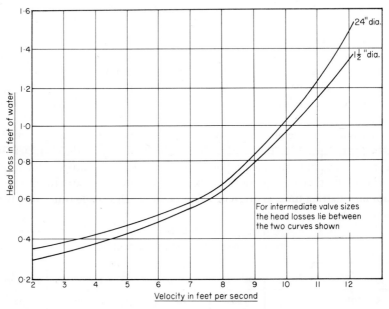

FIG. 52—Loss of head through reflux valves
By courtesy of Glenfield & Kennedy, Ltd.

Dealing first with the rising main, the circumstances suggest the installation of three duty pumps each of 150 cubic feet a minute capacity. This arrangement will give a maximum rate of flow which is three times the normal rate of flow. The nearest diameter of pipe that will give an economic velocity of not less than $2\frac{1}{4}$ feet a second with one pump running and not more than 7 feet a second at the maximum rate of flow is 14 inches. According to Vol. I, Fig. 68, the hydraulic gradient will be 1 in 60 when the flow is 450 cubic feet a minute, which gives a frictional loss of 16·7 feet in the rising main.

Let us next take the station losses. Accurate data of losses through fittings such as valves, etc., cannot be guaranteed; but a number of tables is in use, and probably Table 29 is as good as any (an alternative is given in Vol. I,

SEWAGE PUMPING

TABLE 29
APPROXIMATE FRICTIONAL RESISTANCE OF FITTINGS
(See also Vol. I, Chapter XII.)

Type of fitting	Equivalent length in feet of straight pipe of equal diameter
Sluice valve	6 × Pipe diameter in feet
Reflux valve	50 × ,, ,, ,, ,,
Bend (radius equal to 3 to 5 diameters) .	14 × ,, ,, ,, ,,
Round elbows	30 × ,, ,, ,, ,,
Sharp elbows and T-connections . .	90 × ,, ,, ,, ,,

Chapter XII). This gives the equivalent length of straight pipe that would give the same loss as the appropriate fittings of equal diameter.

Working from Table 29, the losses are as follows:

$$
\begin{aligned}
\text{2 sluice-valves} &= 6 \times 2 = 12 \text{ diameters of pipe} \\
\text{1 reflux valve} &= 50 \times 1 = 50 \text{ ,, ,, ,,} \\
\text{3 medium bends} &= 14 \times 3 = 42 \text{ ,, ,, ,,} \\
\text{1 T-connection} &= 90 \times 1 = 90 \text{ ,, ,, ,,} \\
\text{Total} &= 194
\end{aligned}
$$

Referring to Vol. I, Table 15, the economic diameter is 10 inches, giving an economic velocity of about 5 feet a second. Then, 194 diameters of 10 inches = 153 feet equivalent length of pipe, which added to the 7 feet of straight pipe gives a total equivalent length for the branch of 160 feet. According to Vol. I, Fig. 68, a discharge of 150 cubic feet a minute through one branch requires a gradient of 1 in 90. Therefore loss of head in branch is $\frac{160}{90} = 1.8$ feet.

Thus, the manometric head is made up as follows:

$$
\begin{aligned}
\text{Dead lift} & \quad 28.5 \text{ feet} \\
\text{Loss in rising main} & \quad 16.7 \text{ ,,} \\
\text{Loss in branch} & \quad 1.8 \text{ ,,} \\
\text{Velocity head} & \quad 0.4 \text{ ,,} \\
\text{Total} & \quad 47.4
\end{aligned}
$$

Assuming a pump efficiency of 70 per cent:

$$
\text{Pump horse-power} = \frac{47.4 \times 150 \times 62.3}{33,000} \times \frac{100}{70} = 19.2
$$

which, multiplied by 0.7457, gives 14.3 kilowatts.

The metric equivalent of the above horse-power calculation is as follows:

$$\text{Kilowatts} = \frac{Q \times H \times 100}{6 \cdot 118\dot{3} \times \% \text{ efficiency}}$$

$$= \frac{4 \cdot 25 \times 14 \cdot 4 \times 100}{6 \cdot 118\dot{3} \times 70}$$

$$= 14 \cdot 3$$

where: $Q =$ pump discharge in cubic metres per minute
$H =$ manometric head in metres.

Next, referring to Table 28, the starting time for any one pumping set can be taken as approximately 20 seconds.

The difference between any two cut-in levels should be equal to the amount pumped by one pump in 20 seconds, which is:

$$\frac{150 \times 20}{60} = 50 \text{ cubic feet.}$$

This should be accommodated in not less than 6 inches depth of suction-well, and therefore the required surface area of suction-well is 100 superficial feet. The capacity between cut-in levels can, of course, be increased as desired, if a matter of convenience, but we will assume in the present instance that 100 superficial feet and the net capacity as calculated is satisfactory. It is now necessary to find the depth between the cut-in and cut-out levels of any one pump.

Allowing for a maximum of fifteen starts an hour, or one start every 4 minutes, required capacity of suction-well between cut-in and cut-out levels of any one pump equals 1 minute's pumping rate of one pump—i.e. 150 cubic feet. Then, as the surface area is 100 superficial feet the depth between cut-in and cut-out levels is 1·5 feet.

From the foregoing it can be seen that taking the lowest cut-out level as assumed datum or 0·00, the cut-in and cut-out levels for the three pumps are as in Table 30.

TABLE 30

EXAMPLE OF CUT-IN AND CUT-OUT LEVELS

No. 3 cuts in at 2·5 feet above assumed datum
No. 2 cuts in at 2·0 ,, ,, ,, ,,
No. 1 cuts in at 1·5 ,, ,, ,, ,,
No. 3 cuts out at 1·0 ,, ,, ,, ,,
No. 2 cuts out at 0·5 ,, ,, ,, ,,
No. 1 cuts out at 0·0 ,, ,, ,, ,,

This is a very typical example, and it will be seen that the cut-in and cut-out levels form a simple series occupying in all a depth of 2 feet 6 inches in the well and an overall capacity of 250 cubic feet. Allowing for, say, an additional 50 cubic feet capacity below lowest cut-out level for submersion of suction, etc., the well will not be more than 300 cubic feet working capacity, which is very moderate and easily arranged to be self-cleansing.

Pumping station without suction-well. Where very large sewers and flows are concerned it may be practicable and desirable not to construct a suction-well of sufficient capacity to determine the starting times of the pump, but to rely on the storage capacity of the sewer itself. For example, if the incoming sewer is of 10 feet diameter and is laid at a gradient of 1 in 2000 it is possible for the cut-in and cut-out levels to be arranged at suitable positions not below the invert but between the invert and the crown of the sewer at the pumping station. Then, between any two levels there will be an appreciable storage capacity in the sewer, depending, of course, not only on the diameter and gradient of the sewer but also on the relation of those levels to the invert.

The storage in this type of case can be calculated near enough for practical purposes by finding the space occupied in the sewer by the sewage, assuming the water to be resting level in the sewer at cut-in level of the pump concerned and deducting from this the space occupied when the water at the outlet end is at cut-out level. The simplest case is when it is assumed that cut-out level for any particular pump is the level to which the sewer is filled when the flow in the sewer is at the rate which would cause the most frequent starting and stopping of that pump, assuming that no allowance is made for drawdown in the sewer.

In practice, it may be convenient to arrange the cut-out level below the theoretical level of the water in the sewer. But draw-down is best excluded from the calculation and no extra storage capacity allowed for this and, for calculation purposes, the cut-out level should be taken as being the same as the water-level in the sewer running partly full.

It will be seen that in calculations of this kind, it is necessary to determine the volume of a slice taken through a cylinder at an angle to the axis and deduct from this volume a smaller slice taken parallel to the axis.

In the ideal case the cut-out level of the first pump to cut-in should be at the level of the water in the sewer at the pumping station when the flow is one-half the delivery of the pump. This level can be found from Table 1. For example, if the sewer were a 10-foot diameter sewer laid at a gradient of 1 in 2000 and discharging 24,068 cubic feet a minute when running full and if pump discharge is 4200 cubic feet a minute, half the discharge of the pump will be 2100 cubic feet a minute: then the proportional discharge will be

$$\frac{2100}{24,068} = 0\cdot0876 \text{ approx.}$$ At this discharge the proportional depth will be 0·20

or 2 feet above invert (see Table 1). Now suppose the cut-in level were set 1 foot higher. This would be 3 feet above invert, or 0·30 of the total depth.

The storage in the sewer could then be found by the following formula, which is based on a modification of Simpson's rule:

$$C = \frac{\pi D^2}{4} \times \frac{A_1 + 4A_2 - 5A_3}{6} \times (H_1 - H_2)i$$

where: C = storage capacity in cubic feet
D = diameter of sewer in feet
A_1 = proportional area applicable to H_1 (see Table 1)
A_2 = proportional area applicable to $\frac{H_1 + H_2}{2}$
A_3 = proportional area applicable to H_2
H_1 = depth over invert in feet at cut-in level
H_2 = depth over invert in feet at cut-out level
i = inclination or length in feet in which sewer falls 1 foot.

Then, in the example given and referring to Table 1:

$$C = \frac{\pi \times 10^2}{4} \times \frac{0 \cdot 2523 + (4 \times 0 \cdot 1955) - (5 \times 0 \cdot 1424) \times (3 - 2) \times 2000}{6}$$
$$= 8440 \text{ cubic feet}$$

which is the quantity the pump could pump out in 2 minutes. Thus, in the conditions which have been set in the example, the pump would start not more frequently than once every 8 minutes.

Working on this method, the suitable distance between cut-in and cut-out levels can be found by trial and error. The calculation is, of course, complicated if the storage extends beyond the end of that part of the sewer which is of constant diameter and gradient.

In the case of second and subsequent pumps to cut-in, the cut-in level should be arranged to correspond with the water-level in the sewer when the flow is equal to the total discharge of the pumps already running plus half the discharge of the pump to which the calculation relates.

It is possible to obtain a closer approximation to the truth, and thereby obtain some greater theoretical storage capacity, by allowing for draw-down at the time when the pump is just cutting out and a "back-water" curve when the pump is just cutting in, working out the hydraulic gradients over short lengths starting from the pumping station and continuing upstream until the draw-down or back-water curve has died out. While this method makes a closer estimate of the actual storage available, it introduces uncertainties and for which reason it is recommended that the former method, without allowance for draw-down, is generally to be preferred.

Because of a case in which surge in the incoming sewer had been blamed (probably unjustly) for the cause of failure of a pumping station, the author made model experiments and found that it was impossible to cause surge in an incoming sewer by the starting and stopping of pumps to which it discharged provided that at no time was the sewer permitted to run full. On this basis he designed the sewage pumping station of the Crossness sewage-treatment works, probably the largest of its type in the world at that time. This station has now been running for several years drawing direct from one high-level and two low-level sewers of 11 feet 6 inches diameter, and there has been no surge interfering with automatic pump operation. On the other hand, a case was brought to his notice of a pumping station where the precaution of not allowing the sewer to surcharge had not been taken and violent surging did result.

Storage of storm water. In Chapter IV it is described how allowance can be made for the storage of storm water in the suction-wells of pumping stations so that peak flows exceeding the maximum pumping rate may be permitted. When storage of storm water is allowed for in the well of a pumping station the capacity as calculated should occupy space in the well above the highest cut-in level of any pump and not include that capacity required for determining starting and stopping times. It is also generally desirable for this storage capacity to be below the invert of the incoming sewer.

As in English sewerage practice it is usual to design sewers so as to accommodate the most intense run-off liable to occur say once in three years* and to let surcharge and other adventitious factors take care of all less-frequent greater storms, it would appear reasonable that, where it is found advisable to allow for storage in the suction-wells of storm-water pumping stations, the storage between highest cut-in level and invert of incoming sewer should be the maximum taken up by the design storm, and that greater storms of less-frequent occurrence should be accommodated in that part of the suction-well which lies above the invert of the incoming sewer but below the point at which overflow would occur.

There is a point to be borne in mind when applying the storage formulae to pumping stations which deliver both surface water and storm water, and that is, that the rate of outgo P is not necessarily the maximum pumping rate, but the maximum pumping rate less any constant rate of flow likely to be entering at the time, e.g. dry-weather flow or even peak rate of dry-weather flow during the period of pumping. For whereas in most storm water calculations dry-weather flow is negligible by comparison with peak rate of storm-water run-off, it is by no means negligible in comparison with pumping rate when small pumps discharge over a long period the storm water that came down in a few minutes.

Automatic starting of diesel engines. Automatic operation may be applied to diesel-driven pumps, but this method is not generally popular. It has the advantage of rendering a pumping station independent of outside electric supply and is more economical of power than the use of electrically

* Or once a year if no allowance is made for change of run-off coefficient during storm.

driven pumps and an independent power-station. The last-mentioned method is sometimes of utility where electricity is used for many purposes—for example, at large sewage-disposal works.

The automatic starting of diesel engines requires a few safety devices. For example, if the engine fails to start in 30 seconds, the starting motor should cut-out; also the engine should be automatically stopped if any failure in the lubricating system should occur. In order that starting batteries should

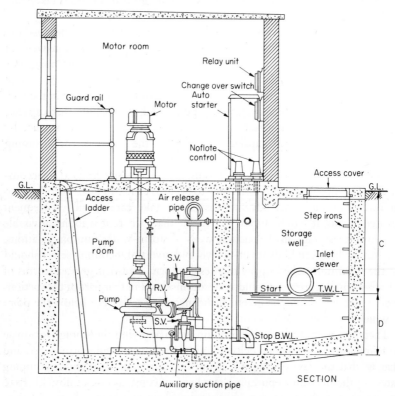

Fig. 53—"Dry-well" pumping station
By courtesy of Adams-Hydraulics, Ltd.

be kept fully charged, the diesel engines should run for not less than, say, 15 minutes at a time—the approximate battery-charging period—which means that the suction-well should hold not less than 15 minutes' pumping capacity of the largest pump installed. This means that the pumps would, under working conditions, never start more frequently than once an hour.

Automatic priming. The most popular method of priming for automatic pumps is the placing of the pumps in a "dry-well" below the level of the sewage in the suction-well, so that the pumps are always full of sewage when there is any to be pumped. This does away with all possibility of failure to prime, but adds to cost of construction by requiring a pump-chamber below ground-level (see Fig. 53).

Pumps which are submerged in the suction-well are also used. A pumping station incorporating these costs less than a "dry-well" pumping station. On the other hand, should a stoppage occur the pump can be cleared only

FIG. 54—"Barrington" self-priming sewage pump
By courtesy of Tuke and Bell, Ltd.

with difficulty: either a man must work in thigh boots in the suction-well itself or else the whole of the installation must be hoisted into the motor-room. To permit the last operation, the motor-house must have sufficient headroom to accommodate the longest individual portion of the pumping plant plus the minimum length of the chain-hoist or other lifting gear. In most instances such pumps should be avoided.

Modern self-priming sewage pumps such as Tuke and Bell's "Barrington" pump do away with the necessity for the pumps being placed in a pump-well below sewage level. This pump is a centrifugal pump set below a priming vessel which holds sewage between times of pumping and primes the pump. Pump, priming vessel and electric motor are all set at ground-level, the suction being carried down to the suction-well (see Figs. 54 and 55).

Screens. In larger installations, particularly those which have to deal with storm water, it is an advantage to use high-efficiency pumps: however, this necessitates screening of the sewage because small particles—for example, corks—can put out of action large centrifugal pumps of the high-efficiency type. Where screens are installed in a large station they should be close-spaced: any screens with clear openings of more than 1 inch may prove unsatisfactory, and spaces of $\frac{1}{2}$ inch or $\frac{5}{8}$ inch are generally preferable.

To avoid added labour and the necessity of removing objectionable screenings, the screens may be automatically raked by a mechanism which is float controlled, so as to start when there is a differential head on each side of the screen. The screenings removed by the tines may be washed off by a sparge supplied by a small pump, and discharged into a disintegrator pump. This last pump should deliver the disintegrated screenings upstream of the screens. In a large station the disintegrator pump or another unchokable pump may be arranged completely to empty the well of sewage, scum and detritus at least once a day. (For more details see Chapter XXIV.)

An arrangement which the writer has found satisfactory* for the protection of small unchokable pumps for stations of minimum size is one or more vertical slots 2 inches wide and about 2 feet high and spaced at least 18 inches apart. These should be in a wall of a manhole discharging to the suction-well. Such slots serve the purpose of a screen in intercepting large solids, but they do not choke to any extent because there are no bars round which fibrous matter can fold and build up. Manually raked screens should never be used at small automatic pumping stations because they are liable to become choked with large quantities of rag and paper, the disposal of which is inconvenient.

Ventilation. The ventilation of a pumping station is of importance. The rising and falling of the sewage in a well has a bellows action which blows air in and out. The air must have free access and egress, and this necessitates the arrangement of a vent column or other ventilator connecting direct to the well and placed where it will not be a nuisance. Ventilation of the sewer is not sufficient, as the sewer may become water-sealed. The suction-well should be completely isolated from the pump-chamber, and any leads from floats should pass through a ventilated chamber before entering the pump chamber or motor-house. The last two should also be separately ventilated. Where diesel or other internal-combustion engines are installed, forced ventilation is necessary, because leakage of exhaust pipes is extremely difficult to prevent, and there is a great danger of carbon monoxide collecting in the pump-chamber.

---

* The arrangement is satisfactory with normal usage, where the slots can be cleared once a day. At one site visited once a month slots had to be removed because they became stopped by stacks of whole oranges which rested vertically one above another and which, with other material, closed the waterway. The pushing of objects through the trap of a water-closet is not uncommon at children's institutions: at some sites (boarding-schools for young children) drains leading to treatment works were choked by rubber ducks. In one instance (a restaurant) a considerable quantity of cutlery was found in a drain and could only have got there by being pushed through the trap of a closet.

Reduction of noise. Pumping stations are very seldom silent, although the amount of noise which can be heard outside a station with closed doors and windows is usually negligible in places where there is much traffic. Where it is necessary, owing to the locality, to be certain that there is no noise, there are two means by which it may be reduced. In the first place, the engines or motors should be the most silent of their kind, and secondly, the

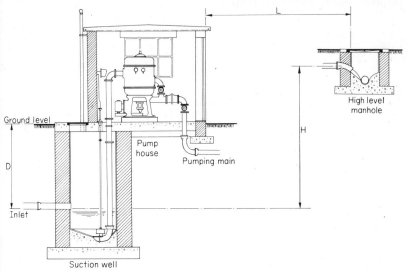

FIG. 55—"Barrington" self-priming sewage pump
By courtesy of Tuke and Bell, Ltd.

buildings should be so constructed as to keep the noise inside. Thick walls and not over-large windows, particularly those with either double-glazing or leaded glass, and roofs of reinforced concrete, are preferable to buildings with large windows lightly glazed and ordinary tiled roofs. Sound insulation depends more on the weight of the material in the walls or partitions than on the kind of material, but cavities greatly help: double windows and doors may be used in extreme cases.

Fuel store. Apart from economy, the type of plant used depends on the importance of maintaining the station in running order at all times. An electrically operated station is dependent on electricity supply, failure of which would mean failure of the station. In some cases more than one supply may be available, reducing this danger. Alternative supplies may be essential in some instances. At the Crossness pumping station mentioned above, there are two supplies from the National grid plus one from the sewage-treatment works power-house.

In a large station involving prime movers the running of the plant is not dependent to the same extent on external agencies. Transport or other strikes, however, can interfere with the supply of fuel, and for this reason the storage of oil or coal should be ample so as to cover against the possibility of a prolonged interruption of supply. When an existing steam power-station is converted to diesel-oil power, the boilers may with advantage be converted to oil tanks and give several weeks' storage. Where only small stores are concerned, the size of the tank used may best be based on the size of the tank-vehicle which delivers the supply, the minimum storage of the station being, say, two vehicle loads in two tanks.

CHAPTER XIV

PNEUMATIC EJECTORS

A PNEUMATIC ejector (Fig. 56) consists of a cast-iron vessel into which the unscreened sewage flows by gravitation, and from which it is blown out by compressed air. As soon as the ejector is full, a float rises, actuating a valve through which the air is admitted. There is a reflux valve on the inlet pipe which prevents the return of the sewage which is forced by the pressure of the air up to the height required, and a reflux valve on the rising main to prevent the sewage from gravitating back into the ejectors. When the last

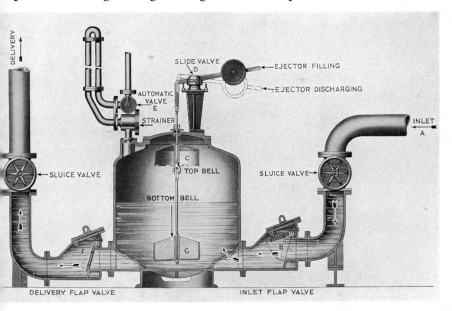

FIG. 56—"Shone" ejector
By courtesy of Hughes and Lancaster, Ltd.

of the sewage has been lifted, the float drops, bringing back the air valve to its original position, whereupon the supply of compressed air is cut off and the air in the ejector released via a pipe which discharges to a manhole on the incoming sewer. These operations go on automatically.

Wherever possible, the storage capacity of the rising main should be less than that of the ejector, so that, should the reflux valve on the delivery leak, the ejector will not go on working when there is no incoming flow. Ejectors are usually installed in duplicate and their air valves so interconnected that one ejector may fill while the sewage is being delivered from the other; but two ejectors, so working, discharge only a little more than one ejector working alone on the same air supply.

Most ejectors have to be placed below the invert of the incoming sewer, but Tuke and Bell's "Lift and force ejectors" (see Fig. 57) do away with the necessity for a below-ground chamber. These ejectors are placed at ground-level, and are controlled by a pneumatic device actuated by the sewage in a small suction-well. They alternately suck sewage out of the well and eject it up the rising main.

The quantity of sewage pumped by an ejector depends on the volume of compressed air supplied for its operation. This has to be slightly more than the quantity of sewage pumped so as to allow for losses by solution, etc. The

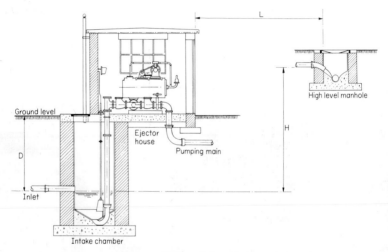

FIG. 57—"Lift and force" ejectors
By courtesy of Tuke and Bell, Ltd.

rate of discharge up the rising main is not the same as the quantity of air supplied: it has to be greater for two reasons. First, in the case of a single ejector, while the ejector is filling there is no flow up the rising main: therefore, when discharge takes place, it must be at a higher rate than the flow of incoming sewage.

The second reason is that the velocity of flow in the rising main must always be sufficient to give self-cleansing conditions. This velocity depends on the manometric head and the pressure at which air is supplied.

The difference between the rate at which air is supplied by the compressor and the rate at which it is used by the ejector is taken up by an air vessel which should be of sufficient capacity to give a flush that will empty the ejector at the self-cleansing velocity of the rising main. The air vessel is also the means by which the compressor is started and stopped, being fitted with pressure-operated switches.

The quantity of air required for a sewage ejector can be found by the formula:

$$C = \frac{Q(H + 34)}{34}$$

where: C = cubic feet of free air per minute
H = total head in feet
Q = discharge of sewage in cubic feet per minute.

Pressure should be sufficient to ensure self-cleansing conditions and always provided at 40 per cent higher than the theoretical manometric head.

Emergency or temporary ejector. During the Second World War the author had to devise a means of pumping chemical-closet sewage without using machinery more than absolutely necessary.* For this purpose he devised a plant which worked on the combined principle of the sewage ejector and air-lift and, in the main, involved nothing more than a steel pressure cylinder, a screwed-barrel rising main, an inlet hopper with sluice-valve and a pressure gauge. This plant had to be operated from an existing air supply liable to fall to a pressure of 45 lb per square inch and the maximum quantity available per installation was that which would pass through a $\frac{1}{8}$-inch diameter orifice. (Over 100 were installed and used with satisfaction.)

A similar installation could be used by an engineering contractor for delivering chemical-closet or water-closet sewage from a site of works at considerably less cost than normal pumping or ejector plant and, as compressed air is almost invariably available on an engineering site, this installation should be worthy of description in this connection.

The plant consisted of a 200-gallon cylindrical pressure tank with an inlet hopper controlled by a sluice-valve on the top side. The sluice-valve was arranged horizontally so as to be less liable to have trouble with solids. On the top side also was an air-main inlet, and air-release outlet. These had plug-type valves linked in such a manner that one was open when the other was shut and vice versa. The tank had also a pressure gauge. The rising main was taken from the bottom of the tank and delivered to a manhole on a drain or sewer.

The principle of operation was as follows. The tank was filled from time to time until it was full. Then the sluice valve was closed and the air turned on. As the pressure developed in the cylinder the sewage was forced up the rising main into the manhole until, when the cylinder was almost empty, air entered the rising main reducing the weight of the column of mixed air and water so that the compressed air in the cylinder violently blew out all that remained. The completion of ejection was indicated by drop of pressure on the pressure gauge, after which remaining air pressure was released and the

* "The Sanitation of Tube Railway Stations used for Air-raid Shelters" (Ingenuity prize paper, Institution of Civil Engineers, 1941).

sluice valve opened ready for the next refilling. It was necessary to discharge to a well-vented manhole or into a break tank for release of air otherwise there was the danger of blowing the traps of sanitary appliances of near-by properties.

On an engineering site an old boiler could be used as ejector and a connection made from a portable compressor. If the ejection were to be made once a day only, the tank would need to have a capacity of at least $\frac{1}{3}$ gallon per head per day in the case of chemical-closet sewage or at least 10 gallons per head per day when taking the flow from water-closets, lavatories and canteens.

Air compression. The pressure of the air must be sufficient not only to lift the sewage to the desired height, but also to overcome the frictional loss in the rising main, the air-main valves, etc. In calculating efficiency of installation and quantity of air required, allowance must be made for the reduction in volume of air due to compression. The volume of air is inversely proportional to its absolute pressure, which is equal to the gauge pressure + 14·7 lb per square inch. If, therefore, air is required at a gauge pressure of P_1 lb per square inch (P_1 + 14·7 absolute), $\dfrac{P_1 + 14\cdot 7}{14\cdot 7}$ cubic feet of free air must be compressed to furnish every cubic foot of compressed air.

One cubic foot of free air at a temperature of 60 °F. and at atmospheric pressure weighs 0·0764 lb, and 13·09 cubic feet weigh 1 lb. At any other pressure, P_1 (on the gauge), 1 lb of air is equal to $\dfrac{14\cdot 7 \times 13\cdot 09}{P_1 + 14\cdot 7}$ cubic feet.

In the compression of air a part of the power expended is converted into heat, which tends to increase the volume of the air, and consequently the amount of power required to compress it. In order to minimise the loss of power thus occasioned, as much of this heat as possible is carried off by a water-jacket which surrounds the cylinder of the compressor. If all the heat could be carried off as fast as it is produced, the temperature of the air would remain constant, and the compression would be "isothermal." If, on the other hand, none of the heat is removed, the compression is "adiabatic."

If W is the horse-power required for the isothermal compression of 1 cubic foot of air per minute, W' the power required for adiabatic compression, Pa the atmospheric pressure in lb per square inch, and P_1 the absolute pressure to which the air is compressed (or the gauge pressure + 14·7) then

$$W = 0\cdot 1477 \log \frac{P_1}{Pa}, \text{ and}$$

$$W' = 0\cdot 2214 \left[\left(\frac{P_1}{Pa} \right)^{0\cdot 29} - 1 \right].$$

The ratio $\dfrac{W}{W'}$, is called the "thermal efficiency" of the compression.

PNEUMATIC EJECTORS

In arriving at the power required to drive a compressor, allowance must be made not only for the thermal loss of power, but also for the mechanical losses in the compressor. A good compressor of moderate size should give an overall efficiency of at least 50 per cent.

Table 31 shows the pressures corresponding to various heights of lift, the

TABLE 31

PRESSURE AND VOLUME OF AIR AND POWER REQUIRED FOR COMPRESSION
(*Efficiency of compression 50 per cent*)

Height of lift (ft)	Gauge pressure (lb per sq in)	Cu ft of free air per cu ft compressed	B.H.P. per 100 cu ft of free air per minute
2	0·87	1·06	0·735
3	1·30	1·09	1·087
4	1·73	1·12	1·43
5	2·17	1·15	1·76
6	2·60	1·18	2·09
7	3·03	1·21	2·41
8	3·47	1·24	2·72
9	3·90	1·27	3·02
10	4·34	1·30	3·32
11	4·77	1·32	3·60
12	5·20	1·35	3·89
13	5·64	1·38	4·16
14	6·07	1·41	4·43
15	6·50	1·44	4·70
16	6·94	1·47	4·96
17	7·37	1·50	5·21
18	7·80	1·53	5·46
19	8·24	1·56	5·71
20	8·67	1·59	5·95
25	10·84	1·74	7·09
30	13·01	1·89	8·13
35	15·17	2·03	9·10
40	17·34	2·18	10·00
45	19·51	2·33	10·84
50	21·68	2·48	11·62
55	23·84	2·62	12·37
60	26·01	2·77	13·07
65	28·18	2·92	13·73
70	30·35	3·07	14·37
75	32·51	3·21	14·97
80	34·68	3·36	15·54
85	36·85	3·51	16·10
90	39·02	3·65	16·62
95	41·18	3·80	17·13
100	43·35	3·96	17·62

TABLE 32

LOSS OF PRESSURE IN AIR MAINS

Cubic feet of free air per minute at 60 °F

Diam.	1 in	1¼ in	1½ in	2 in	3 in	4 in	5 in	6 in	7 in	8 in	9 in	10 in	11 in	12 in
A	2·7	5·1	8·5	19·3	59·9	132	243	397	601	857	1,171	1,546	1,986	2,494
B	2·8	5·4	9·1	20·6	63·9	141	259	424	641	914	1,249	1,649	2,118	2,660
C	3·0	5·7	9·7	21·8	67·9	150	275	451	681	971	1,327	1,752	2,250	2,827
D	3·2	6·1	10·2	23·1	71·9	159	292	477	721	1,028	1,405	1,855	2,383	2,993
E	3·4	6·4	10·8	24·4	75·9	168	308	503	761	1,085	1,483	1,958	2,515	3,159
F	3·5	6·7	11·4	25·7	79·9	176	324	530	801	1,143	1,561	2,061	2,648	3,325
G	4·4	8·4	14·2	32·1	99·8	221	405	662	1,001	1,428	1,951	2,576	3,309	4,157
H	5·3	10·1	17·0	38·5	120·0	265	486	795	1,201	1,714	2,341	3,091	3,971	4,988
I	6·2	11·8	19·9	45·0	140·0	309	567	927	1,401	2,000	2,732	3,607	4,633	5,819
J	7·1	13·5	22·7	51·4	160·0	353	648	1,060	1,602	2,285	3,122	4,122	5,295	6,651
K	8·0	15·2	25·6	57·8	180·0	397	729	1,192	1,802	2,571	3,512	4,637	5,957	7,482
L	8·9	16·8	28·4	64·2	200·0	441	810	1,325	2,002	2,856	3,902	5,152	6,619	8,313
M	9·7	18·5	31·2	70·7	220·0	485	891	1,457	2,202	3,142	4,293	5,668	7,281	9,145
N	10·6	20·2	34·1	77·1	240·0	529	972	1,590	2,402	3,428	4,683	6,183	7,943	9,976
O	11·5	21·9	36·9	83·5	260·0	573	1,053	1,722	2,603	3,713	5,073	6,698	8,604	10,807
P	12·4	23·6	39·8	89·9	280·0	618	1,134	1,855	2,803	3,999	5,463	7,213	9,267	11,639
Q	13·3	25·3	42·6	96·4	300·0	662	1,215	1,987	3,003	4,285	5,854	7,729	9,929	12,470
R	14·2	27·0	45·4	103·0	320·0	706	1,296	2,120	3,203	4,570	6,244	8,244	10,590	13,301
S	15·1	28·6	48·3	109·0	339·0	750	1,377	2,252	3,403	4,856	6,634	8,759	11,252	14,133
T	15·9	30·3	51·1	116·0	359·0	794	1,458	2,385	3,604	5,142	7,024	9,274	11,914	14,964
U	16·8	32·0	53·9	122·0	379·0	838	1,539	2,517	3,804	5,427	7,415	9,790	12,576	15,795
V	17·7	33·7	56·8	128·0	399·0	882	1,620	2,650	4,004	5,713	7,805	10,305	13,238	16,627
W	18·6	35·4	59·6	135·0	419·0	926	1,701	2,783	4,204	5,999	8,195	10,820	13,900	17,458
X	19·5	37·1	62·5	141·0	439·0	970	1,782	2,915	4,404	6,284	8,585	11,366	14,562	18,289
Y	20·4	38·8	65·3	148·0	459·0	1,015	1,863	3,048	4,605	6,570	8,976	11,851	15,224	19,121
Z	21·3	40·4	68·1	154·0	479·0	1,059	1,944	3,180	4,805	6,856	9,366	12,366	15,886	19,952

TABLE 33
LOSS OF PRESSURE IN AIR MAINS
Loss of pressure: lb per square inch per 100 feet of pipe

	Gauge pressure: lb per square inch															
	5	6	7	8	9	10	12	14	16	18	20	25	30	35	40	45	50
A	0·0265	0·0252	0·0240	0·0230	0·0220	0·0211	0·0195	0·0182	0·0170	0·0159	0·0150	0·0131	0·0117	0·0150	0·00953	0·00873	0·00806
B	0·0301	0·0287	0·0273	0·0261	0·0250	0·0240	0·0222	0·0207	0·0193	0·0181	0·0171	0·0149	0·0133	0·0119	0·0108	0·00994	0·00917
C	0·0340	0·0324	0·0309	0·0295	0·0283	0·0271	0·0251	0·0233	0·0218	0·0205	0·0193	0·0169	0·0150	0·0135	0·0122	0·0112	0·0104
D	0·0381	0·0363	0·0346	0·0331	0·0317	0·0304	0·0281	0·0262	0·0245	0·0230	0·0216	0·0189	0·0168	0·0151	0·0137	0·0126	0·0116
E	0·0425	0·0404	0·0386	0·0368	0·0353	0·0339	0·0313	0·0292	0·0273	0·0256	0·0241	0·0211	0·0187	0·0168	0·0153	0·0140	0·0129
F	0·0470	0·0448	0·0427	0·0408	0·0391	0·0375	0·0347	0·0324	0·0302	0·0283	0·0267	0·0233	0·0207	0·0187	0·0169	0·0155	0·0143
G	0·0735	0·0700	0·0667	0·0638	0·0611	0·0586	0·0542	0·0505	0·0472	0·0443	0·0417	0·0365	0·0324	0·0291	0·0265	0·0243	0·0224
H	0·106	0·101	0·0961	0·0919	0·0880	0·0844	0·0780	0·0727	0·0679	0·0638	0·0600	0·0525	0·0467	0·0420	0·0381	0·0349	0·0322
I	0·144	0·137	0·131	0·125	0·120	0·115	0·106	0·0989	0·0925	0·0868	0·0816	0·0715	0·0635	0·0571	0·0519	0·0475	0·0439
J	0·188	0·179	0·171	0·163	0·156	0·150	0·139	0·129	0·121	0·113	0·107	0·0934	0·0829	0·0746	0·0678	0·0621	0·0573
K	0·238	0·227	0·216	0·207	0·198	0·190	0·176	0·164	0·153	0·144	0·135	0·118	0·105	0·0944	0·0858	0·0786	0·0725
L	0·294	0·280	0·267	0·255	0·244	0·235	0·217	0·202	0·189	0·177	0·167	0·146	0·130	0·117	0·106	0·0970	0·0895
M	0·356	0·339	0·323	0·309	0·296	0·284	0·263	0·244	0·228	0·214	0·202	0·177	0·157	0·141	0·128	0·117	0·108
N	0·424	0·403	0·384	0·367	0·352	0·338	0·312	0·291	0·272	0·255	0·240	0·210	0·187	0·168	0·153	0·140	0·129
O	0·497	0·473	0·451	0·431	0·413	0·396	0·367	0·341	0·319	0·299	0·282	0·247	0·219	0·197	0·179	0·164	0·151
P	0·576	0·548	0·523	0·500	0·479	0·460	0·425	0·396	0·370	0·347	0·327	0·286	0·254	0·229	0·208	0·190	0·176
Q	0·662	0·630	0·601	0·574	0·550	0·528	0·488	0·454	0·425	0·399	0·375	0·328	0·292	0·262	0·238	0·218	0·202
R	0·753	0·716	0·683	0·653	0·626	0·600	0·555	0·517	0·483	0·454	0·426	0·374	0·332	0·298	0·271	0·248	0·229
S	0·850	0·809	0·771	0·737	0·706	0·678	0·627	0·583	0·545	0·512	0·481	0·422	0·375	0·337	0·306	0·280	0·259
T	0·953	0·907	0·865	0·827	0·792	0·760	0·703	0·654	0·611	0·574	0·540	0·473	0·420	0·378	0·343	0·314	0·290
U	1·06	1·01	0·964	0·921	0·882	0·847	0·783	0·729	0·681	0·640	0·601	0·527	0·468	0·421	0·382	0·350	0·323
V	1·18	1·12	1·068	1·021	0·978	0·938	0·868	0·807	0·755	0·709	0·668	0·584	0·518	0·466	0·424	0·389	0·358
W	1·30	1·23	1·18	1·13	1·078	1·034	0·957	0·890	0·832	0·781	0·735	0·643	0·572	0·514	0·467	0·428	0·395
X	1·42	1·35	1·29	1·24	1·18	1·14	1·05	0·977	0·913	0·857	0·808	0·706	0·627	0·564	0·513	0·470	0·433
Y	1·56	1·48	1·41	1·35	1·29	1·24	1·15	1·07	0·999	0·937	0·883	0·772	0·686	0·617	0·560	0·513	0·474
Z	1·69	1·61	1·54	1·47	1·41	1·35	1·25	1·16	1·09	1·02	0·962	0·840	0·746	0·670	0·610	0·559	0·516

volumes of free air required to furnish 1 cubic foot of compressed air at the different pressures, and the power required to compress 100 cubic feet of free air per minute.

Loss of pressure in air mains. The drop in pressure which takes place when air flows through a pipe may be calculated by any of the usual formulae or from Tables 32 and 33.

In the preparation of these tables a difficulty presented itself which does not occur in connection with similar tables for water or sewage. In those cases we are dealing with a fluid of constant density, and have therefore only three variables to take into account. With air, on the other hand, we have a fourth variable, for the density of air varies with every change in its temperature or pressure. This difficulty has been overcome by having two tables, each line in the first showing the quantity of air (measured as free air) which can be carried by pipes of different diameters with the loss of pressure shown in that line in the second part of the table which bears the same reference letter.

In using these tables, the quantity of free air nearest to that required must be found in Table 32, and the loss of pressure, for the working pressure adopted, in the corresponding line in Table 33. If the loss of pressure in the pipe, as compared with the working pressure, is considerable, the latter should be taken as the mean of the initial and final pressures, but when the loss of pressure is very great, a special formula should be used.

CHAPTER XV

PROBLEMS PECULIAR TO THE SEWERAGE OF COASTAL TOWNS

INLAND towns are dependent on natural watercourses or occasionally soakage into the ground for the discharge of sewage effluent, and consequently their effluents usually have to be treated to the standard recommended by the Royal Commission on Sewage Disposal (see Chapter XX). In this respect the sewerage of seaside towns differs, for it is possible for sewage to be discharged into the sea without treatment or with partial treatment only, according to circumstances. On the other hand, discharge into tidal waters sets up fresh problems equally difficult, and, on the whole, generally less familiar to most engineers than those of sewage treatment. For sea outfalls have to be constructed to withstand storms and erosion of the foreshore, and often there is the problem of storage during tide-lock or for the purpose of taking advantage of favourable tidal streams so as to avoid pollution of the foreshore.

Standards of effluents. Sewage cannot always be discharged into the sea without precautions, for a number of reasons. Crude sewage contains solids, some of which float, others that settle, and, were haphazard discharge permitted, banks of sludge would collect at the ends of many outfalls,* and solids would wash up on to bathing beaches and in other places where their presence would be objectionable. As yet, little is known of the degree of the actual risk involved in bathing in contaminated sea-water. Both of these unpleasant occurrences are common and, for obvious reasons, it is desirable for improved methods of discharge into the sea to be employed wherever nuisance would otherwise be caused.

The prevention of pollution of shell-fish beds has also to be considered, but in this connection it should be remembered that the usual methods of sewage treatment do not ensure destruction of the typhoid bacillus, and the only satisfactory means of preserving shell-fish beds free from infection is to keep sewage effluent away from them entirely.

Among the methods in use for preventing complaints of nuisance are:

1. Discharge of sewage at such states of the tide that currents will carry the sewage out to sea. This method is restricted to those circumstances in which there is a good tidal stream capable of carrying away the suspended solids.

2. Disintegration of solids, with the aid of Comminutors or disintegrating pumps or machinery, and their discharge with the sewage. These have largely been installed for the purpose of disguising the solids so that

* They usually do.

bathers do not recognise them for what they are. But the solids are still there.

3. Partial treatment of sewage by screening and sedimentation, the sludge being removed and disposed of on land. This method reduces obvious nuisance, for when it is employed no visible particles of sewage are found in the sea, but it does not prevent bathers from unconsciously bathing in tank effluent should they come in contact with it before it has been diluted.

4. Full treatment of the sewage so as to produce a "Royal Commission" effluent.

In every case the degree of treatment desired should be considered in relation to cost. Full treatment is the ideal, but is not always necessary; discharge without treatment involves the least maintenance costs, but cannot always be permitted. Often some compromise has to be found. For example, in a particular instance it might be found advisable to treat the sewage by screening and sedimentation and, in addition, to discharge tank effluent at a state of the tide at which there would be little risk of it approaching the foreshore. (See also the section on "Sea outfalls" in Chapter XXI.)

Effect of sewerage. In the past, when little thought was given to treatment or storage of foul sewage discharged to the sea, many coastal towns were sewered on the combined system. This set up problems for the engineers who later had to arrange for treatment or storage. In a new coastal sewerage system there would be every reason to prefer separate sewerage because surface water can be discharged to soakaways, any stream or merely on to the foreshore, without an expensive outfall pipe, thus permitting the use of smaller pumping plant, treatment works, storage tank or outfall for foul sewage than would otherwise be required.

The possibility of inland disposal should never be overlooked. The fall of the land, more often than not, is other than towards the sea, and often it may be found most economical to sewer the town so as to discharge to some inland watercourse, in which case treatment works would be necessary. It should be mentioned that pumping is more often involved in coastal sewerage schemes than elsewhere, because of the tendency for some seaside towns to straggle along the coast regardless of contours, together with the necessity, on occasion, to pump against tide-lock.

Tidal experiments. When a sea outfall is proposed, experiments have to be made to determine the effects of the tide, so as to decide the most suitable position for the outfall and the times at which discharge may be permitted. Seaside local authorities usually have good information on the tidal ranges, and predicted tides can be found for the principal coastal towns in the Admiralty Tide Tables. Should, however, an outfall be proposed in a part of the world where such information is not available, records would have to be made by automatic tide recorder, over a period of not less than three months. From such records tidal curves for normal spring and neap tides can be plotted.

The time for the tide to rise and fall is approximately $12\frac{1}{2}$ hours, and on most parts of the coast of Great Britain the curve of height against time is very nearly a true sine curve. In many calculations a sine curve may be assumed, but not all: for example, there is a double tide at Southampton Water and a noticeable lag on the ebb in the tidal reaches of the large rivers.

Position of outfall. There are positions where, at suitable states of the tide, discharged sewage will be carried out to sea with little risk of its return to the foreshore. Headlands jutting out to sea are particularly favourable not only because they carry the outfall farther away from land than elsewhere but also they tend to direct lateral currents out to sea sometimes at both peak-ebb and peak-flow times. A similar effect is produced at ebb tide at the mouth of a harbour or natural haven. The value of these probably suitable points of discharge has to be ascertained by float experiments.

Float experiments. The behaviour of tidal streams or currents is found by putting floats in the water and plotting their movements. The floats need not be elaborate: provided they are easily visible, float deep in the water so as not to be unduly affected by the wind and yet are not liable to sink, they serve well. Occasionally floats are expensively made; but all that is necessary is a billet of wood about 8 feet long, weighted so that it floats upright with about one-sixth of its length out of the water and painted bright red or orange. Half a dozen of these simple floats are cheaper and easier to handle, and far less liable to be damaged, than one elaborate float heavily loaded and fitted with fins and miniature flag-pole.

On those parts of the coast where there is a prevailing wind towards the shore it may be considered necessary to make additional experiments with shallow floats. Painted blocks of wood loaded with iron serve this purpose: large oranges have been used because they float almost submerged but are easy to see.

Procedure in making tidal experiments. The behaviour of the currents at all states of the tide must be known, and this means that a continued series of experiments must be carried out. The procedure is on the following lines. One or more floats are placed in the water at the far end of the proposed point of outfall, and are followed by boat, and their courses plotted at 15-minute intervals until they wash ashore, have gone right out to sea and are obviously continuing in that direction, or have been in the water for at least 2 hours and shown no sign of approaching the foreshore. As soon as the courses of the floats have been observed in accordance with the above requirements, the floats are taken back to the point of outfall and the experiment repeated. This work is continued day after day until the floats have been placed in the water at the point of outfall every hour after high tide over a period of one tide or, say, 12 hours, and the experiment has been made for both spring and neap tides.

There are several methods of surveying the courses of the floats. One of the simplest is for the whole of the work to be done from the boat (which, incidentally should be a fair-sized power-launch), the surveyors working with nautical sextants and observing points on the shore for determining their positions. This entails some preliminary work in establishing shore landmarks and, where necessary, erecting easily distinguishable poles, the positions of which must be plotted on the Ordnance map. The survey book is then set out in the form shown in Table 34.

TABLE 34

EXAMPLE OF SURVEY BOOK

Date	Time	Hours after high water	Float number	Stations left right	ϕ_1 (left)	Stations left right	ϕ_2 (right)	Remarks (weather wind, etc.)
4th June 1946	4.35 p.m.	4	2	P$_4$ P$_5$	31° 20′ 45″	P$_5$ P$_6$	30° 8′ 30″	Light wind N.E.

Plotting the survey. The position of the boat at the time of taking the two angle readings on to three points of known position is found geometrically by the three-point problem, which can be found in most surveying textbooks, (see Fig. 58).

In practice the survey is plotted with a station-finder, an instrument with three arms, two of which are movable, connected to a protractor; or else an improvised station-finder drawn on tracing-paper as follows. The two adjacent angles ϕ_1 and ϕ_2 are drawn, and the piece of tracing-paper is moved

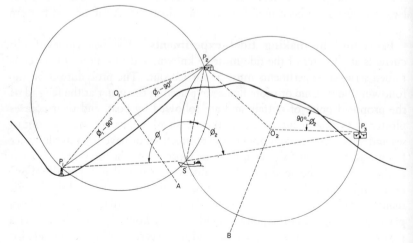

FIG. 58—Three-point problem

over the Ordnance sheet until the three lines forming the correct angles pass through the three shore stations exactly. Except in very rare circumstances, which need not be described, it will be found that there is only one position of the station-finder in which this can occur. The point of intersection is then pricked through onto the Ordnance map, and denotes the position of the boat and float. This work is continued until the courses of all floats have been plotted, when the points are joined by lines showing the directions of movement of all the floats. From the final plan so prepared the engineer can decide without much difficulty at what states of the tides and over what periods of time discharge of sewage may be permitted.

Calculating discharge of sea outfalls. The simplest calculation relating to sea outfall problems is the determination of the required diameter of outfall to accommodate the maximum flow of sewage at peak discharges from the town under the conditions of minimum head—i.e. at the time when the tide is at its highest. The hydraulic gradient is measured from the crown of the sewer at the point where it connects with the outfall (an allowance being made for loss of head through the tide flap at this point) to sea-level at the bottom end of the outfall (a second allowance being made for the difference in specific gravity of sea-water and sewage). In the first place, the required size of outfall can be found from "Escritts' Tables of Metric Hydraulic Flow," "Crimp and Bruges' Tables," or other tables or diagrams preferred by the designer, after which the accuracy of the calculation is checked and allowance made for the difference in specific gravity of sea-water and sewage by substituting for the value of H (head in feet), in the formula selected, the following expression:

$$H - \frac{h}{36}$$

where: H = the height of maximum sewage-level in tide-flap chamber above highest sea-level in feet

h = height in feet of highest tide-level above centre of outlet-end of outfall.

(In the more complex calculations which follow this, it may prove too difficult to make this allowance throughout and, instead, rule-of-thumb adjustment can be made to the whole calculation throughout according to the above formula. For example, suppose at high tide the end of the outfall was submerged 8 feet, for the sake of convenience it may be assumed throughout the calculation that the tide is $\frac{8}{36}$ feet higher than its true level.)

Storage of sewage. The more difficult problems are those relating to the discharge of sewage from a storage tank when allowance has to be made for incoming flow of sewage and the continual variation of tide-level.

Sewage is stored for two reasons, for apart from the storage required when it is intended that discharge shall take place only at certain states of the tide so as to avoid nuisance, it is not uncommon for land near the sea to be low-lying, with the result that sewers are tide-locked over high tide, and as discharge cannot take place, storage is necessary if pumping is to be avoided.

The calculations for finding size of outfall that will permit a tank to empty between periods of tide-lock are of varying complexity according to circumstances. The simplest case is when a tank with vertical sides and flat bottom discharges through an outfall the outlet of which is above high-tide level, there being no inflow at the time. This case can be solved by the following formula:

$$C = \frac{2A\sqrt{L}\left(\sqrt{H_1} - \sqrt{H_2}\right)}{t}$$

where: t = time to empty tank in minutes
L = length of outfall pipe in feet
H_1 = maximum head from top water-level to centre of pipe at outlet end
H_2 = minimum head from invert of tank (low end) to centre line of pipe at outlet end
A = surface area of tank in feet super
D = diameter of outfall in inches.
$C = 3 \cdot 575 D^{2\frac{1}{8}}$

The case is somewhat more complicated when, other conditions remaining the same, a constant rate of inflow has to be allowed for. Then the diameter of the pipe can be found by the following formula, which, however, involves trial and error:

$$C = \frac{2A\sqrt{L}(\sqrt{H_1} - \sqrt{H_2})}{t} + \left[\frac{2ALQ}{Ct} \times 2 \cdot 3026 \log_{10} \frac{\sqrt{H_1} - \frac{Q\sqrt{L}}{C}}{\sqrt{H_2} - \frac{Q\sqrt{L}}{C}} \right]$$

The notation is as for the previous formula, with the addition:

Q = rate of flow of sewage entering tank in cubic feet per minute.

Table 35 gives the values of C for various sizes of pipe. It should be noted that where there are two or more outfall pipes in parallel, the value of C in the formula is the sum of the values of C of all the pipes used.

To reduce the amount of experiment required for the above formula, the writer has devised the following formula, which, in most cases, gives a reasonable approximation to the correct answer:

$$C = 2\sqrt{L} \times (\sqrt{H_1} - H_2) \times \left[\frac{A}{t} + \frac{Q}{(H_1 - H_2)} \right]$$

TABLE 35
VALUE OF $3\cdot 575\ D^{\frac{21}{8}}$ FOR CIRCULAR PIPES

Diameter in inches	Coefficient "C"	Diameter in inches	Coefficient "C"
6	394	46	82,797
7	591	47	87,605
8	839	48	92,583
9	1143	49	97,732
10	1507	50	103,055
11	1936	51	108,554
12	2433	54	126,127
13	3001	57	145,360
14	3646	60	166,311
15	4370	63	189,036
16	5177	66	213,589
17	6070	69	240,024
18	7052	72	268,394
19	8128	75	298,752
20	9300	78	331,149
21	10,570	81	365,635
22	11,943	84	402,262
23	13,421	87	441,077
24	15,008	90	482,128
25	16,705	93	525,465
26	18,517	96	571,134
27	20,445	99	619,182
28	22,494	102	669,656
29	24,664	105	722,600
30	26,960	108	778,061
31	29,383	111	836,082
32	31.937	114	896,709
33	34,623	117	959,984
34	37,446	120	1,025,952
35	40,406	123	1,094,655
36	43,508	126	1,166,135
37	46,752	129	1,240,435
38	50,142	132	1,317,597
39	53,681	135	1,397,662
40	57,369	138	1,480,672
41	61,211	141	1,566,666
42	65,208	144	1,655,685
43	69,363	147	1,744,769
44	73,678	150	1,842,959
45	78,155	153	1,941,293

The stage-by-stage calculation. The above calculations, unfortunately, serve only in the minority of instances, for in most cases discharge is into tidal waters; in many cases the storage tanks or tank-sewers are complicated in form, having arched crowns and arched or sloping inverts; and in some, storage may be in a natural contoured basin. Consequently, a method applicable to all such variations is necessary. In the first place, calculation of the discharge from a simple tank through an outfall which discharges below high-tide level is a problem for which, so far, a mathematical solution suitable for practical use has not been found, although J. R. Daymond has devised a graphical method based on model experiments.[58] As a result, most engineers prefer to deal with these problems by stage-by-stage approximations.

In brief, a stage-by-stage method is to consider the total time of discharge in short periods of a few minutes. Commencing with the first, say, 5 minutes, the discharge under the initial conditions of head is found, from this is deducted any inflow, and the change of level in the storage tank at the end of 5 minutes and the change of tide-level are calculated, giving the new hydraulic head. The procedure is repeated for the second period of 5 minutes, and so on, until the condition is reached in which the tank is empty. For any degree of accuracy of the calculation, periods in excess of 5 minutes should not be adopted.

The full procedure is as follows. The engineer has to make a shrewd guess as to the size of outfall pipe required or, assuming an average head, find the approximate size by the last formula given. The stage-by-stage calculation is then made for the purpose of checking the accuracy of this guess. Should it be found that the outfall is not sufficiently large to empty the tank in the available time, or should the outfall be found to be unnecessarily large, a new size must be chosen and the calculation repeated.

It is usual to work with the aid of a diagram on which are plotted the levels of sewage in the tank and water in the sea at all times after high tide. The first of these to be drawn is either a recorded tide curve or a tide curve assumed to be a sine curve, which is prepared by dividing the circumference of a semicircle into twenty-five equal portions representing 15 minutes and drawing the curves for rising and falling tide. Next, related to the same datum, a graph is drawn on the same paper, showing the surface area of the sewage in the storage tank at all levels from top water-level to lowest invert. Then from these is prepared the diagram showing level of sewage in tank at all times after high tide.

The calculation is carried out as shown in Table 36, which gives the commencement of the calculation for determining the time of emptying a small storage tank, the data of the example chosen being as follows:

Rate of inflow is constantly at 10 cubic feet per minute.
Tank (with vertical sides) has a surface area of 570 feet super.
Outfall is 9 inches diameter and 1000 feet in length.

TABLE 36
EXAMPLE CALCULATION

Time (mins)	Level of sewage in tank (ft above datum)	Sea-level (ft above datum)	Head (ft)	Hydraulic gradient (1 in:)	Discharge of outfall (cu ft min)	Inflow from sewage system (cu ft min)	Excess of outflow over inflow (cu ft min)*	Surface area of sewage in tank (ft sup)	Fall of sewage level in 5 min (ft)
0	12·50	4·70	7·80	128	94·4	10	84·4	570	0·74
5	11·76	4·52	7·24	138	91·0	10	81·0	570	0·71
10	11·05	4·30	6·75	148	87·9	10	77·9	570	0·68

* When inflow exceeds outflow, of course the difference is a negative value and raises the level of sewage in the tank.

The example is worked as follows:

Taking zero time as the time of commencement of discharge, at that moment the level of sewage in tank is 12·50 feet above datum, the level of sea is 4·70 feet above datum, which gives a head of 7·80 feet. The length of outfall divided by the head gives the hydraulic gradient, which is 1 in 128. Then, according to Vol. I, Fig. 68, a 9-inch pipe at a gradient of 1 in 128 discharges 94·4 cubic feet per minute. From this figure is deducted the rate of inflow (10 cubic feet per minute). The remainder multiplied by 5 minutes and divided by the surface area of sewage in the tank gives the fall of level during the 5-minute interval. This gives the new level of sewage in the tank after 5 minutes. With this new level, together with the new sea-level for the same time, the new head is found, and the calculations repeated, so as to find the new level in the tank after 10 minutes from zero time, and so on.

Allowances have to be made for all special conditions. For example, should the end of the outfall become exposed, the fall of the tide will no longer affect the calculation, and from then onwards tide-level should be taken as being constant at level of centre of outlet of outfall, until the calculation is completed, or the tide rises again to submerge the outfall. Should the tide rise above the level of sewage in the tank there will be no outflow from the tank at that time, assuming a tide flap is installed, or if there is no tide flap, flow of sea-water into the tank would have to be assumed.

Storage of storm water in tanks is dealt with in Chapter IV.

CHAPTER XVI

CONSTRUCTION OF STORAGE TANKS AND SEA OUTFALLS

THE main items of construction peculiar to the sewerage of coastal towns are sea outfalls and storage tanks (or tank-sewers, as such tanks are called when constructed in the form of lengths of sewer the diameters of which are extra large so as to permit storage).

Sea outfalls. Sea outfalls are frequently very expensive, and even hazardous to construct, for they have to be laid in or on the foreshore within the range of the tide and often below low-tide level, and they are constructed on all coasts, not uncommonly in difficult positions, or in places where the action of the sea is violent. Thus it is that losses of plant and materials during construction occur from time to time, and expensive works are damaged or destroyed by the sea before they have served a useful life. It follows that the designer of a sea outfall needs to have a wide experience, and he should design all works according to local conditions.

In protected waters the simplest forms of sea outfalls can be constructed of lines of iron pipes laid down the foreshore, either on the surface or buried. A very inexpensive form of construction is welded steel tube, which will keep its position in spite of moderate wave action, which does not become distorted if it is undermined, and which can be laid in deep water very easily, as will be described. Steel is not so resistant to corrosion as cast-iron, for which reason the latter material is more commonly employed when works that will be difficult to replace have to be constructed. But steel outfalls, in favourable circumstances, are easily constructed and can be bitumen coated: they should always be kept in mind. The method of laying is to weld up the full length of outfall along the foreshore, temporarily plug the ends, roll the outfall into the sea, where it will float, tow it into position and then, removing the plugs or caps, sink it, after which it can be connected to the sewerage system and put to use (see Fig. 59).

When outfalls are laid in or on foreshores of loose material they frequently have to be secured in position by piles and frames driven deep into the foreshore. For this class of work heavy cast-iron pipes with turned and bored joints are commonly used. Where the work is below low tide and divers have to be employed, cast-iron pipes with turned and bored joints, the faces turned to a curve so as to allow a "ball-and-socket-joint" movement, and lugs for bolting the pipes together, have advantages. These pipes can be lowered into position, but, as divers are working under difficulties, they cannot be expected to lay the pipes with the comparative ease with which pipes are laid above sea-level. The lugs and bolts, however, make it possible for

CONSTRUCTION OF STORAGE TANKS AND SEA OUTFALLS 183

the divers to draw the spigots into the sockets and ease the pipes into position. The lugs should not be arranged in such positions that any will come too near the ground to permit the use of a spanner.

When an outfall has to be laid on a rocky foreshore the construction is different. Piles cannot be driven into rocks; on the other hand, a trench can

Fig. 59—Steel outfall being floated into position
By courtesy of Stewarts & Lloyds, Ltd.

be excavated and the outfall set solid in concrete therein. Alternatively, if the outfall is above the surface of the rock it can be surrounded in concrete, the concrete being let into a shallow trench, or else the pipe can be tied down with iron straps and rag-bolts grouted into the rock.

All except the smallest cast-iron pipes, when empty, are lighter than water, and would float were there no flow in them and were the sea excluded. For this reason a tide flap is very seldom placed at the lower end of any outfall that runs down the foreshore into the water, and never in those circumstances in which it would be possible for the outfall to become empty and consequently floated out of position.

Generally it is not necessary for a tide flap to be so provided, even were it possible, for the flow of sewage keeps the outfall clear of silt; therefore only in the case of such storm-water outfalls as need to be carried to low-water mark—e.g. relief sewers for combined sewage—is a tide flap used, in rare circumstances, with special precautions against flotation of the outfall and the danger of the flap becoming buried and prevented from opening. It is, however, usual practice to install a tide flap in a chamber at the top end of an outfall between it and the sewerage system which discharges to it.

The ends of outfalls have to be marked by buoys or beacons with top-marks to the requirements of Trinity House and any other controlling authority.

Submission of plans to the Department of the Environment. Plans for all works constructed below High Water Mark Ordinary Spring Tide have to be submitted for the approval of the Department of the Environment. These plan must include Ordnance Survey maps, which must not be prints or cuttings, but full-size original maps submitted in duplicate. The maps should be to the scale required by the Department. It is required that all land occupied shall be indicated in red, and all dredging or excavation shown in blue. Full details should also be submitted showing longitudinal cross-section of works and methods and materials of construction. High- and low-water levels should be shown, their values related to Ordnance or Newlyn Datum. Detail drawings should be true-to-scale prints on opaque linen. These papers should be submitted in duplicate together with a statement as to whether an application is being made for a loan sanction.

Tank-sewers. The choice as to whether a tank-sewer or a rectangular storage tank shall be used depends on the site. A long tank-sewer capable of storing the required quantity is in some respects the best solution, for besides giving storage it does away with a length of connecting sewer which would otherwise be required, and it is most easily designed so as to be self-cleansing. But all circumstances do not permit its construction.

The capacity of the tank-sewer is determined by length and diameter, which are partly problems of economy. The cost of a comparatively short tank-sewer of large diameter, with a length of approach sewer of the diameter required for accommodating the flow of sewage, can be balanced against a longer length of smaller-diameter tank-sewer without so great a length of approach sewer. Estimates for alternatives should be compared. The cross-section of a tank-sewer should be such that any sludge that settles during storage will gravitate to a central grip or narrow invert which will be adequately flushed by the normal flow of sewage when the tank is emptied. Opinion on choice of cross-section varies, and the choice in any particular instance must depend on the circumstances in which construction will take place. A fair example is that in which the section does not vary much from the circular, but the sides near the invert slope at 1 in 6 to an invert of small radius, which is calculated to ensure a self-cleansing velocity of at least 2½ feet a second at the average rate of flow of sewage.

At the top end of the tank-sewer the incoming sewer should enter with its invert no lower than top water-level, and a backdrop or cascade should be constructed to carry the sewage to invert level of the tank-sewer. At the lower end of the tank-sewer should be constructed the installation for controlling the flow by penstock, the tide flap to prevent back-flow and the connection to outfall. The crown of the outfall should preferably be below the invert of the tank-sewer to permit complete emptying of the latter, even when the outfall is flowing full. The outfall may consist of one or a number of pipes, according to the flow and the practicability of constructing the outfall of one large-diameter pipe.

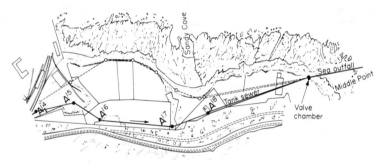

FIG. 60—Plan of sea outfall
By courtesy of John Taylor & Sons, Consulting Engineers

Access should be given to the tank-sewer at both ends and at intervals not exceeding 360 feet, preferably by side-entrance manholes with ladders and safety chains leading to the invert. The tank-sewer should be very adequately ventilated.

Control of discharge. The flow out of the tank must be controlled by a penstock, which can be opened and closed quickly at the predetermined states of the tide. This means that if the penstock is to be manually controlled it must be visited by men regularly four times in the twenty-four hours, at most inconvenient times day and night, and the work must be performed without fail. This requires careful organisation and an appreciable labour cost; therefore it is preferable that the penstock should be automatically controlled wherever electricity is available.

Electrically operated penstocks can normally be obtained without difficulty. The motors can be controlled by a "lunar clock"—i.e. a clock the cycle of which is exactly equal to the cycle of a tide, nevertheless having an adjustable twenty-four-hour dial. The clock should be electrically wound. With this arrangement the labour of controlling the penstock is greatly reduced: it is necessary only for men to make periodic visits to check the clock and look to the maintenance of the installation. (Power penstocks close at about 2 ft per min.)

Electrically operated penstocks constrolled by floats may also be used, but are not so popular as clock controls. Manual operation of the penstocks should be provided in all cases, to allow for breakdown of the electrical installation, and a weir should be arranged above high-tide level to permit overflow of the tank-sewer should, owing to any mischance, the sewer become surcharged.

Storage tanks. Rectangular storage tanks may be economically constructed on convenient sites out of reinforced or mass concrete, or brickwork with concrete backing, in the same manner as reservoirs for water, but it is important that they should be so detailed that they are self-cleansing, as otherwise they are liable to become very foul. An effective means of ensuring self-cleansing conditions is to divide the tank into long compartments leading one from another, and to design the inverts as in a tank-sewer: the tank becomes, in fact, a tank-sewer folded upon itself backwards and forwards. With this arrangement the flow goes right through the tank, and when the tank is empty the invert is scoured clean.

A problem which usually arises when storage tanks are constructed on the foreshore is their complete disguise in order that the public shall not recognise them for what they are and imagine foul odours. The flat tops of such tanks are often used as car parks or open promenades, the sewer-vents are disguised as flagstaffs, and shelters or seats provided, making it appear that the constructions are for amenity purposes only.

When storage tanks are intended to serve the dual purpose of storage and sedimentation tanks, they should be provided with either hopper-bottoms, for removal of sludge under hydrostatic head, or mechanical sludge-removal gear. The latter is to be preferred when construction of deep pyramids is impracticable. Sedimentation tanks so constructed are usually different from the sedimentation tanks at sewage works, in that they are completely covered over, and this makes design and working more difficult than in the case of either ordinary storage tanks or ordinary sedimentation tanks.

Protection from the sea. When storage tanks are constructed on the coast, full precautions should be taken against damage by wave action. Consequently, the outer walls, if not protected by a sea wall, should be of heavier construction than the sea wall normally designed for the same locality, and should be similar in detail, for in these matters local experience is of great value.

The coastline is subjected to far more rapid denudation or erosion than is usual inland. In some places the coastline has moved inland at rates in the order of 15 feet per annum, with the result that buildings, and even villages, have been undermined and destroyed.

Geological factors affect the rate of erosion. The sea acts most readily on soft beds which slope down to the foreshore, but it also undermines cliffs of hard rock, such as limestone. When formations dip towards the sea there is frequently a tendency for landslides to occur with an increased rate of erosion. Springs also increase the rate at which the rocks break down.

Coast-defence works are constructed to prevent loss of land and to protect buildings and structures. From the point of view of the sanitary engineer, they are required for the protection of sewerage works. Coastal protection works can be extremely expensive, but it is false economy to construct works which will not last long.

Sea walls. Protection of high cliffs is usually not practicable. When they are constructed of hard rocks there is little need of protective works, because erosion is slow, and the fallen material itself defends the base. But cliffs of sandstone are easily eroded at the base, and when undermined they fall, and if the sandstone is not of a hard variety it breaks up to form sand and is carried away. Softer materials are protected by sea walls where it is considered that the expense is justified.

Sea walls are constructed in heavy masonry, stone-faced concrete or concrete. For the best works the largest masonry blocks that can be moved are used, and these are bedded with cement joints in order that the sea may not loosen and remove the wall piece by piece. The walls are battered, curved or stepped so that they do not receive the full force of the sea, and at the tops they may be turned outwards, or provided with a cornice, to throw back the waves to prevent their falling on the promenade: practice varies considerably. The foundations of the walls are deep, and protective aprons or sheet piles are provided, on the seaward side, to prevent scour from undermining them.

Where the action of the sea is not severe, less expensive works are constructed, consisting of reinforced concrete, timber and stone pitching.

While sea walls immediately protect the coast from the action of the waves their presence may add to the lateral scour which removes the material from the foreshore, and therefore when a sea wall is constructed groynes may also be required.

Groynes. The purpose of groynes is to prevent the scour and removal of existing foreshore material and to cause deposition. The rate of accretion is dependent upon the quantity of material travelling along the coast. If there is not much material in suspension little can be deposited, even when numerous groynes are constructed, but when there is much material in motion the construction of groynes causes deposition. They are not altogether effective in that scour occurs due to the turbulence caused by flow over them and this increases the erosion at their shore ends.

Groynes are constructed mostly out of timber piles and sheeting, or out of mass concrete, but there are many special designs less frequently employed. Their lateral spacing can be found by experiment only, a number being constructed and intermediate groynes added if found to be necessary. Generally, the spacing should be about one to one-and-a-half times the length of the groyne. Timber groynes are used where the foreshore is of soft material and piles have to be driven. But where there is a firm, rocky bed below the beach a more expensive concrete groyne is worth its cost because of its durability. Pebbly beaches may be protected in some cases by groynes, but if these are not found to produce a deposit of sand, some form of longitudinal defence work may be required.

CHAPTER XVII
UNDER-DRAINAGE

ALTHOUGH under-drainage is closely bound up with sanitation, it is not familiar in all its forms to sanitary engineers. For example, field drainage for agricultural purposes could generally be said to be outside the sphere of both everyday civil engineering and sanitary engineering. And in this respect under-drainage is similar to river maintenance, which has become a specialised study.

On the other hand, under-drainage is a necessary part of sewage treatment by land filtration; under-drainage for structural purposes is civil engineering work and drainage of the subsoil is important to the public health engineer, particularly in relation to site development, because saturation of the soil and subsoil not only militates against the amenity of the site, but it has bearing on health. Wet ground is notoriously unhealthy to live on. According to Parkes' "Hygiene":

"A moist soil is cold, and is generally believed to predispose to rheumatism, catarrh and neuralgia. It is a matter of general experience that most persons feel healthier on a dry soil. In some way which is not clear, a moist soil produces an unfavourable effect on the lungs; at least in a number of English towns which have been sewered, and in which the ground has been rendered much drier, Buchan has shown that there has been a diminution in the number of deaths from phthisis."

There is not, however, very much reliable information on the effects on health of the vapours that are given off by polluted soil or made ground, perhaps containing tipped rubbish.* The modern requirements of a sealing layer of concrete under the ground floor of a house should make the type of soil less important from this point of view. Nevertheless, there is evidence of the importance of site in relation to health.

Under-drainage of building sites. Under-drainage for building development is frequently constructed, with undue confidence, for the purpose of rendering the site around the buildings dry on the surface. Very frequently this work is unnecessary, and the fact that is usually overlooked is that subsoil drains do not remain in working condition for ever; they are liable to silt and, particularly in built-up areas, they are very liable to be rendered ineffective by sections being broken out during excavations for new drains and services.

Land drains for building sites include French drains—i.e. trenches filled with large stones but not any pipes—and agricultural tiles in trench, the

---

* The author encountered one instance of lethal quantities of carbon dioxide given off from an old tip used as a housing site.

trench partly filled with stones or brick rubble. Generally, the methods of drainage applicable to agricultural drainage serve for building sites. Subsoil drains connected to drainage systems for surface water or soil sewage must be isolated by a gully with trap or a reverse-action disconnector, so arranged that the subsoil drain is ventilated at its lower end.

Agricultural drainage. For nearly all agricultural purposes the soil must not be waterlogged. It should contain a reasonable proportion of moisture, but the level of saturation should at all times of the year be lower than the level of the roots of the crops that are being cultivated, leaving an adequate proportion of air spaces in the soil. Land that is at all waterlogged can be improved by agricultural drainage.

About 1763 Joseph Elkington showed that land could be drained by pipes laid in suitable positions. These early land drains were laid in low parts of the ground to which water naturally gravitated, and this system still has its uses. In 1823 James Smith introduced the parallel system, which is the most common method now used. In this system agricultural tiles are laid in parallel straight lines, more or less in the direction of the fall of the land, and connected into larger drains, which in turn discharge into ditches. The distance apart of these parallel drains depends on the type of the soil. For a permeable sandy soil a distance of 75 feet apart has been suggested, while 50 feet has been proposed for a sandy clay, 30 feet for heavy soil and 20 feet for stiff clay. One variation of the parallel system is the herring-bone lay-out in which main drains are laid in low points, while the parallel drains discharge into them diagonally from both sides.

There has been some variation of opinion on the desirable depth of land drains. James Smith considered a depth of 2 feet 6 inches to invert satisfactory. Later practice tended towards greater depths, such as 3 or 4 feet, but more recently agriculturists have expressed opinions in favour of the original shallow depths; while Basil Furneaux, in a letter to the author, suggested that 18 inches cover is sufficient to give protection to land drains, and stressed the importance of arranging the depth as far as possible in accordance with the depth of the water-table and the condition of the subsoil. For example, if the soil to be drained is a permeable material overlying stiff clay, the latter being only 2 or 3 feet deep, the agricultural drains should be just in, and not below, the surface of the impermeable material unless elaborate precautions are taken to prevent their becoming covered over by the impermeable material.

Drains should not be laid so shallow as to be disturbed by cultivation. On the other hand, it is not always easy to lay them to the regular falls required by good practice across an undulating surface without rendering them deep. If in such a case the drain has to be laid below the level of an impervious stratum, it is advisable to fill that portion of the trench which is within the impervious material with faggots, large stones or other material which will assist the percolation of water and at the same time keep the agricultural tiles free from silt.

UNDER-DRAINAGE

Land drains most commonly consist of earthenware agricultural tiles, having at least $2\frac{1}{2}$-inches internal diameter. Horseshoe tiles of ordinary brick-earth were formerly largely used for land drainage, but they have long since given place to circular pipes of the same material. Porous concrete pipes are also obtainable, but cost more than earthenware and deteriorate if sulphates are present. Porosity offers no advantage, for water can enter an ordinary land drain quite freely through the joints.

For the main drains, sizes of 6-inches internal diameter upwards may be used, according to the estimated flows. The laterals or parallel drains discharge into the main drains by connecting to them at the crown, the pipes being cut and the intersections covered in such a manner as to prevent earth from entering at the junctions.

The bottoms of the trenches for agricultural drains should not be cut flat as in the case of sewers, but should be formed with a narrow, round-pointed tool to the size and shape of the pipes, which should then be lowered into place with a pipe-hook and *butted tightly together*. The pipes are generally covered with inverted sods, or a little grass or similar material to keep out silt, and the trenches are then filled and well rammed. Care must be taken to keep the soil and the subsoil separate, and to return them in their proper order.

TABLE 37
MINIMUM GRADIENTS FOR LAND DRAINS

Diameter of drain in inches	*Gradients for drains laid in normal soils. 1 in:*	*Gradients for drains laid in running sand. 1 in:*
2	500	125
3	860	215
4	1260	315
5	1680	420
6	2160	540
7	2680	670
8	3200	800
9	3720	930

As far as practicable, under-drains should be laid to self-cleansing gradients similar to those employed for sewers, because under-drains must receive some silt from the soil, and this will inevitably tend to accumulate unless there is an adequate flow of water at a self-cleansing velocity. Nevertheless, it is frequently impossible for such ideal gradients to be maintained and, in fact, the practice of land drainage has been to lay under-drains to gradients which are very flat compared with those of sewers and which, theoretically, must lead to silting. But the fact remains that authorities have stated that these slack gradients have proved satisfactory, and there is reasonable agreement as to what the minimum gradients should be. The average opinion appears to be that generally the drains should be laid in most soils at gradients which will ensure velocities of not less than $\frac{2}{3}$ feet per second and, in running sand, at gradients which will give velocities of about twice that figure. Table 37 gives the gradients which will produce these velocities as calculated by Crimp and Bruges' formula.

Catchpits. Catchpits should be placed at the junctions of main drains. They are generally built of open-jointed brickwork, half a brick thick, about 3 feet in diameter inside, and domed in at the top to receive a flat circular cover. The bottom of the catchpit should be 18 inches or more below the invert of the outgoing pipe, depending on the nature of the ground. No floor is required, but a course of headers should be laid on the ground to carry the wall.

Where the main drains are laid at flat gradients, or the ground contains much iron or fine sand, where practicable provision should be made for flushing them. Water for the purpose may sometimes be obtained from a stream.

The ends of drains where they discharge into ditches require support, otherwise the falling in of the sides of ditches may have the effect of closing them. Concreting the pipes will serve in some cases, but where the ends of main drains enter ditches or streams in which there is appreciable flow, concrete support should be carried well below the invert of the bed of the stream. The ends of all pipes should be closed against rodents by gratings or light flap valves.

Roots. The worst enemies of land drains are roots, especially those of water-loving trees, such as willows and alders. Elm trees, too, are particularly obnoxious: oaks are less dangerous. Where a drain crosses a hedge, or passes near a tree which cannot be cut down, socket pipes with cemented joints should be used. The joints must be made with the greatest care, for a fine rootlet can thred its way into any cracks in joints, and, eventually growing into a dense mass of roots, will completely block the pipe. Where it is desired to preserve a tree near the line of a drain the drain should be surrounded with concrete or constructed of cast-iron.

Brush drainage. The Board of Agriculture and Fisheries, in their Food Production Leaflet No. 62, issued in 1919, refer to faggot or brush drainage as an ancient practice, still in use in several localities, particularly in the Eastern Counties. In view of the shortage of drain-pipes then existing, they felt

> "that the practice might be extended, especially in connection with mole drainage on heavy clay soils. Faggot drains may be expected to last as long as the mole drains which they serve.
>
> The object aimed at in the construction of a brush drain is to secure an underground passage for water. This object is gained by laying brushwood and other material in the bottom of a trench in such a way as to support the covering soil without either itself impeding the flow or allowing the soil to drop through and choke the drain. Later, when the brushwood rots, the passage becomes clearer.
>
> Care is necessary at each stage of the work.

The trench should be dug with steeply sloping sides and narrowed to a width of not more than 3 inches at the bottom. Used as a feeder, it should not be more than 27 inches in depth."

The wood should be used green, and should be placed in the trench with the bushy ends on top, pointing away from the outfall.

Mole drainage. Where the subsoil is a stiff clay, pipes are sometimes dispensed with, and the drains consist merely of holes bored through the ground by a bullet-shaped tool fixed to the bottom of a knife-edged coulter and drawn across the field by a powerful tractor, or by a cable worked by a fixed engine. The drains, which are usually $3\frac{1}{2}$ inches in diameter, are about 2 feet deep and 5 yards apart in medium clay soils, and from 18 to 20 inches deep and about 3 yards apart in very heavy land. The lengths of the branches should not exceed 220 yards. The main drains may be formed in the same way, and may be up to 6 inches in diameter, but either piped or French drains are more satisfactory. In suitable conditions, mole drains will remain in service for a few years before the work has to be renewed.

Drainage of playing-fields. The surface of a playing-field must always be kept firm and dry. To bring this about, two requirements must be met. The soil must be so open in texture that it will absorb immediately the greatest quantity of rain which may fall on it, and there must be sufficient depth of dry subsoil to accommodate the rain until it can be carried off by the drains.

Ditches. Ditches must be laid to regular falls, proportioned according to the flow and maintained in good condition. The top water-level in a ditch should be below the incoming drains at all times; the depth of flow should be deep—i.e. about one-half the width of the water at the surface.

For forest land mole drains and tile drains are impossible. The former cannot be constructed, and the latter quickly becomes useless owing to their being rapidly choked by roots. This necessitates that forests requiring drainage shall be drained by open ditches excavated parallel, about 60 feet apart, the depth being about 2 feet. The parallel drains discharge into main ditches 3 or 4 feet deep according to the estimated flow, and these main ditches in turn deliver into natural watercourses.

Drainage of aerodromes. Drainage of aerodromes has been evolved from land-drainage practice, but the modern tendency is to adopt new methods, including surface drainage. If the soil of the aerodrome is very permeable and drainage is necessary only because of a high water-table, methods somewhat similar to those employed for agricultural drainage are satisfactory, except at the edges of concrete runways. But when the soil is impermeable, the drainage has to be so arranged that it will keep the *surface* free from water, in addition to draining the subsoil. It is this necessity for keeping the surface dry that leads to differences of broad outline and of detail. For, while agricultural drains usually *follow the slope of the land* and are buried in earth, aerodrome drains are laid in trenches which are filled with loose rubble, in some cases topped with porous tar-macadam, and the drains are laid more or less *parallel to the contours*. This ensures that the water running over the surface downhill has to travel only a short distance to reach a porous trench into which it may fall, and thence find its way to the drains. It follows that the run-off from aerodromes is much more rapid than the run-off from a field, because in the latter case time is taken by the percolation of water through the soil, while in the case of an aerodrome on impervious soil there is only a short period of time required for water to run over the surface and trickle down through porous material to the drains. This makes it necessary to calculate the sizes of the main drains by a time of concentration method (described in Chapter III). The once-in-five-years storm is sometimes allowed for, and it may be advisable to base run-off on the rainfall without any allowance for change of impermeability during storm.

At the most recent aerodromes, runways have been drained by shallow, wide, dished concrete channels laid at the sides of the runways and connected to special gullies. The former practice of constructing drainage trenches with surfaces of porous tar-macadam or other porous material was found to be less effective. When this type of drainage is provided, the Lloyd-Davies method of calculating run-off should be used.

Drainage to stabilise earthworks, etc. Drainage works, sometimes of an elaborate nature, are carried out for the purpose of preventing subsidence, landslides and the movement of cuttings and embankments. These always necessitate study of the local geological conditions, and are designed in accordance therewith. Sub-drainage of basements can relieve water pressure on walls, floors and often obviate tanking.

CHAPTER XVIII
RIVER MAINTENANCE

River maintenance and flood control form a branch of surface drainage, and therefore are closely related to the work of the public health engineer. To some extent public health engineers need to have knowledge of river works, because this branch of engineering is not unconnected with either water supply or sewerage. Rivers form sources of water and accept the effluents from surface-water sewers and disposal works.

Nevertheless, not every public health engineer gains wide experience in river works, and not infrequently public health engineers have approached catchment problems by the methods applicable to sewerage systems, instead of adopting the practice built up by engineers who specialise in river maintenance and flood control.

There are marked differences between river works and surface-water sewerage which are due to the facts that the majority of natural watercourses have natural beds not protected against scour, and therefore not permanent as is the invert of a surface-water sewer and that flooding of water-meadows, etc., affects flow records, estimates of flow and design.

Run-offs of small catchments or parts of catchments have been estimated (with doubtful accuracy) by the Lloyd-Davies method. But this method necessitates a reasonable estimate of the time of concentration and, in the case of river catchments, this may amount to a time-lag of several days, for it is influenced by not only the fall of the land and the nature of the surface and vegetation, but also by the discharge of under-drains and springs.

For this reason, such an approach to the run-off of natural catchments involves the possibility of considerable error, and therefore the alternative of *observations and calculations based on records* is more advisable.

Another main difference between river works and surface-water sewerage —i.e. it is impracticable for all except the smaller tributaries of rivers to be lined with concrete or other durable material, and that in most instances the bed and banks must be the natural ground—means that the average velocity of flow of the river must be maintained, otherwise either silting or erosion will take place.

The line of the bed of a river or *thalweg* forms a logarithmic curve to the horizontal which is always traceable, although distorted by variations in the geological structure. Near where the river rises in the mountains, the gradients of the *thalweg* and the hydraulic gradient of the stream are steep, the velocity of flow is rapid, and consequently the bed of the stream consists of large stones, all small particles being washed away by the rapid current. Farther downstream the slopes of the *thalweg* and the hydraulic gradient are slacker, the velocity lower and the size of particles forming the bed smaller. Finally, in the lower reaches of the river the gradient is very slack, is reversed by the tide at high tide; velocity of flow is low, and fine mud and silts are deposited.

The scour due to the flow of water and the movement of suspended particles is always cutting down the bed of the river in the upper reaches, where the energy dissipated by the fall of water under gravity is greatest in proportion to the surface area of the bed. Lower down the river, where the energy is less, the rate of scour is less. In the lower reaches the scour of the bed is small, sometimes absent—in fact, where the river flows through permeable strata the bed may tend to *rise* as a result of deposition until the river is flowing between natural banks actually *above* the ground-level of the main part of its own valley.

Ancient rivers and streams flowing at slack gradients in their own valleys, while they do not cut down their beds erode their banks and always in an irregular manner, producing meanders or sharp bends from side to side. Wherever the slightest bend has been produced in the line of the river, scour becomes greatest on the outside of the bend and deposition occurs on the inside, so that the degree of curvature increases until finally the river *meets itself* and locally straightens its course once more.

Effects of new works. When works which have the effect of altering the average velocity of a river are executed, the stability of the bed of the river is upset. If a high velocity is produced by straightening the course, and thereby steepening the hydraulic gradient, scour is to be expected. If, to increase capacity of flow, the bed of the river is enlarged without the hydraulic gradient being increased, the velocity of flow is reduced and, as a consequence, silt deposited, which in time may make re-excavation necessary.

It is these considerations that make it advisable for all river works to be executed on the assumption that they will be of temporary value only unless:

    1. They are so designed as not to upset the stability of the river.

    2. While stability is admittedly upset by *increasing* the velocity, protection is given to the bed or banks.

Estimating run-off. The most satisfactory method of estimating the run-off of a catchment or part of a catchment so as to calculate the size of waterway required with the available hydraulic gradient is to make extensive records over a period of several years of flows at various points on the river system, together with records of rainfall and of flood-levels. With the aid of such information it is possible to plot curves showing the maximum run-off per acre likely to be expected from areas of different sizes and in different localities. In the absence of actual records for the catchment concerned, run-off may be estimated by comparison with another catchment of similar character for which records are available, or by comparison with information from various sources, but this always involves risk of error.

Rates of run-off from upland catchments are very high because of the steep gradients, comparatively impermeable surfaces and small catchment areas. Intense storms of short duration have the most effect on them. Short storms have little effect on lowland catchments and catchments of large size generally; these are affected mostly by prolonged wet periods.

The run-off from upland catchments has already been discussed with relation to water supply and the design of spill-ways of impounding reservoirs. In the design of spill-ways for reservoirs, allowance has to be made for the greatest run-off likely to occur during the life of the reservoir, because failure of a dam could mean disaster, and the construction of a spill-way to take the greatest possible run-off is not, in the circumstances, an extravagance. In the design of river-works, it is also usual to assume that the most intense run-offs will be catered for by the waterways of major magnitude, apart from such flows as can be accommodated by flooding land of low value. These practices contrast with the practice of engineers who design surface-water sewers which, theoretically, are filled to crown level about once in three years and at less frequent intervals are surcharged.

On run-off from large areas in England, far less information is available than is desirable, and consequently very crude rules of thumb have been applied in making run-off estimates. F. Johnstone-Taylor, in "River Engineering,"[135] gave a table which related gross area of various drainage areas in England to maximum run-off according to flood records. From this table it could be seen that where acres of land varied from about 100,000 to 200,000,000 acres or more the run-off per acre varied roughly between the limits of ½–1⅛ cubic feet per minute, but a considerably larger run-off was observed for an area of a few thousand acres. Roughly, the figures suggested the following formula:

$$Q = \frac{25}{\sqrt[4]{A}}$$

where: Q = cubic feet per minute per acre
A = area of catchment in acres.

It will be seen that this conforms with Dickens' formula, which is mentioned in the following paragraph.

Close accuracy is seldom possible in river-works calculations and the above figures may be considered to show *comparatively* little variation. Thus, in English practice, a run-off figure of ⅓ inch of rainfall per day has been adopted in many cases. Other authorities have preferred to use as a rule-of-thumb guide a daily run-off equal to 1 per cent of the total annual rainfall; while in the Fen District a rainfall of ¼ inch per day was assumed. R. D. Walker[246] gave a curve showing the proportion of daily run-off to be expected in accordance with mean annual rainfall that agrees with the above figures for English rainfall intensities. But for small areas the run-off per acre is greater, and for this reason a number of empiric formulae have been devised to express the relation between size of area and run-off. Some of these may be expressed as follows:

$$Q = C.M.^n$$

where: Q = the discharge in cubic feet per second
M = area in square miles
n = an index, which in Ryves' formula is $\frac{2}{3}$ and in Dickens' formula is $\frac{3}{4}$
C = a constant.

The value of the constant "C" is dependent on the country, impermeability of the soil, shape of the catchment, slope of the land, vegetation and the rainfall. It is determined for any particular area by the flow records.

Maximum surface run-off from small areas may be estimated by a time-of-concentration method similar to the Lloyd-Davies or the rational method described in Chapter III. In 1897 George Chamier[38] wrote with regard to the run-off of water-supply catchment areas:

"The general rule to be followed is to take the greatest rainfall to be anticipated, from previous records, for such a time as is required for the flood-water to reach the outlet from the furthest extremities of the catchment area."

G. Bransby Williams[256] gave the following formula for estimating the time of concentration for circular areas:

$$t = \sqrt[5]{\frac{A^2}{h}}$$

where: t = time of concentration in hours
A = area of catchment in square miles
h = average fall in feet per 100 feet, from the watershed to the point of discharge.

Where the shape of the catchment area is irregular the time of concentration is multiplied by $\frac{e}{d}$, where e = length of the area draining to the point of outlet and d = diameter of a circle of area equal to that of the drainage area.

When the time of concentration has been found, the maximum rainfall likely to occur during that period of time is converted from inches to cubic feet per minute per square mile or other suitable unit, and a suitable percentage of run-off estimated.

The amount of rain likely to fall during any period of time up to 48 hours may be estimated by Bilham's formula given in Chapter III. A rough approximation to the percentage run-off was given by Bransby Williams, although it must be understood that the figures referred to Indian country, and in any case could only be taken as rough approximations. They were as follows: flat land to gentle slope, 10–20 per cent; moderate slope or undulating, 20–40 per cent; wooded hill slopes, 40–60 per cent; steep mountainous and rocky country, 60–80 per cent. Generally estimates of impermeable area of natural land and estimates of time of concentration may be suspected of tending towards underestimations, on both counts more or less balancing each other out.

A final method of checking the maximum discharge of a river is to measure the cross-sectional area below maximum flood-level at a number of points on a straight length, find the hydraulic gradient and by flow formula calculate the flow that the hydraulic radius and gradient should permit and check the result against Lacey's formula, described herein.

Rainfall allowance on river catchment: areas covered by storms. While sewers are usually designed to run full when taking the most intense rainfall likely to occur in, say, one to three years, additional flows being accommodated by surcharge, rivers are designed on their "bank-full" capacity during the most intense rainfall for which allowance can reasonably be made. It is here that the danger arises of the engineer allowing for some extremely rare rainfall such as the once-in-100-years storm and over-designing his works. For while such rainfalls do, of course, occur on the average according to Bilham's formula, it is a matter of experience that the run-off from a large catchment due to such an infrequent rainfall is never observed.

The reason for this is that storms of great magnitude in intensity and total volume, including those which have been recorded as causing widespread and disastrous floods, cover limited areas that can be smaller than the catchment and the effect of this fact is that, in the case of any particular large catchment, there is some optimum frequency of occurrence which produces a greater run-off than any storm of greater magnitude but lesser frequency of occurrence. In one instance it was found that the maximum run-off in 30 years was produced by the once-in-10-years Bilham storm.

There is at present little information on this important matter of areas covered by storms in Great Britain. A useful investigation was made by Dr. Glasspoole in which the thirty-three "most disastrous" storms occurring in the whole of Great Britain during a period of 40 years were examined. The writer, plotting Dr. Glasspoole's figures, found that there was a reasonable degree of regularity in the form of the curves for storms lasting from $1\frac{1}{2}$ to 144 hours, and that these approximated to the formula:

$$R = \frac{5 \log_{10} T + 2 - 1\cdot 5 \log_{10} A}{T}$$

where: R = inches of rainfall per hour
T = duration of rainfall in hours
A = area covered by that rainfall in square miles.

This gave reasonable agreement with the data for 1 to 500 square miles. It should be noted also that the data did not include any falls totalling less than 1 in of rain during storm, and the formula is not applicable where $RT < 1$.

If this formula is plotted to logarithmic ordinates and abscissae, a number of curved lines are produced, the maximum values on which can be enveloped by a straight-line curve, the formula of which is:

$$R = \frac{2}{A^{0.3}}$$

This formula gives the maximum intensity of rainfall likely to cover any area regardless of duration. Storms of greater intensity have occurred during and since the period observed by Dr. Glasspoole, but they are of extreme rarity and should not need to be taken into account in ordinary river-work calculations.

It is interesting to compare this enveloping curve with the "normal maximum enveloping curve" adopted by the Institution of Civil Engineers Committee on Floods in Relation to Reservoir Practice, which, to the above notation, is about:

$$R = \frac{1.54}{A^{0.4}}$$

The curves differ in that the former is for rainfall and the latter for run-off—i.e. rainfall reduced by permeability and other losses. The difference of slope of the curves, indicated by the power of A, is probably due to change of impermeability during rainfall. The author's formula, $R = \frac{2}{A^{0.3}}$, assumes that the area of the storm falls within that of the catchment, but the shapes of the areas covered by storms and of catchment vary and they seldom coincide. For this reason it is suggested that, if the method is used in design, A in the formula should be taken as meaning the area of a circle totally enclosing the catchment and not that of the catchment itself. Further, except in the design of reservoir spill-ways, where no flooding can be permitted in any circumstances, it will be reasonable to make a considerable reduction in the flow calculated on the basis of the rainfall found by the above.

Hydrograph methods. For many purposes it is necessary to ascertain the maximum peak rate of flow only in the river at a selected maximum rainfall, which should generally be the rainfall of the greatest magnitude likely to occur in a period of many years and having a duration equal to the time of concentration of the river system above the point at which flow is desired to be known. It will thus be seen that the methods used for design purposes have a broad similarity to those applicable to surface-water sewerage, differing mainly by reason of the greater uncertainties that apply to river work. In some circumstances it may be advantageous to know not only the peak rates of discharge but also the form of the hydrograph—i.e. the curve of run-off intensity resulting from any particular rainfall.

American hydrograph methods, particularly the unit hydrograph method,[79] appear to be gaining popularity in the British Isles.[193] There are several varieties or modifications of the unit hydrograph method, but broadly the procedure is as follows.

Particular attention is given to important rainfalls lasting for a pre-selected unit period of time which is short compared with the time of concentration of the catchment—e.g. one-quarter of the time of concentration and the run-offs during all such storms are observed. From these are deducted the steady run-off that was occurring prior to the rainfall as a result of soakage from the subsoil, so that a fair estimate of the surface run-off due to the rainfall may be arrived at.

As run-off from a catchment must continue for a time not less than duration of rainfall plus time of concentration, and as time of concentration is considerably longer than the unit time of rainfall, the run-off will continue for several unit periods. The run-off can then be set out as given in column (5) of Table 38 and converted to percentages of the total run-off. The working is generally as given in Table 38. This procedure should be carried out for several storms and the average results found.

TABLE 38
CALCULATION FOR UNIT HYDROGRAPH

(1) Period No.	(2) Rainfall (inches)	(3) Total run-off (cusecs)	(4) Run-off by soakage from subsoil (cusecs)	(5) Surface run-off (3)–(4) (cusecs)	(6) Surface run-off as a per cent of total = distribution ratio
1	0·75	1976	450	1526	21·8
2	—	4510	450	4060	58·2
3	—	1320	450	870	12·5
4	—	950	450	500	7·2
5	—	470	450	20	0·3
6	—	450	450	0	0·0
			Totals	6976	100·0

The data of rainfall and run-off have then to be studied so as to determine the percentage of rainfall that runs off the surface. This, as observed in the case of surface-water sewers (see Chapter IV), may be expected to vary with duration and quantity of rainfall. The values for all conditions should be determined.

With the information now obtained, it is possible to make an estimate of the amount of run-off at any time during and after rainfall, and plot the hydrograph of run-off for any storm or series of storms occurring in an aggregate time consisting of several unit periods. The procedure is as set out in Table 39: suppose 2 inches of rain fall in the first unit period and 0·34 in the third, and the percentages of run-off as ascertained from records for these periods are 25 and 30 per cent respectively, then the run-off *into* the river system will be 0·5 inch of rainfall for the first period and 0·1 for the third. Column (5) of Table 39 is taken from column (6) of Table 38, but normally would be an average of such figures made from several tables similar to Table 38, based on several storms. Column (6) of Table 39 gives the run-off

TABLE 39
APPLICATION OF UNIT HYDROGRAPH

(1) Period No.	(2) Rainfall (inches)	(3) Run-off per cent	(4) Run-off (inches)	(5) Distribution ratio per cent	(6) (7) Distributed run-off in inches for rainfall in:		(8) Total (inches)	(9) Total (cusecs)
					Period 1	Period 3		
1	2·0	25	0·5	21·8	0·109	—	0·109	5,860
2	—	—	—	58·2	0·291	—	0·291	15,700
3	0·34	30	0·1	12·5	0·0625	0·0218	0·0843	4,540
4	—	—	—	7·2	0·036	0·0582	0·0942	5,070
5	—	—	—	0·3	0·0015	0·0125	0·014	753
6	—	—	—	—	—	0·0072	0·0072	388
7	—	—	—	—	—	0·0003	0·0003	16

for each unit period of time for the rainfall that occurred in period 1 and column (7) for the rainfall in period 3. Column (8) gives the total for all rainfalls in inches. This is converted from inches rainfall to cusecs by multiplying by the gross area of catchment in superficial feet and dividing by twelve and by the length of the unit time-period in seconds. The figures given in column (9) are based on the assumption that the catchment has an area of 500 square miles and the unit period is 6 hours.

Flow records. In order to obtain records on which to base future calculations for design purposes, automatic recorders should be installed in a number of key positions on the river system. These should not all be located on the main flows, but in well-considered positions, in order that the run-offs from tributaries and small areas should be learnt as well as the discharges of the various parts of the main river and of the whole of the catchment. The records can be taken on large waterways, where weirs are already inserted for navigation purposes. But it may be necessary to construct gauging weirs across quite large spans of water purely for the purpose, in which case the effect of the weir on the flow of the river must be kept in mind, as otherwise silting and flooding may result. Small streams can be gauged by standing-wave flumes and rectangular or V-notch plate weirs, or composite weirs, according to the size and seasonal variation of the flow concerned. Autographic rain-gauges should also be installed in order that run-off may be compared with rainfall.

Stable channels. The recognised procedure in producing a more adequate natural watercourse is to vary the cross-sectional areas and gradients of the different parts of the stream in such a manner as to increase discharge and yet maintain stability. This means that at the available new gradient, after straightening the course of the river, the cross-section is such that the maximum flow is accommodated, and at the same time the limits of velocity are such that scour at maximum flow and deposition at other times practically balance out.

Textbooks gave what were at one time believed to be the velocities of flow at which various deposits, from coarse gravel to fine silt, commence to scour. Too much reliance should not be laid on these figures, because it is known that the velocity at which a grade of particle or material scours varies, being higher for large, deep rivers. Kennedy[140] showed that stable channels could be constructed for rivers flowing in their own alluvium, and Lacey,[155] elaborating Kennedy's theory, showed that the scouring velocity depended on the hydraulic radius of the watercourse.

Lacey derived the formulae:

$$P = 2 \cdot 67 \, Q^{\frac{1}{2}}$$
$$V = 1 \cdot 151 \sqrt{fR}$$
$$S = 0 \cdot 000547 \frac{f^{\frac{5}{3}}}{Q^{\frac{1}{6}}}$$

where: P = the wetted perimeter in feet
f = the silt factor
R = the hydraulic mean depth in feet
V = the velocity in feet per second
Q = the discharge in cusecs
S = the inclination, or fall divided by length.

These formulae related to stable channels in alluvium, i.e. channels of more or less constant flow which did not tend either to scour or silt up to any great extent. In the case of wide channels the wetted perimeter P was taken as equal to the width. The cross-section was considered semi-elliptical and of area dependent on silt factor or coarseness of the grains of sediment.

In the design of English river work consideration has to be given to watercourses in which the flow varies between dry-weather flow and maximum flood flow. Marshal Nixon examined a collection of data on British rivers and, working from Lacey's formulae, gave the following formulae as applicable to stable river channels running bank-full:

$$W = 1 \cdot 65 \, Q_b^{\frac{1}{2}}$$
$$d = 0 \cdot 545 \, Q_b^{\frac{1}{3}}$$
$$V = 1 \cdot 12 \, Q_b^{\frac{1}{6}}$$
$$A = 0 \cdot 9 \, Q_b^{\frac{5}{6}}$$

where: Q_b = "bank-full" discharge in cusecs
W = the water-surface width in feet
d = the average depth in feet
V = the velocity in feet per second
A = the cross-sectional area in square feet.

The "bank-full" discharge means the discharge when the river is running almost to overflow and at the rate of flow liable to occur 0·6 per cent of the time. For shorter periods of time during the year the river must be considered to spill over and flood, otherwise it must be made deeper or be provided with flood banks to accommodate the design discharge.

The most important factor in stable channel design is the width. If this is not correctly proportioned the channel will not be stable and will tend to scour or silt.

The method of using Nixon's formulae is as follows:

1. Plot a flow-duration curve in which cusecs are plotted against time as a percentage time of the total.
2. From this curve read off the discharge when flow takes place for 0·6 per cent of the time: this is the "bank-full" discharge.
3. Calculate by the formulae the required width and depth.
4. Calculate from the data so obtained the required hydraulic gradient to pass the "bank-full" discharge.
5. If flooding is not permissible, increase the depth of the channel by providing flood banks or cutting deeper so as to accommodate the flood flow greater than the "bank-full" flow, *but do not alter the width*.

Consideration of the system as a whole. It is desirable that, when river-works are being designed, the whole of the river system should be considered throughout, in order that works executed in one section do not lead to trouble elsewhere. For this reason a survey of the whole of the catchment area should be made, the run-offs of all parts of the catchment determined as far as possible by the methods already described, and finally sections plotted showing the bed-levels, top of bank-levels and top water-levels recorded on different occasions.

Such sections show where flooding occurs and in what circumstances, and indicate to the engineer where works are necessary, and what types of works are advisable: where the plotted hydraulic gradient or top water-level rises above bank-level, flooding has occurred. From study of the sections and maps of the surrounding land, the engineer can decide whether or not to permit the flooding to continue in order to avoid trouble elsewhere, to raise the banks at the point of flooding, or to make the length of river downstream more adequate so that it will be able to take the maximum flow.

Methods of flood prevention. Several methods of flood prevention are possible. These include:

1. Clearing the river or stream of weeds, etc.
2. Enlarging the cross-sectional area.
3. Increasing the hydraulic gradient by straightening the course.
4. Raising the banks.
5. Permitting or, by impounding, causing flooding of low-value land so as to provide storage over the period of peak flow and thereby reducing flow downstream.
6. Removing, where permissible, any disused mill dams or disused navigational dams.
7. A combination of the above methods.

By far the greater number of cases of flooding and waterlogging of lands are due to neglect of watercourses and their obstruction by weeds, deposited rubbish and ill-designed structures. Before large-scale works are put in hand to make improvement to the river, the engineer should therefore ascertain that the courses are clear of rubbish, vegetation and the silt which vegetation induces.

As has already been mentioned, increasing the cross-sectional area of a natural watercourse may, in most instances, be considered a temporary measure only, because silting will eventually reduce the size of the stream to what it was originally. The method has its applications, but they are limited.

Straightening the course of a river, that has developed tendencies to flood because of meandering and a rising bed-level, is effective. It may produce some scour, but this does not necessarily matter very much unless it results in trouble due to deposition farther downstream. Such straightening of the course of a river is one of the cheapest and most effective methods, and the easiest to apply.

Raising the banks of a river is a method which is applicable to the prevention of local flooding and to the reclamation of land—for example, in tidal reaches. When banks are raised, the flood-water is kept in the river, but at the same time the run-off from adjacent lands to the river is prevented. It is obvious that not only the banks of the main river must be raised, but so must those of any tributaries joining it in the area concerned. The water in ditches which drain to the river and tributaries can be excluded temporarily, provided there is sufficient storage, but must be dealt with eventually. This water can discharge to the river system at other than flood-time through flap valves, the valves preventing back-flow to the low-lying lands in times of flood. Alternatively, the water may have to be pumped, as is the practice in the Fen District.

Some flooding of developed areas and valuable land is prevented by the reduction of run-off due to the time-lag by storage of flood-water in water meadows. Similarly, rate of run-off can be reduced by impounding upland catchments.

Embankments. Embankments need to be of sound construction, because the bursting of an embankment during flood-time can have serious results. Failure can be due to inadequate proportions, wrong materials of construction, or damage by plants and animals.

Embankments should be sloped at inclinations dependent on the earth of which they are formed. In no case should a bank be steeper than 1 in $1\frac{1}{3}$, and batters of 1 in 2, 1 in 3, 1 in 4 and 1 in 5 are given by some authorities as suitable in various circumstances. The embankments must be watertight, and if the earth excavated locally is of a pervious nature, clay puddle cores may be necessary.

Where there is not much scour, the internal surface of an embankment may be of earth; but on the outside of bends, and wherever high velocities of flow are to be expected, protection must be given. One of the simplest methods of protecting the face of an embankment is to bury faggots 3 or 4 feet long bundled with tarred yarn, the brush ends pointing towards the waterway. As these brush ends become exposed by scour, they have the effect of locally reducing velocity close to the surface of the earth and prevent further scour from taking place.

A more expensive method, but one which is preferable in built-up areas and often justified in other circumstances, is to face the embankments with pitched large stones or else spalls (split stones), the last bedded in the earth with the smooth (split) surface exposed to the flow of water. Pitched stones or spalls may also be bedded in cement mortar or concrete.

The backs of embankments should be protected by vegetation, in the form of grass or small bushes, to prevent the earth from falling away: large trees should not be permitted to grow on them because the roots can break through clay puddle and because the falling of a tree during a storm can lead to rupture of the bank. Banks should be inspected from time to time and precautions taken against their being damaged by rabbits. One rabbit-hole through the bank below flood-level could mean disaster.

Embankments are not always placed directly in contact with the water's edge; where space is available they may be set well back. This serves the twofold purpose of protecting the embankment from scour and providing storage which reduces the required size of waterway downstream.

In many instances construction of the waterway between limited boundaries necessitates vertical sides: similarly, vertical sides are desirable if the bed of the river needs to be deep throughout to permit navigation. In these circumstances sheet piling with timber, concrete or steel sheet piles is usually the most effective method. Comparatively steep banks can be built of concrete in bags and occasionally, but rarely with economy, retaining walls of concrete or brick work may be constructed. The difficulty of constructing these works is that they necessitate "work in the dry." They are not difficult to construct in those circumstances where the river is being diverted to a new course. They may be essential where bridges are built: elsewhere they are usually out of the question because they involve temporary sheet piling so as to permit the work to be built.

The bed of a river can be protected by inverting with concrete where it is possible to dry out the bed while the work is in progress, by concrete in bags or by trenches dug in the bed in the form of chevrons pointing downstream and filled with heavy material such as large stones or concrete bags.

Pumping. Pumping of surface water most commonly involves the lifting of large quantities against very small heads. For this purpose axial-flow centrifugal pumps are particularly suitable (see Vol. I, Fig. 79). Centrifugal pumps should be protected against damage from floating debris by the installation of screens (see Chapter XIII).

It is not practicable in all circumstances to install pumps that will deal with rainfall run-off as fast as it accumulates in ditch systems. Usually these ditch systems can be kept comparatively empty in dry weather; they then provide storage which balances the difference between pumping rate and peak run-off. This storage also gives the time-lag necessary to permit manual operation of pumping plant where electric power is not available (see Chapter IV).

Tidal reaches. In parts of the world where there is little or no tide, deltas are formed where rivers discharge into the sea, because of deposition where gradients are slack. In Great Britain tides are so wide in range that the scour produced by the flow of tidal water in and out of estuaries keeps them comparatively free from silt. In view of this, work in tidal reaches should be so designed that the flow of the tide is assisted, not prevented: the engineer should not fall into the error of constructing a barrage with tidal gates excluding high tide, and thereby reducing the flow of the stream at high water. This would most probably temporarily improve matters, but eventually the reduced current would permit deposition of sediment, and the consequent reduction of the waterway would set up a new problem.

The most effective way of maintaining a clear channel in a tidal estuary is to straighten the course of the river and to form it in the shape of a trumpet widening towards the sea. This makes the most of the scouring effect of the tide.

Model experiments. Where estuarine works are of sufficient magnitude to justify considerable expenditure on research, it is advisable for model experiments to be made so as to determine the effects which new works will have on the natural formation of the waterway. It should be said at once that this class of experimental work is not for the engineer to undertake, but that it should be passed to scientists specialising in this particular branch of physics, for it is too easy for misleading results to be obtained by errors of technique.

There is extensive literature on the subject of hydraulic scale models, most of which refers to experimental work with river models. Perhaps the clearest exposition of the basic theory is that of Professor A. H. Gibson[91] who writes:

The ratio of the corresponding speeds depends upon the type of model. Whereever flow takes place under the action of gravity, as, for example, in the discharge over a weir, or where surface waves, whose form depends essentially on the force of gravity, are formed, corresponding speeds must be in the ratio of the square roots of corresponding dimensions.

There are, however, cases in which forces due to fluid friction are all-important, and in which gravity has no effect on the phenomenon. For example, the lines of flow around the hull and also the resistance of an airship or a deeply submerged submarine are independent of the force of gravity. So is the resistance to flow through a pipe and the distribution of velocities over its cross-section. In such cases the density ρ and viscosity μ of the fluid are the only potent factors, and theory shows that corresponding speeds must be such as to make the product $vl\rho/\mu$ the same for original and model. Here l denotes some definite linear dimension, such as the diameter of the pipe or the length of the submarine. The expression $vl\rho/\mu$ or vl/v* is known as the "Reynolds number." If the same fluid is used in the model as in the original, the first type of model (in which gravity-effects are important) requires

$$\frac{v}{V} = \sqrt{\frac{l}{L}},$$ while the second type (in which viscosity-effects are important) requires $$\frac{v}{V} = \frac{L}{l}.$$

It is evidently impossible to satisfy both these requirements simultaneously, so that where, as is often the case in practice, both gravitational and viscous forces are involved, it becomes impossible to choose corresponding speeds which will give exact similarity as regards all the forces called into play.

In any kind of model, if the scale-effect is not to be serious the type of motion must be the same as in the original. As is well known, two distinct types of fluid motion are possible—laminar or streamline, and turbulent.

The first of these tends to occur at low speeds, with fluids of high viscosity and in flow through channels of small dimensions—that is, at low values of the Reynolds number. In it the fluid forces are entirely governed by the viscosity of the fluid. Turbulent motion sets in at some definite value of the Reynolds number and persists at all higher values. In such flow inertia forces predominate, and viscous forces become less and less important as the turbulence increases until, when the resistances become proportional to v^2, viscosity ceases to have any further modifying effect on the lines of flow, or on the resistance.

* v is called the "kinematic viscosity" of the fluid.

CHAPTER XIX

SAFETY OF SEWERAGE OPERATIVES

THE dangers to sewerage operatives and the men employed at sewage-treatment works are greater than is generally appreciated, and while accidents and cases of infection seldom receive publicity, the fact remains that these occupations are comparatively hazardous. In "Occupational Hazards in the Operation of Sewage Works"[80] it is pointed out that in the United States the number of accidental deaths per thousand employed at sewage-treatment works exceeds that for machine shops by 215 per cent, and that in several States the insurance rates for sewage-disposal operatives are considerably greater than for machine-shop workers.

The risks to which workers in sewers, sewage-pumping stations and treatment works are exposed include:

1. Drowning as a result of flooding or of falling into tanks, etc.
2. Physical injury by contact with moving machinery.
3. Various physical injuries of major or minor character due to mishaps such as falling from ladders (the most frequent cause).
4. Physical injury as a result of explosion of gas in sewers or tanks.
5. Asphyxiation as a result of oxygen deficiency or excess of irrespirable gas.
6. Poisoning by gases or vapours.
7. Infection.

The means of minimising accidents in sewers or at treatment works are:

1. The proper design of works, so as to prevent mishaps from taking place, as far as is practicable, making full allowance for carelessness and mismanagement.
2. Efficient administration and education and control of staff, so as to prevent unsafe practices, and to ensure proper use of safety equipment provided.
3. The provision of safety equipment.

The various dangers liable to be encountered and the means of obviating them will be considered from the above points of view.

Drowning due to rapid filling of sewers. It is unsafe to rely on weather reports for deciding whether or not there is any risk of sudden filling of sewers as a result of rainfall run-off and, similarly, full reliance cannot be placed on rules and regulations for preventing sudden discharge of waste from premises. It is therefore always necessary to take adequate precautions when men enter sewers in which they could be swept away by the flow. If the sewers receive storm water, men posted at the surface should watch for the approach of storms, so as to be able to give warning.

Structural precautions to be taken include the provisions of safety bars or chains at manholes (see Chapter XII). But of prime importance is the proper spacing of manholes, in order that, on receiving a warning, men shall not have to run long distances laden with equipment, and perhaps up to the thighs in water. As has been described in Chapter XII, the Ministry of Housing and Local Government restricted the distance between manholes on sewers of small size to a maximum of 360 feet. Previously this restriction applied to all sizes of sewers, but was later relaxed for no apparent reason, and many large sewers have manhole spacings in excess of this figure. American practice would appear to be somewhat similar to English practice in this respect, for "Manual of Practice" No. 18[0] recommends that manholes shall be not more than 500 feet apart and preferably 300 feet apart.

All tanks into which there is any danger of men falling should be adequately fenced or covered. This is far from being common practice at present. Fences or handrails of inadequate height have been the cause of fatal accidents: the not uncommon height of 3 feet is not enough.

Dangerous machinery. Local authorities' machinery, such as pumping plant, not being liable to examination by factory inspectors, is very often faulty from the point of view of safety. For example, fast-running shafting is considered particularly dangerous and, in a factory, must be given careful protection, even when it can be reached only by ladder. But it is not uncommon for the high-speed shafts that connect from electric motors to sewage pumps to be totally unprotected and to be in close proximity to access ladders. It is recommended that designers of sewerage and sewage-treatment works should attempt to render their machinery at least as safe as would be required by a factory inspector.

Accidental injury. Design plays a large part in either the cause or prevention of accidental injury to operatives. For example, in the design of manholes the arrangements of ladders or step-irons and the proportioning of the component parts of the manholes, such as benchings, are of importance. Ladders are generally to be preferred to step-irons in all deep manholes, and wide step-irons placed one above the other are much to be preferred to the narrow, staggered step-irons that are in vogue at present. Step-irons should always be equally spaced vertically and lineable vertically. The spacing of step-irons and of rungs of ladders depends on usage: men wearing thigh boots prefer closer spacings than men not so encumbered. Step-irons should be of non-corrodible material or else heavily galvanised. They should be inspected from time to time and, if found to be faulty, replaced immediately.

Means of giving warning against hidden dangers, as described in Chapter XII, should not be overlooked. Benchings should be adequately wide, and sloped at reasonable but not excessive falls, and should not be rounded.

Dangers from traffic should not be overlooked. When manholes are in the road, men entering and leaving them should be protected by a vehicle being halted against them: this is the usual practice where sewer-gangs are mechanised. An alternative is the use of "manhole cages."

Gases and vapours. Gases and vapours, being liable to cause explosions as well as asphyxiation or poisoning, deserve special attention. They arise from a number of causes, including decomposition of sewage, leakage of coal-gas into sewers, evaporation of petrol spilt from road vehicles and the discharge of trade wastes. Natural gas from the soil may also be encountered in sewers.

The principal gases liable to be found in sewers, their sources, physiological effects and the means of their detection are as follows:

Hydrogen sulphide. Hydrogen sulphide is a normal product of decomposition, and is always present in septic sewers. In a concentration of 0·2 per cent it is fatal in a few minutes, having a paralysing effect on the respiratory centre. Less than half this concentration causes acute poisoning in a short time. Although hydrogen sulphide has a strong offensive odour, it impairs the sense of smell as the concentration increases, and therefore it is not safe to rely on its odour as the means of detection. The usual method of detection is exposure of lead acetate paper. Filter-paper is moistened with 5 per cent lead acetate solution for five minutes; in the presence of hydrogen sulphide in low concentration the paper changes in colour to grey or brown.

Methane. Methane is the main constituent of sludge gas, such as is produced in sludge-digestion tanks. It is also an important constituent of coal-gas and of natural gas. Mixed with air it is explosive, and consequently it has been the cause of a number of explosions in sewers and at sewage-treatment works. Physiologically it is not poisonous, but when in high concentration it causes asphyxiation by excluding oxygen.

Methane has no odour. It is detected either by a combustible-gas indicator or an oxygen-deficiency indicator.

Carbon dioxide. Carbon dioxide in sewers can be a product of decomposition, or combustion, or of the reaction of some trade wastes. It is also found in the soil in some circumstances. (Dangerous quantities of carbon dioxide have been encountered in headings driven in chalk.) A concentration of about 10 per cent acts on the respiratory centre, and cannot be endured for more than a few minutes. It is considered to be odourless, but in strong concentration has an acid taste. It is usual to test for carbon dioxide with an oxygen-deficiency indicator.

Carbon monoxide. Carbon monoxide is produced by incomplete combustion, such as the explosion of sewer-gas or the exhaust of internal-combustion engines. It is also a constituent of coal-gas. About 0·2 per cent concentration produces unconsciousness in thirty minutes, as little as 0·1 per cent is fatal in four hours. The gas is completely odourless. It is detected with the aid of an electric carbon monoxide indicator, a chemical detector or carbon monoxide ampoules.

Hydrogen. Decomposition produces hydrogen, which, in sufficient concentration, is explosive, and can also have a similar physiological effect to methane in preventing the absorption of oxygen. It can be detected either by combustible-gas indicator or oxygen-deficiency indicator.

Petrol vapour. Petrol vapour mostly finds its way to sewers from garages and certain trade premises, such as dry-cleaning establishments, but it is not unknown for it to be discharged from private dwelling-houses. In this country explosions due to petrol vapour are rare, but in America very many have been traced to this cause. Besides being explosive, petrol vapour has anaesthetic effects, and is rapidly fatal when in 2·4 per cent concentration, and one-half this concentration is very dangerous. It can be detected by means of combustible-gas indicator or by oxygen-deficiency indicator.

Oxygen deficiency. Oxygen deficiency in sewers is usually due to the depletion of the oxygen in the air as a result of the oxidation of the organics in sewage. It can also be due to combustion of explosive gases. When the oxygen in the air is reduced to about one-half of its normal concentration of about 21 per cent the condition is dangerous to life, and if it is reduced to about one-quarter of the normal concentration it is almost certain to be fatal. The usual test is by oxygen-deficiency indicator.

Precautions. The obvious precaution to be taken in the design of sewers so as to prevent the dangers of explosion, asphyxiation or poisoning by gases or vapours is to provide adequate ventilation. In addition, the decomposition in sewers that produces the more dangerous gases, such as hydrogen sulphide, is mostly of sediment, the accumulation of which can be prevented by laying sewers to adequate self-cleansing gradients. Other precautions that can be taken are the insistence on petrol interceptors and the exclusion or treatment of dangerous trade wastes.

The normal precautions to be taken when entering sewers are mainly aimed at avoiding the dangers due to gas concentrations or oxygen deficiency. These are set out in "Accidents in Sewers: Report on the Precautions Necessary for the Safety of Persons Entering Sewers and Sewage Tanks,"[183] and briefly are as follow. Before a man enters a sewer or a tank it should be ventilated, in the case of a sewer by the removal of the manhole cover and the covers of the manholes immediately adjacent, both upstream and downstream at least half an hour before the man enters the sewer. Then tests should be made for hydrogen sulphide, for inflammable gases and for asphyxiating conditions. The first man to enter the sewer should be secured by life-line, and throughout the period of work in the sewer there should not be fewer than two men at the surface ready for rescue work.

The Greater London Council have their own code which is still more stringent. It requires, *inter alia*, that before any person enters a sewer the position of the nearest telephone shall be ascertained, manholes both upstream and downstream shall be opened and no entrances closed while men are working between them. In no circumstances is a man permitted to travel through a sewer alone. The council uses the "Spiralarm" lamp which is lowered into the manhole or chamber and left there for two minutes while, at the same time, a moistened lead acetate paper is lowered in a wire cage. The lead acetate paper is drawn up after one minute's exposure and its colour compared with that of an unused paper.

The Spiralarm lamp indicates the presence of explosive gases and oxygen deficiency. Explosive gases cause a red light to operate and oxygen deficiency causes the oil flame of the lamp to diminish or go out. The lamps are tested in a gas chamber once a month. The lead acetate papers have to be moistened but not soaked in a solution of glycerol in distilled water.

In the event of the Spiralarm flame indicating oxygen deficiency or the paper showing the presence of hydrogen sulphide no person is permitted to enter the manhole or chamber until the atmosphere has been cleared and satisfactory tests obtained unless, in an emergency, he is wearing breathing apparatus. The council requires that one, or where practicable, two men shall be stationed at the surface and keep in touch with the men in the chamber or sewer by calling or signalling at frequent intervals and give warning if rain should fall. Throughout operations the men in the sewer and at the surface shall have lighted Spiralarm lamps and lead acetate paper.

No light or fire should be exposed within 10 feet to the entrance of any sewer. No smoking should be permitted or other than safety lamps used within a sewer. Electric lamps should not be switched on or off within a sewer or near its entrance.

Rescue kit should be carried by every travelling sewer gang, and this should include not fewer than two life-lines of adequate length, a breathing apparatus (a recommendation has been made by the Institution of Civil Engineers regarding breathing apparatus for use in sewers), and this kit should be regularly examined and maintained in efficient condition.

In the event of an accident the casualty should be brought to the surface without delay, care being taken not to augment injury. First-aid should be given before leaving the sewer if practicable. Disinfectant should be applied to all open wounds and the casualty should be seen by a doctor on the same day. Should a man be overcome by contaminated air the men at the surface must be warned and, if he cannot be removed immediately, the men in the sewer must leave before they themselves are overcome, when rescue should be made with the aid of breathing apparatus.

In cases where breathing has ceased artificial respiration (in which all staff should be trained) should be applied, during which the casualty should be kept warm and his limbs rubbed upwards to help circulation.

Work in compressed air. Men should not be permitted to work in compressed air unless they have been medically examined and found fit. The rules and precautions applicable are covered by a Ministry of Labour leaflet "Workers in Compressed Air" (Form 754).

Radioactive contamination. While, in most instances, discharge of radioactive material to sewers is very unlikely the possibility as a result of fire or other accident should always be allowed for in sewerage areas where there are premises on which such materials are used. In the event of such circumstances the police should be informed immediately and men be withdrawn from all localities where they might be affected. Any men who could have been affected should be instrumentally checked and, if found to be contaminated, treated by the appropriate medical authority by arrangements made in conjunction with the police. All premises likely to have been contaminated should not be re-entered until instrumentally checked and found to be safe.

Infection. It is often said that sewer men are particularly healthy and robust, and this in spite of their exposure to all manner of diseases. Whether or not this is true, every precaution should be taken against infection, in view of the high concentration in normal sewage of bacteria, many of which are pathogenic. In this connection mention should be made of *Spirochaetosis Icterohoemorrhagica* or Weil's disease. This disease is, fortunately, rare, but sewer workers are particularly exposed to it. Rats are carriers of the disease, and infection can be from their urine.

One of the troubles with Weil's disease is that, owing to its rarity, it is not always recognised by medical practitioners and has been mistaken for influenza, pneumonia, tonsillitis, rheumatic fever and other disorders in the early stages. For this reason the Greater London Council provide all staff who are liable to come in contact with sewage with cards giving the symptoms of this disorder and the instruction that they must show this card whenever going to a hospital or doctor on account of illness.

The precautions to be taken against this and other diseases are the avoidance of contamination of food and drink or of the infection of cuts and abrasions. Sewer men should be taught not to touch their faces during work or until they have washed, and adequate washing accommodation should be allowed, including the provision of surgeon's taps which can be operated by means of the wrist or elbow or pedal-operated taps or spray nozzles. All cuts and abrasions should be given immediate treatment, however slight they may be, and gangs should be provided with the necessary first-aid equipment.

PART II
SEWAGE TREATMENT AND DISPOSAL

CHAPTER XX

RATES OF FLOW AND CHEMISTRY OF SEWAGE

DECOMPOSITION of organic matter is continually taking place. The soil itself is made up of a mixture of mineral detritus and decaying organic matter. When dead organic material falls on the ground it rots and becomes part of the humus of the soil. Similarly, decomposition takes place in ponds, rivers and the sea.

Decomposition of organic matter under natural conditions is not offensive or deleterious to health and, in the main, it goes on unnoticed because the aggregate of material at one place is seldom great. In these circumstances the material is oxidised with the aid of aerobic bacteria and other organisms.

Civilised conditions, with or without sanitary services, have led to the accumulation of large quantities of organic waste, including faeces, liquid sewage and solid refuse. Much of this material is in itself offensive in bulk, and if allowed to decompose in large masses becomes septic and emits foul odours. Such organic matter left upon the ground encourages vermin, in particular flies and rats.

If solid or liquid organic waste is discharged into watercourses, first, the oxygen in the water is absorbed by the bacteria, which live on the sewage, and as a result any fish in the streams die. Second, once the water has become free of oxygen, anaerobic bacteria set up offensive conditions. It is for these reasons that sewage treatment becomes necessary. Water supply has led to large quantities of water being used. Sewerage systems collect this water when it has been sullied: the foul sewage must be discharged eventually to natural waters of some kind—i.e. most frequently rivers or streams, the sea in the case of coastal towns, more rarely underground waters in those places where there are no watercourses available and soakage into the ground is possible and permissible.

In practically all cases of discharge to natural watercourses, sewage must be so treated that it cannot impair the quality of the water of the stream. When sewage is discharged to the sea a lower standard of purification is usually permissible, because of the high degree of dilution that is available, and the standard adopted depends on circumstances. When sewage is soaked into the ground the main consideration is the possibility of pollution of water supplies from wells.

The sizes of the units of sewage-treatment works are based on two main factors: the rate of flow to be treated and the strength of the sewage. These multiplied together give the load on the works or the work to be done in purification, but they also have to be considered separately. Furthermore, there are hour-to-hour variations of flow, proportion of storm water, peculiarity of chemical content, etc., which require special consideration.

Rates of flow. The *average* rate of flow per day is the total flow during the year divided by the number of days in the year. This is often only slightly more than the dry-weather flow; but the difference should not be overlooked. The average rate of flow is used in calculations of total quantities of detritus, screenings and sludge to be disposed of, total sludge gas available, and general estimation of pumping and other power requirements.

The *dry-weather* flow has been defined as the average daily rate of flow as recorded on days when not more than $\frac{1}{10}$ inch of rain has fallen in the twenty-four hours. This figure is of importance because it is usual to design treatment works to treat up to three times dry-weather flow.

The difference between dry-weather flow and average daily flow is *not* storm water. Storm water is sewage discharged without full treatment, and is less in quantity than the above difference.

The rate of flow of sewage during the day varies from a minimum at about 6 a.m., rises to a peak at about 10.30 a.m. or later and then falls off fairly regularly, apart from a secondary peak in the middle afternoon. For small catchment areas, where the time of concentration is short, the daily flow curve takes a somewhat triangular form and may reach a peak of as much as twice average flow, but where the catchment area is large and the time of concentration lasts for several hours, the curve is rounded so that it tends to approach a sine curve and the peak occurs late: for example, at London's Crossness Works the peak occurs between 4 and 5 p.m. and is only $1\frac{1}{3}$ times average daily flow. In all cases, however, the rise of intensity in the morning is sharper than the tail-off in the afternoon. In a few instances the minimum flow at night is as much as half the average rate of flow, but in most new sewerage systems it is much less, for night flow is mostly due to infiltration of subsoil water.

From the curve of hour-to-hour variation of flow the degree of infiltration can be estimated approximately, particularly if there is a period during the night when flow remains constant for some hours; this can be taken as being infiltration. Although infiltration must vary from time to time, it is often assumed to be a constant figure throughout the day and from day to day.

There are the various industrial wastes discharged to the sewers in addition to domestic flow, infiltration and storm water. These are not easy to estimate, but can be taken as being at least the known trade wastes discharged to the sewers under agreement or at least equal to the metered trade water supply.

In addition to hour-to-hour variation there is the variation throughout the week. The total variation due to both day-to-day variation and hour-to-hour variation gives a weekly peak that may be in the region of three times dry-weather flow.

The foregoing rates of flow should be, as far as practicable, obtained from instrumental records. In the absence of such records and of the possibility of obtaining them, it is necessary to make assumptions based on comparison with other drainage areas. In the first place, it can be assumed that the dry-weather flow due to trade waste and domestic demands is equal to, or slightly greater than, the public water supply and that the peak flows for the day and week are almost as marked as the peaks of water demands at the same time. In all cases the trends for water supply to increase from year to year should be studied and the future dry-weather flow estimated in accordance with these trends.

Storm water. At existing works flows of storm water should be measured by recorders in order that minimum rates of flow and total flows throughout the year can be known. These figures are required for the design of storm overflow weirs and storm-water pumping plant and, in some circumstances, for storm-water treatment plant.

In the case of proposed sewerage systems or where appreciable extensions of the sewerage system are expected, peak storm flows can be estimated by one of the time-of-concentration methods described in Chapter III. This gives the maximum rate of flow likely to occur on the average once in three years. At less-frequent intervals greater rates are to be expected, and in some circumstances it may be necessary to design those parts of the works taking storm water to accommodate flows of about twice or thrice the calculated value.

The total run-off per year may be approximately estimated by the formula:

$$C = 22{,}687\, Apr$$

where: C = gallons per year
  r = total inches rainfall per year
  Ap = contributing *impervious* area in acres.

or: $C_1 = 10\, Ap_1\, r_1$

where: C_1 = cubic metres per year
  r_1 = total millimetres rainfall per year
  Ap_1 = contributory impervious area in hectares.

Chemistry of sewage. More than nine hundred and ninety-nine parts out of every thousand of an ordinary sewage are merely water. The foreign matter which is present in each cubic foot of sewage is usually less than a cubic inch, and consists for the most parts of harmless salts, such as lime and magnesia, derived from the water supply.

Organic pollution is present in three different forms:

1. Suspended solids, such as faeces, fat and vegetable matter.
2. Colloidal or finely divided suspended solids.
3. Organic matter in solution.

The composition of a sewage can be ascertained by means of a chemical analysis, but such an analysis is usually confined to a few of its constituents.

Chlorine, combined with sodium as common salt (NaCl), is always present in sewage, being a constituent of urine, and affords a useful measure of its strength. The amount of chlorine is not affected by any process of purification, so it should be the same in the purified effluent as in the crude sewage. A difference in the chlorine figures for the two liquids shows either that the effluent from which the sample was taken does not correspond with the sewage, or possibly that the sewage has become diluted on its way through the works. In arriving at the amount of purification which has been effected, it is necessary to be sure that the same liquid is being sampled in both cases, and that the effluent from a weak night sewage is not being compared with the strong sewage which comes down during the day or vice versa.

Organic carbon (or oxygen absorbed). Carbon is an essential constituent of all organic matter, and the organic matter present in a sewage was formerly recorded in terms of carbon content. Nowadays, instead of measuring the carbon directly, the chemist ascertains the amount of oxygen required to oxidise it. This oxygen may be obtained from an acidified solution of potassium permanganate, a measured quantity of which is added to the sample under examination and kept at a temperature of 80 °F. for four hours. A determination is sometimes made of the oxygen absorbed in three minutes also. The latter test indicates the amount of readily oxidisable matter present, some of which, however, may not be of a polluting character.

The most important test applied to sewage or treated effluent is the test for biochemical oxygen demand, which is usually referred to as B.O.D. value. This is the amount of oxygen absorbed from an incubated sample over a period of days, the period frequently selected being five days. The oxygen for oxidising the organic matter is derived from tap-water saturated with oxygen, a given volume of which is mixed with the sample and kept at a constant temperature for the incubation period.

Nitrogen. Nitrogen is an essential constituent of protein. It is present in most organic materials, and in the course of their decomposition and the subsequent purification of the sewage it undergoes a series of characteristic changes. As the result of decomposition the complex organic molecules are broken down into a number of simpler substances, including ammonia, a typical constituent of stale sewage. In the later stages of purification the ammonia is brought into combination with oxygen, forming first nitrous acid (HNO_2) and afterwards nitric acid (HNO_3). These acids react with the alkaline substances present, with the formation of nitrites and nitrates. Albuminoid ammonia, and to a less extent free or saline ammonia, are measures of impurity. Low ammonias and high nitrates indicate a well-purified and stable effluent.

The nitrogen in its various combinations was formerly determined as ammonia. Results expressed as ammonia may be converted into their nitrogen equivalents by multiplying them by 14 and dividing by 17.

Dissolved oxygen. Fresh sewage contains little or no dissolved oxygen, and stale sewage none, any free oxygen which it may have possessed having gone to oxidise some of the products of decomposition. A purified effluent, on the other hand, should contain an appreciable amount of dissolved oxygen.

Stability. Sewage is an unstable liquid. In other words, it tends to decompose. A purified effluent, on the other hand, is comparatively stable in composition. In an American analysis of a sewage effluent its "relative stability" is often expressed numerically.

Alkalinity and acidity. Domestic sewage is usually alkaline, by reason of the soda, potash and ammonia which are always present in it, but the sewage of a manufacturing town, containing, for instance, large quantities of waste water from electro-plating works, may be strongly acid. The alkalinity is frequently expressed in terms of pH value, a high value denoting an alkaline sewage, a value of 7 denoting a neutral sewage and a low value denoting an acid sewage. pH value is the measure of the hydrogen-ion concentration (being the logarithm to the base 10 of the reciprocal of the hydrogen-ion activity) and hence of the acidity or alkalinity of a solution, expressed on a scale of numbers ranging from 0 for a solution containing 1 gram-ion of hydrogen ions per litre, corresponding to extreme acidity, to 14 for a solution containing 1 gram-ion of hydroxyl ions per litre. As the scale is logarithmic, the average of a number of values can be found only by using logarithm tables. The method is to find the anti-logarithm of each figure and then find the logarithm of the average of the anti-logarithms.

Detergents. About the year 1949 the domestic use of synthetic detergents had become of sufficient magnitude to involve sewage-treatment problems. The effect of these surface-active agents at sewage works was to interfere with sewage treatment, particularly the activated sludge process, by reducing the rate of transfer of oxygen from the air to the water. They also cause very severe foaming on aeration tanks and at other parts of the works. They are only partially removed by treatment.

By far the most common synthetic detergents are the alkyl benzene sulphonates, which have the above-mentioned effects. A method for their estimation has been recommended by the Ministry of Housing and Local Government. This depends on the formation by the anionic surface-active agent of a complex with methylene blue which can be extracted by chloroform and measured colorimetrically.

To a small extent non-ionic detergents are used for domestic purposes, and these and cationic detergents are used in industry. No satisfactory methods have been found for estimating the quantities of these types of surface-active substances in the concentrations in which they are likely to be found in sewage.

Strength. The strength of sewage varies greatly. It is sometimes said that every user of the sewers contributes about the same amount of polluting matter, and that the strength of sewage is therefore proportional to the number of gallons per head. This is not necessarily the case, for, in addition to excreta and urine, sewage contains a great deal of soap, etc., from lavatories and laundries, scraps of meat and vegetables, grease from sculleries and other organic matter. Other things being equal, however, the sewage from a town using 40 gallons of water per head will be only about half as strong as that from one which uses 20 gallons.

The Local Government Board used to classify sewages on the basis of the amount of oxygen which they absorbed from potassium permanganate in four hours. A sewage which absorbed from 17 to 25 parts of oxygen was called a strong sewage; a sewage which absorbed from 10 to 12 parts an average sewage; and one which absorbed only 7 or 8 parts per 100,000 a weak sewage.

A classification based solely on oxygen absorbed was felt to be unsatisfactory, and the Chemists to the Royal Commission on Sewage Disposal carried out a large number of experiments with a view to arriving at a better method, which should be applicable to sewages and tank effluent alike. As the result of these experiments they proposed alternative formulae for expressing the strengths of sewages and tank effluents. The formulae are given in "Methods of Chemical Analysis as Applied to Sewage and Sewage Effluents."[185]

Formulae (McGowan's formulae) were given for sewages and septic tank liquors and for precipitation liquors. The latest publication of "Methods of Chemical Analysis as Applied to Sewage and Sewage Effluents" gives that for sewage only as follows:

Strength of Sewage = $4 \cdot 5 \times$ (ammoniacal nitrogen + organic nitrogen) in parts per 100,000 + $6 \cdot 5 \times$ permanganate value using N/8 permanganate in parts per 100,000.

The factor of $4 \cdot 5$ represents the amount of oxygen required to convert the nitrogen to nitrate and is correct provided that all the nitrogen is originally present as ammonia or one of its substitution products. The factor $6 \cdot 5$, however, is empirical and depends upon the proportion of the oxidisable carbonaceous matter which reacts with permanganate under stated conditions, and this proportion varies very widely with different trade wastes. The McGowan formula suffers from the disadvantage already mentioned, that it is based solely on oxygen requirements, although it has proved useful in the absence of appreciable concentrations of trade wastes.

It should be noted the McGowan strength is calculated from the permanganate value using N/8 permanganate and not N/80 permanganate. . . . The relation between the two is no doubt variable. A figure of $1 \cdot 6$ has been given as the general ratio between the N/8 test carried out at $25 \cdot 7$ °C and the N/80 test carried out at $18 \cdot 3$ °C. The ratio must be smaller when the N/80 test is carried out at 27 °C . . . and probably does not exceed $1 \cdot 2$. This figure could be used to obtain an approximate figure for the McGowan strength, but analysts still using it are recommended to adhere to the original method of determination. . . .

It should also be borne in mind that when McGowan's formulae were devised, results of analyses were expressed in parts per 100,000, whereas now, following American practice, it is usual to express them in parts per million (or milligrammes per litre).

McGowan's formulae were aimed at obtaining an estimate of the total biochemical oxygen requirement of the sewage or virtually a three months' B.O.D. value in parts per 100,000.

At first "strength (McGowan)" as determined by the above formula was used as the means of estimating the strength of sewage and the basis of determining required capacity of sewage-treatment works. But this applied to English practice only. Abroad (particularly in America), B.O.D. value has for many years been accepted as the most useful method of determining strength, and now in Great Britain the same practice is followed.

There is no direct correlation between strength (McGowan) and B.O.D.; two sewages of the same B.O.D. may have strikingly different strengths as calculated by McGowan's formulae. But, as both methods are used for the same purpose ultimately (the calculation of size of works), it is not unreasonable to compare one with the other *on the average*. As far as can be ascertained with the data at present available, the B.O.D. of an average-strength sewage, of strength (McGowan) 100, is in the region of 350 parts per million after five days incubation.

TABLE 40

B.O.D. AND STRENGTH OF SEWAGE

Approximate strength (McGowan) p.p. 100,000	Biochemical oxygen demand, pp. 1,000,000
240	840
230	805
220	770
210	735
200	700
190	665
180	630
170	595
160	560
150	525
140	490
130	455
120	420
110	385
100	350
90	315
80	280
70	245
60	210
50	175
40	140

Opinion now is that there are three most important tests for sewage:

1. Ammoniacal and organic nitrogen.
2. Organic carbon or another test for carbonaceous matter.
3. Biochemical oxygen demand.

The first two give broadly a measure of the organic content requiring oxidation; the third, when considered in relation to the first two, gives an indication of the speed at which oxidation takes place.

The Royal Commission gave as an example of an average sewage the sewage of Oswestry, the analysis of which was as follows:

Ammoniacal nitrogen	3·53 parts per 100,000
Albuminoid nitrogen	0·91 ,, ,, ,,
Total organic nitrogen	2·25 ,, ,, ,,
Oxidised nitrogen	trace
Total nitrogen	5·85 ,, ,, ,,
Oxygen absorbed at 27 °C (80 °F) in 4 hours	11·27 ,, ,, ,,
Chlorine	9·16 ,, ,, ,,
Solids in suspension	29·40 ,, ,, ,,

The flow of sewage at Oswestry was at the time 35·7 gallons per head per day.

The night-soil content of sewage. The following information, gleaned from various sources, is seldom to be found in textbooks on sanitation; its uses are mostly in relation to conservancy sanitation. The average amount of faecal matter has been stated to be 135 grammes per person per day at 75 per cent moisture content. The quantity of urine has been stated as being, for the average man, 1500 grammes of water per day, containing 72 grammes of solid matter. According to some experiments of limited extent, it was found that closets were used for defaecation one-and-a-half times per day by each person in an establishment and four times as often for urination only.

Sampling. Great care should be taken when sampling sewage or sewage effluent, because careless sampling can result in misleading analyses. All samples should be taken quietly, so that they are not aerated, in half Winchester quart bottles, filled to the top.

Samples for determination of dissolved oxygen must be taken so that the contents of the vessel is displaced several times yet in such a manner that there is no agitation or inclusion of air bubbles. The samples should be analysed at once or chemically fixed before transport to the laboratory. Generally, samples for other determinations should be examined as soon as practicable and, if stored, should be kept in the dark at a temperature of about 5 °C. Fresh sewage should not be examined for nitrogenous compounds until about a day after the sample has been taken, so as to permit the hydrolysis of urea, for urea interferes with the determination of the various forms of nitrogen content.

Analyses required for determining the capacities of new works should always be twenty-four-hour average "weighted" samples. The procedure in taking such samples is to collect twenty-four samples at hourly intervals throughout the day and night. Each sample should be taken in a wide-necked bottle of about 300 cc capacity, stoppered and set aside in a cool place. At the end of the twenty-four hours a quantity should be taken from each of the sub-samples in accordance with the rate of flow of sewage gauged at the time each sample was taken, and these quantities mixed to form the final sample for analysis. Another method is to read the flow gauge and measure off a sample according to the rate of flow at the time. Each measured sample should be tipped into a carboy or other suitable vessel surrounded with ice. If samples are not weighted the result is that the aggregate sample is lower in strength than the true average sample for the day, because high rate of flow during the day generally coincides with high strength. Samples should be kept on ice and analysed within twenty-four hours of being collected.

In the taking of sewage samples, detritus and large solids, such as screenings, are liable to be excluded to some extent. The heavier particles of detritus move along the invert of the sewer and are not picked up, while the large screenings are best omitted from the sample, as their inclusion would produce very irregular results. For these reasons, quantities of detritus and screenings cannot be determined from sewage analyses as normally made and have to be estimated from measurements of detritus and screenings formerly collected or by comparison with existing works of similar character. Both are very variable figures, depending much on the type of the sewerage system.

Automatic sampling. Automatic samplers for sewage, treated effluent, etc., can be installed at sewage-treatment works to take weighted samples for routine analyses. These can work on various principles. One machine has a rotating arm which dips out a sample of effluent, the sample being automatically sized according to the rate of flow. This arrangement is not altogether satisfactory for sampling crude sewage, because it does not take an average sample in so far as sewage near the bottom of a channel may be denser in sludge and detritus content and that at the top denser in scum content. Another method is to take a sample of constant size by means of a dredger bucket which is set in motion every, say, 100,000 or million gallons of sewage arriving at the works on an impulse sent out by an integrator on the sewage-flow meter.

An automatic weighted sampler designed by the author for use at Crossness Sewage-treatment Works involves no moving parts except a suitable pump. The sampler (see Figs. 61 and 62) works on the leap-weir principle. Every time X gallons are clocked on the main instrument panel, an electric impulse sets the pump running and, by time mechanism, sewage or effluent is pumped for long enough to clear the pipework. The pumping rate and the sizing of the gunmetal nozzle inside the sampler ensure that the velocity of injection is sufficient to throw the flow over a weir inside the sampler and down the waste-pipe. Should the flow exceed the discharge of the nozzle, overflow takes place via the overflow-pipe at the top between the rising main and the waste-pipe. The vertical air vent at the top prevents siphonage. On cessation of pumping the velocity falls and the contents of the Z-shaped pipe connecting to the nozzle falls into the sample bottle.

The requirements in the design of this sampler are:

1. The rising main must be of a suitable diameter to be free from chokage with the kind of liquid concerned—e.g. $1\frac{1}{4}$ inches for final effluent or settled sewage.

2. The pump must have a delivery which will ensure a self-cleansing velocity in the rising main.

3. The nozzle must have a diameter that, under the maximum head permitted by the overflow-pipe, will throw the liquid over the dam into the waste-pipe.

4. The Z-shaped sampling pipe must have a capacity equal to the required individual sample between the nozzle and the point on its vertical component at which the surface of the water will stand when the trajectory of the jet throws the liquid just inside the dam and flow starts to fall into the sample bottle.

5. The vessel receiving the samples should be large enough to receive all the samples taken on the day of maximum rate of flow, preferably it should be housed in a refrigerator to prevent fermentation of the sample and it should be thoroughly cleaned before re-use.

Accurate automatic sampling is not easy, for imperfections in the design of the machine and the bad handling of the samples such as using imperfectly cleaned vessels can falsify the results. Apart from ensuring that the pipework is thoroughly cleared of sediment before the sample is taken, the sampler should be such that there is no segregation of solids-content during sampling. In one design, the sample was not delivered to the sampler by pump but sucked up by the sampler itself. This caused gas to be given off by the sewage and the rising bubbles carried grease and solids to the surface. The sampler then took a sample of the first part of the flow containing an excess of grease and other solids and threw away the rest. This made all samples tend to have a much higher B.O.D. value and solids-content than the sewage. Each component sample was then poured into a vessel that was part of the machine and could not be removed for cleaning except as part of a major maintenance operation. In this vessel settlement took place on a hopper bottom of inadequate slope, from which periodically sludge fell when the vessel was emptied to a sample bottle. At the same time, fermentation took place in the dirty vessel. The overall effect was that, in addition to the samples being stronger than the sewage, the strength and solids-content varied from day to day in the extreme and the sample showed evidence of organic action having taken place.

Even with the most perfect automatic sampling, the analysis of an individual sample is of comparatively little value for there are always the effects of individual differences. As a general rule, the design of new works and all research work should be based on the average of samples taken over a long period, preferably a year.

Instrument panel. At very large sewage works several flows have to be measured, including total flow of sewage arriving at the works including storm water, and total flow separated and delivered for full treatment. Alternatively flow to treatment and storm flow may be measured separately, and totalled to give total flow to works. In large activated sludge works, it is usually advantageous to split the plant into a number of complete independent units to which flow should be measured separately.

It is not altogether good policy to install flow meters as a means of distributing flow to individual sedimentation tanks, aeration tanks, etc. if this can be achieved by other and simpler means: also records should not be taken and filed unless a useful purpose can be served by them.

Fig. 61—Automatic sewage–effluent sampler, Crossness sewage-treatment works

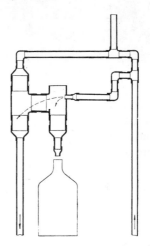
Fig. 62—Principle of automatic sampler

Total, but not individual flows of raw sludge should be measured, total flow of surplus activated sludge, and total flow of liquor from digestion tanks, etc. in order that a day-to-day check may be kept on the way that the works are operated. Quantity and quality of sludge gas from digestion tanks should be measured.

At small works, where there are few records to be taken or instruments inspected, local instruments are best. At large works, the converse is true. There should be at least one control centre in which are located the instrument boards and such remote controls as exist.

The instrument boards are best so arranged that the relations of the various instruments to the various units can be seen at a glance. In one arrangement, the instruments are set out on a conventionalised plan of the works, each in their correct position. On the same board as the instruments are mounted,

Fig. 63—Control room, Crossness sewage-treatment works

coloured lights show where motors are running or where emergency conditions have arisen (see Fig. 63). Another arrangement is to place the instruments on the board in a convenient compact manner, but to paint on the board in various colours the several flows of sewage, sludge, liquor, etc., so that the purpose of each instrument is immediately obvious.

Standards of effluents. It is impossible, without going to altogether unreasonable expense, to purify sewage completely. The utmost that can be done is to carry the purification to such a point that the effluent can safely be discharged into the stream which is to receive it.

The Royal Commission on Sewage Disposal went very closely into the question of fixing legal standards of purity. Most of the witnesses whom they examined on this point were of opinion that it was undesirable and impracticable to do so. Recognising, however, the need for some definition of a good effluent, the Commissioners suggested that such an effluent discharged into a flow of natural water that diluted it by eight to one should not contain more than three parts of suspended matter per 100,000, nor absorb in five days more than two parts per 100,000 of dissolved oxygen. This standard, though it has no legal sanction, is generally considered a reasonable one.

Under common law a riparian owner is entitled to receive river water in its natural state without *any* pollution. But to make a sewage effluent absolutely pure would be impracticable, and therefore it is felt that some statutory effluent would protect the interests of local authorities and others who discharge treated effluents to watercourses. River Boards now state their requirements in this respect.

The Analyst to the Thames Conservancy regards an effluent as infringing the Thames Conservancy Act if:

1. The albuminoid ammonia exceeds 2 parts per million.
2. The suspended matter exceeds 30 parts per million.
3. The dissolved atmospheric oxygen absorbed in five days exceeds 20 parts per million.

Effluents are judged by the analyses of samples taken from time to time, and are expected to pass all three tests.

While the Royal Commission effluent was intended to be applied to circumstances in which the discharged effluent would be diluted by eight times its volume of river water, practice over the years had been to require Royal Commission effluents whenever discharge is to inland water, or even, in the absence of watercourses, soaked into the ground, no matter how high the dilution. The Royal Commission standard has also been accepted in many cases where dilution was below eight to one. Recently, however, there have been instances where a higher standard than that called for by the Royal Commission has been insisted on, and the Ministry of Housing and Local Government found it necessary to issue Circular No. 37/66 together with a pamphlet, "Technical Problems of River Authorities and Sewage Disposal Authorities in Laying-Down and Complying With Limits of Quality for Effluents more Restrictive than those of the Royal Commission." The Ministry considered that to ask for a standard more restrictive than that of the Royal Commission, as a general rule without reference to particular conditions, could lead to diversion of resources from more important tasks. It was suggested that where a river authority required an effluent better than Royal Commission standard they should explain their reasons in detail to the local authority or, in the event of a local investigation, to the Ministry's inspector. But where stricter standards were demonstrably needed the local authority would have to pay the extra cost that this involved.

It is particularly the removal of suspended solids that has to be considered where higher standards are required, for suspended solids can settle in the stream, accumulate and become offensive. Also their removal reduces the B.O.D. value as well as the solids content of the effluent, the B.O.D. reduction being about one-third of the reduction of suspended solids. The Ministry, in the above-mentioned publication, considered that if suspended solids were to be reduced from 30 parts per million to 20 parts per million the B.O.D. could be reduced from 20 parts per million to 17 parts per million, but if the suspended solids were reduced to 10 parts per million this would effect reduction of B.O.D. to 15 parts per million without additional treatment: no further significant reduction would be justifiable on this account even if the suspended solids limit was reduced to 5 parts per million. It was particularly suggested that river authorities should not make a demand for a B.O.D.

limit less than 20 parts per million without first taking into account all the data relating to the river as regards oxygen replenishment etc., which could have bearing on the need for a particularly high standard of sewage treatment.

Some river authorities have called for a minimum concentration of nitrates. On this matter the Ministry considered that such a demand is not appropriate except in the case of certain industrial rivers.

The means of obtaining high standard effluents are discussed in the section on effluent polishing near the end of Chapter XXIX.

Where effluent is discharged into the sea and it may be considered to be diluted by five hundred times its volume of sea or other natural water, the Royal Commission considered that crude sewage could be discharged (see Chapter XXI).

Bacteriological qualities of effluents. All the standards which are now applied to effluents relate solely to their chemical qualities and aim at the prevention of nuisance. None of them affords any security against the chief danger to which sewage may give rise—namely, the transmission of disease.

Sewage is liable to contain the faeces and urine of persons suffering from enteric and paratyphoid fevers, dysentery and diarrhoea, all teeming with the germs by which these diseases are spread. Cholera, too, is conveyed in the same way, but fortunately, during the present century, cholera has been rare in this country. Serious epidemics have resulted from drinking water which has been contaminated by the excreta of cholera or typhoid patients: deaths have been caused by eating shell-fish taken from beds polluted with sewage.

None of the modes of purification in ordinary use can be relied on with absolute certainty to remove or destroy all the germs of disease, and the Royal Commission, in their Interim Report, said that sewage effluents, whether from land or from artificial processes, "usually contain large numbers of organisms, many of which appear to be of intestinal derivation, and some of which are of a kind liable, under certain circumstances at least, to give rise to disease." They were of opinion, therefore, that "such effluents must be regarded as potentially dangerous." "In the present state of knowledge," they said, "and especially of bacteriology, it is difficult to estimate these dangers with any accuracy, and it seems quite possible that they should be either exaggerated or undervalued according to the predisposition of those who have to deal with them." The Commissioners did not see their way to recommend the institution of a bacterial standard of purity.

Sewage effluents and cattle. While even the best-quality sewage effluents, owing to the bacteria which they contain, are dangerous to human beings, cattle appear to be virtually immune to dangers of infection. Cows have been known to drink sewage effluent regularly, and in some instances even crude sewage, without taking any harm. There remains the danger of infection of milk if inadequate precautions are taken.

CHAPTER XXI

DISPOSAL OF SEWAGE BY DILUTION

A TOWN which is situated by the sea or on a great river can often discharge its sewage in a crude state. This mode of disposal is known as "disposal by dilution." Where the discharge takes place under proper conditions the sewage is not merely got rid of, it is also effectively purified.

Clean river-water at a temperature of 60 °F contains about ten parts by weight of dissolved oxygen per million. Salt-water contains rather less. The lower the temperature of the water, the more oxygen it can hold in solution, and vice versa. In the presence of this oxygen the bacteria and other forms of life in the water feed on the impurities of the sewage, and convert them into inoffensive compounds.

Water which contains all the oxygen which it is capable of dissolving is said to be "saturated." The oxidation of polluting matter tends to deoxygenate the water, and if the pollution is gross and the water is stagnant, the latter may be robbed of all its oxygen. The supply of oxygen is, however, constantly being replenished—slowly if the water is still, but more quickly if its surface is broken into ripples or waves, or if it flows over rapids or falls over a weir.

Self-purification of rivers. Many years ago there was a lively controversy as to the self-purifying power of a polluted river. It was asserted, on the one hand, that the organic matter contained in sewage and other polluting liquids was rapidly oxidised during the flow of the river into which it was discharged, and, on the other hand, that there was no river in the United Kingdom long enough to effect the destruction of sewage by oxidation. It is now agreed that the latter opinion is nearer the truth.

Where the conditions are favourable, sewage can undoubtedly be purified by admixture with a sufficient volume of clean, thoroughly oxygenated water, and this is the most effective, the least offensive and by far the cheapest method that can be employed. But these suitable conditions do not occur in Great Britain.

In applying this method certain essential conditions must be observed. There must be sufficient diluting water to supply the requisite amount of oxygen; the sewage and the diluting water must be quickly and thoroughly mixed together, and the currents past the outfall must be strong enough to prevent the deposition of solids. If all these conditions cannot be complied with, the sewage must be subjected to some form of preliminary treatment before its discharge. How far this treatment should be carried depends on the relative volumes of the sewage and the diluting water.

The Royal Commission were of opinion that a "standard filtered effluent" requires 8 volumes of diluting water and an average septic tank effluent 98 parts. With a dilution of more than 500 volumes they considered that all tests might be dispensed with and crude sewage discharged, subject to such conditions as to the provision of screens or detritus tanks as might be called for by the Central Authority which they proposed to set up. It is nearly always necessary to settle or screen out the grosser solids which would otherwise float on the surface of a river or form deposits on its bed.

The question of dilution has also received close consideration from a Committee of American Engineers appointed by the American Public Health Association. They reported that "a stable dilution for municipal sewage may be obtained with from 4 to 6 second feet (240 to 360 cubic feet per minute) of relatively clean water per 1000 people contributing sewage . . . the desirability of removing the coarsest suspended matter being governed by local conditions." With a sewage of 30 gallons per head per day the 500 volumes suggested by our own Royal Commission are equivalent to 1667 cubic feet per minute per thousand people. Their recommendation would appear, therefore, to be framed on conservative lines.

Many cities and towns in Germany and America are situated on the banks of large rivers which are well able to purify their sewage, but the sewage is usually passed through fine screens before its discharge. On the shores of the Great Lakes which divide the United States from Canada there are many important cities which, with a single exception, discharge their sewage into the lakes. In many cases it is merely passed through fine screens; in some it is also sterilised; in few is there any further treatment, and the whole duty of dealing with the dissolved impurities is usually thrown upon the water of the lakes.

Artificial lakes. Sewage may be purified also by admixture with a large body of still water. In a state of rest the heavier suspended matter settles to the bottom and the suspended colloidal matter undergoes a natural flocculation. In the presence of light and the oxygen derived from the atmosphere and from aquatic plants a process of biological self-purification takes place.

A striking demonstration of purification by this means was afforded by the work of the Ruhr Federation. The River Ruhr supplied about a quarter of the water which is consumed in the whole of Germany. It received the sewage from all the towns in the valley and the liquid wastes from the various important industries which are carried on in it. The Ruhr Federation were responsible for keeping the river in such a state that the water from it was clean and fit for use.

The Federation decided to form a series of eight lakes along the lower course of the river in which the natural biological processes should be utilised. By April 1934 three of these lakes had been constructed, the first being at Hengstey, below the confluence of the Ruhr and the Lenne. This lake was nearly three miles long and a quarter of a mile across at the widest part. It contained about 620 million gallons, and at periods of mean flow it afforded a detention period of 34 hours.

The Hengstey lake was kept under close observation. Of the two rivers which flowed into it, the Lenne was the more polluted, and the iron wastes which the Lenne contained had a flocculent action on the domestic sewage and trade wastes in the Ruhr. Sludge was thus precipitated, and settled out in the upper part of the lake. The mechanical purification thus effected amounted to 94·5 per cent. The phenol content of the water was reduced by 54 per cent, the total bacteria by 50 per cent and the *B. coli* by nearly 60 per cent. The fish which were placed in the lake increased in numbers, and the fishing conditions in the river below improved.

Attempts were made, without much success, to dredge the settled sludge from the bottom of the lake. It was later allowed to accumulate for a year, and sent down the river at periods of very high water.

At the outlet from the lake there was a fall of 15 feet, which was utilised in a turbine electrical plant. The current generated, when not required for use elsewhere, was employed to pump water from the lake into a reservoir on the hilltop, some 525 feet higher, from which it ran down to generate current at the hours of greatest demand.

The possibilities of the lake as a pleasure resort and sports ground were fully developed, facilities being provided for bathing, sun-bathing, boating and fishing.

Sea outfalls. We have in this country no large rivers and no lakes comparable with the Great Lakes of North America. The opportunities for disposal by dilution are practically limited, therefore, to towns which are situated on large tidal rivers or on the sea-coast. Even in such cases it is not always wise to discharge the sewage without some form of treatment. In addition to the serious dangers of infecting shell-fish beds, it is manifestly wrong to discharge crude sewage near a bathing-beach; and unless there is a constant current seaward past the outfall, there is always the risk that floating solids will be washed ashore. In most cases, therefore, it is advisable to keep back at least the grosser solids. This may be done by passing the sewage through screens, but it is often difficult to find a suitable place for these, and still more so to remove and dispose of the screenings.

At some coastal towns disintegrators have been installed to break up the larger solids in the sewage. The reason for installing disintegrating plant is not to render the discharge of sewage safe and sanitary, but to disguise it: bathers recognise sewage when they see it floating in the sea, but fail to observe it when it is broken up into fine particles. From the public health aspect, such a method is therefore not one to be recommended.

In a large majority of instances where sewage cannot be discharged in the crude state it suffices for preliminary treatment to be given in the form of screening, detritus settlement and the settlement of organic solids in sedimentation tanks. The tank effluent may then be passed to the sea, where it is diluted.

Full treatment of sewage up to the standards required for discharge into inland waterways is sometimes given even when the discharge is into the sea, but this is very seldom required.

The precautions to be taken when discharging into the sea and the means by which contamination of the foreshore may be prevented have already been discussed in Chapter XV.

CHAPTER XXII

TREATMENT OF SEWAGE ON LAND

The earliest attempts to deal with sewage were by application to land, less, however, with a view to its purification than to the utilisation of its fertilising constituents. For nearly two centuries the sewage from part of the city of Edinburgh was applied to the Craigentinny meadows, producing from four to six luxuriant crops of grass every year on what was formerly barren sands Several large towns have, at some time, had extensive and well-managed sewage farms; but land treatment is now becoming less and less popular.

Land treatment, to be effective, must be carried out intelligently and with interest. The ground should be properly prepared in the first instance and then the sewage distributed over one-third of the area available, while the other two-thirds are being rested, these in turn being brought into use before the quality of the effluent from the irrigated area begins to fall off.

Land treatment should come after sedimentation, for although it is possible to treat on land sewage which has not been settled, a much greater area is then necessary. In the method known as "land filtration" the tank effluent is irrigated into parallel trenches which are dug with level inverts throughout their lengths, so that the sewage does not run down them to the lowest point. Intermediate between these trenches, agricultural tiles are laid a moderate distance below the surface and connected to a main drain that leads to the outfall. The agricultural drains should *not* be laid at right-angles to the irrigation ditches, as is often recommended, for this encourages short-circuiting from the bottom of the trench straight down to the drain through the earth that was disturbed when the drain was laid—an occasional cause of a very bad effluent.

"Broad irrigation" is used in all cases where the ground is not sufficiently porous for land filtration, and sometimes even where land filtration would be possible, because it involves less capital cost. In all circumstances, it is less efficient than land filtration and requires the use of more land. The tank effluent is distributed by a trench which is dug along the contour at the highest part of the land to be used, and provided with hand-stops or with agricultural tiles that can be plugged with clay as a means of distributing the flow on the part of the land to be used at any one time. The ground is ploughed in the direction of the contours so that the tank effluent flows from furrow to furrow, taking a long time to find its way downhill. It is then collected by another ditch, to be redistributed at a lower level and collected again by a third ditch, and this process may be repeated as many times as the site permits.

Area required. It is not possible to state exactly the amount of land required for treatment, for this varies with the nature of the subsoil and the skill of operation. The following recommendations, taken from the Royal Commission Reports and Appendices, are given as general guides.

The Royal Commission stated in their Fifth Report that:

> The total acreage of a farm must be relatively much greater when the sewage is purified by surface irrigation than when the method of filtration is employed, and a larger percentage surplus area is also desirable in the former case. We are not able to lay down any rule as to what the ratio of surplus acreage to total acreage should be but, generally speaking, a large surplus area is advisable.

McGowan, Houston and Kershaw made a special report to the Commissioners on the land treatment of sewage, in which they said:

> To summarise all our results within the limits of a few sentences is impossible, but we may say in conclusion, and speaking in general terms, that we doubt whether even the most suitable kind of soil worked as a filtration farm should be called upon to treat more than 30,000 to 60,000 gallons per acre per 24 hours at a given time (750 to 1500 persons per acre); or more than 10,000 to 20,000 gallons per acre per 24 hours, calculated on the total irrigable area (250 to 500 persons per acre). Further, that soil not well suited for purification purposes, worked as a surface irrigation or as a combined surface irrigation and filtration farm, should not be called upon to treat more than 5,000 to 10,000 gallons per acre per 24 hours at a given time (125 to 250 persons per acre); or more than 1000 to 2000 gallons per acre per 24 hours, calculated on the total irrigable area (25 to 50 persons per acre).

It is doubtful if the very worst kinds of soil are capable of dealing even with this relatively small volume of sewage. The population per acre is calculated on 40 gallons of sewage per head per day. It is here assumed that the sewage is of medium strength, and is mechanically settled before going on to the land.

Table 41, which shows the volume of settled sewage which can be dealt with per acre per day by different classes of land with different methods of working, is based on figures in the Fifth Report of the Royal Commission on Sewage Disposal.

Crops. The best crops for sewage farms are quick-growing plants such as ryegrass, lucerne and mangold-wurzel. Peppermint is also a very useful crop. Wheat and other cereals are generally unsuitable, because the ground has to be kept dry for a long time before the seed is sown and again while the crop is ripening. Potatoes, too, are not to be recommended. Willow trees have done very well in the West Riding of Yorkshire, but have not been so successful elsewhere. Where they are grown, it is difficult to get at the surface of the ground to keep it clean, but this objection may be overcome to some extent by intercepting the solids before the sewage is discharged on to the land. Sewage-grown willows are said to be brittle.

Where crops are grown there is always a tendency to study their interests at the expense of the purification of the sewage, and cropping is accordingly regarded with disfavour except where ample land is available.

Even under the best conditions a sewage-farm is rarely a source of profit, and most farms in this country are carried on at a loss.

TREATMENT OF SEWAGE ON LAND

More attention is given now than formerly to the risks to health that are involved when sewage comes into contact with crops that may be used for human consumption uncooked. In this connection the rules applied in California might be studied by sewage works managers. (See Chapter XXVIII.)

TABLE 41

AREAS OF LAND REQUIRED

Class of soil and subsoil, and method of working	Gallons of settled sewage per acre per day	Area required for sludge disposal per million gallons of sewage per day
Class I. All kinds of good soil and subsoil, e.g. sandy loam overlying gravel and sand:		acres
(a) Filtration with cropping	12,000	5
(b) Filtration with little cropping	25,000	5
(c) Surface irrigation with cropping	7000	5
Class II. Heavy soil overlying clay subsoil:		
Surface irrigation with cropping	5000	12
Class III. Stiff clayey soil overlying dense clay:		
Surface irrigation with cropping	3000	20

Sewage sickness. Land which is over-dosed with sewage is liable to become "sewage sick." Land which has become sick can be restored only by a prolonged rest, but its recovery may be hastened by the growing of suitable crops. Regular periods of rest are advisable even for land in good condition.

CHAPTER XXIII

SITING AND LAY-OUT OF SEWAGE-TREATMENT WORKS

Cost of sewage-treatment works. When making preliminary estimates for sewage-treatment works, one can often work to a figure per head of population, dependent upon the size of the works. On the basis of prices for the year 1959, approximate costs can be estimated by the formula:

$$\text{Cost per head of population} = \frac{\pounds 55}{P^{0 \cdot 16}}$$

where: $P =$ head of population served.

For adjustment for future schemes see Vol. I, Table 3.

Choice of site. The main consideration in the siting of sewage-treatment works is the fall of the land and the consequent direction of the main sewers. The treatment works should be, as far as is practicable, below all properties to be served, in order that the sewage may gravitate to the works without pumping or sewers of inadequate gradient having to be constructed.

The effluent from the sewage works eventually has to pass to natural waters, usually a river or stream. The sewage works are usually placed near this point of outfall, but should be constructed above highest flood-level—that is, the effluent from the works should be able to discharge freely to the river at all times of the year without pumping and without any part of the works becoming flooded.

It is often found that the lowest point on the incoming sewer at the works is not many feet above the highest flood-level. This may influence design considerably. A normal percolating filter scheme, by careful design, can be fitted in between an invert-level of sewer and a flood-level that is as little as 8 feet lower than that invert; but generally it is considered that for a percolating filter scheme it is desirable that the invert of the incoming sewer should be about 12 feet above highest level of the natural water into which treated effluent is discharged.

An activated sludge plant does not use up head to the same extent, but it is often difficult to avoid considerable head losses in large works.

For a percolating filter scheme, land falling at a moderate slope of about 1 in 50 is the most convenient for the setting out of the components of the works one above another in such a manner that the sewage flows from unit to unit without it being necessary for tanks to be stilted up or filters or humus tanks to be dug deep into the ground. A very slightly sloping site is most convenient for an activated sludge plant.

SITING AND LAY-OUT OF SEWAGE-TREATMENT WORKS

The nature of the soil is most important when land treatment is considered. Then a porous soil and a low level of subsoil water (water-table) are desirable. The same conditions are to be preferred at all sewage works, for not infrequently it may be found convenient for small quantities of sewage, sludge liquor, storm water or partially treated effluent to be treated on land, even when other methods are adopted for the treatment of the greater part of the flow.

Consideration should be given to the stability of the subsoil. The sites of sewage-treatment works, being adjacent to streams and rivers, may be on alluvial deposits, which call for special precautions, such as piling of foundations, adding 15 or 20 per cent to the cost of construction.

The site of the works depends very much on the amount of land required and the area available in the preferred location. The areas required for land treatment have already been discussed.

Examination of all twenty sewage-treatment works in Greater London, serving populations from 2,966,856 to 60 persons, showed that an adequate area for works of any type (other than land treatment) could be found by the formula:

$$A = \frac{P^{\frac{2}{3}}}{30}$$

where: A = acres
P = head of population.

(The *average* area occupied by the works was two-thirds of this figure, but included works which had no sludge-drying arrangements.)

The site of sewage works should, as far as practicable, be kept away from building development, particularly housing, but this is not always possible. When sewage works cannot be sited other than near to buildings, the type of works must be selected so as to minimise the risk of nuisance due to flies, odour, foam or the possibility of spreading disease by airborne globules of moisture containing bacteria. Writers on town planning have often stated that sewage works should be located on the leeward side of towns, so as to limit such nuisance: such opinion can be expressed only in the absence of knowledge of the problem. The site of works is determined primarily by the position of the point of outfall as before described, and while a leeward position may perhaps be preferable when it is possible, generally the direction of the wind is not considered, being of far too minor importance.

Regionalisation. As far as is practical and economic, sewerage systems should be arranged so as to gravitate to a few comparatively large treatment works rather than to a large number of small works. In the past there was a tendency to construct large numbers of small works and for sewerage to be the responsibility of each local authority. Now, co-operation between authorities and the design and construction of comprehensive schemes is becoming more usual.

The term "regionalisation" is loosely used to denote either the formation of sewerage boards or the collection of sewage from wide areas to large treatment works. Strictly speaking, however, it should be applied in the administrative sense—i.e. the formation of boards to control the sewerage and sewage disposal of areas the boundaries of which conform to natural drainage boundaries extending beyond local authority boundaries.

An advantage of the formation of sewerage boards is that such large authorities have the finances that make possible the employment of highly skilled technical staff capable of sampling and testing effluents, operating activated sludge works, generally maintaining sewage-treatment works in good working conditions, and thereby ensuring satisfactory effluents. This ideal condition can be contrasted with the multiplicity of small sewage works that may be found all over the country, where each plant is operated by a labourer-manager and the effluent is tested by the catchment authority, if at all.

The size of a regional sewerage scheme depends more on administrative convenience than on technical requirements. While the boundaries of the area should conform with the limits of a catchment, this does not mean that the whole of a large river catchment should come within the jurisdiction of one sewerage board. It should suffice for the catchment of a tributary of moderate dimensions to serve as the sewerage area, no part of which should be an undue travelling distance from headquarters.

It should also be understood that when a regional board is set up this does not mean that the whole of the sewage within the area of administration must necessarily be delivered to one treatment plant at the cost of pumping. The board may administer several works, the sizes and positions of which are determined by sound economic policy.

Typical examples of regional authorities were the London County Council which had, separated by the River Thames, the two catchments of the Beckton and Crossness Outfall Works, the Middlesex County Council which had the two catchments of the Mogden and Deephams Works, the Wandle Valley Main Drainage Authority, the North Surrey Joint Sewage Board and the Richmond Main Sewerage Board. The *London Government Act*, 1963, created a single main-drainage authority, the Greater London Council, which embodied all the authorities mentioned above. It also required the Greater London Council to consider the sewers and sewage-treatment works serving their sewerage area or any part of Greater London (with the exception of that part served by the West Kent Main Sewerage Board) and, if satisfied that such works should be transferred to them, make a declaration of vesting. This made possible the embodiment in one sewerage authority of the main sewers and treatment works of twenty separate catchments with the further possibility that the number of catchments and sewage-treatment works could be reduced to eight or even fewer with both technical and administrative advantages at some time in the future. The exclusion by the Act

SITING AND LAY-OUT OF SEWAGE-TREATMENT WORKS

of that part of Greater London served by the West Kent Main Sewerage Board may seem odd. About 80 per cent of the population served by that Board's works are in Greater London: very considerable expenditure on new treatment works, putting a heavy load on the rates, was envisaged at the time and the local rate payers would probably have been relieved of about 80 per cent of this expenditure had their works become vested in the Greater London Council.

Choice of method of treatment. The method of treatment to be adopted depends on:

1. The system which will involve the minimum of running costs including loan repayment.
2. The extent to which nuisance due to flies or odour will matter.
3. The area of land available and the suitability of the soil or slope of land for each of the methods of treatment.
4. The degree of treatment required.
5. The available fall from incoming sewer to point of outfall.

These together form a common-sense economic problem to which there should be only one correct answer in any particular instance. In fact, however, decisions as to type of works too often depend on the preferences of individual engineers and on what may be considered fashionable at the time.

Land treatment can be least expensive where land is cheap. Both level and falling land can be used and the method can mean very little loss of head through the works. Land treatment also permits the growth of some useful crops. The objections to land treatment are that larger areas of ground are required for this method than any other and that some odour is unavoidable.

The percolating-filter method, which is by far the most popular, requires much less land than land treatment and costs less in capital expenditure in those localities where land is expensive. In fact, a well-designed percolating-filter scheme is often the least expensive of any. Percolating filters produce large quantities of filter flies: there is some odour which, however, permits small works to be located near habitations without being noticeable, provided the maintenance of the other parts of the works is satisfactory.

The activated sludge systems occupy the smallest areas of land and can involve lower capital costs than percolating-filter schemes. No filter flies are produced and the slight tarry odour of the aeration tanks is not unduly offensive. Activated sludge systems involve comparatively little loss of head through the works and, therefore, in some circumstances they obviate pumping, but they do consume considerable power for aeration. All the activated sludge methods, however, require skilful operation: the maximum economies are not achieved except where the plants are large. For these reasons the activated sludge methods are used mostly for treating sewage flows from large populations.

Lay-out of plant. When the site and method have been selected, the designer proceeds to decide the position of the works thereon and the lay-out of the components, fitting each structure to the contours of the land and arranging their positions according to the hydraulic gradients through the works. At the same time he endeavours to economise in the use of land, labour and materials and to arrange a works plan that will give the future operative the shortest journey from point to point during his daily tasks.

This is the ideal, and in good works is well achieved, but unfortunately the lay-out of sewage-treatment works is frequently more open to criticism than many other parts of the design.

What appears to be a far too common procedure in the design of sewage-treatment works is as follows. First, a more than adequate plot of land is selected. The designer then, using the boundary of the land as a frame in which to enclose his lay-out plan, spreads out the components so as to occupy more space than need be, and attempts to economise in excavation only. He orientates the lay-out in accordance with the contours only, and not in accordance with both the contours and the boundaries of the site, and he places the individual components so as to reduce excavation to a minimum.

This procedure tends to a disordered lay-out, which sometimes, during construction, is further disordered by variations decided on the site. Finally, when future extension becomes necessary, the new proposals may in no way resemble those originally contemplated, and space has to be found, wherever it may be available, in between the components of the old works. The result is an untidy and unsightly irregular juxtaposition of filters and tanks of all sizes. Figure 64 shows a sewage works lay-out which is ordered, functionally efficient and aesthetically satisfying.

The amount of space required between and around the components of the works, for purposes of construction and, eventually, operation, is quite small. There should be no need for more space between the boundary of the site and the works (apart, of course, from nearness to habitations, etc.) than would be required by a reasonable constructional easement. The spaces between the components should not need to be more than the walking and working space necessary at the crests and toes of embankments between the earthworks and the tanks, etc., together with the space occupied by the embankments themselves except in the case of a large scheme where wide lanes are necessary to the contractors during construction.

Spreading out works unduly is a very common fault to be found with percolating-filter schemes, but not with activated sludge schemes. It is often excused on the grounds that space is required to make the components easily accessible and to permit the convenient insertion of pipes and channels. But required working spaces are not difficult to determine; they should be well known to the competent designer, and they are usually quite small. Moreover, spacing out the works often introduces problems relating to pipe gradients, etc., that could have been avoided with a more compact lay-out, and adds to length and, therefore, cost of pipework. The skilful designer attempts to make the works as compact as possible, always keeping in mind reasonable economy of cost and ease of construction and of working in the future. Included under the latter head is the avoidance of great stretches of wilderness between the components which cannot be either let out for farming or kept tidy and free from weeds except at a cost.

The correct relative positions of the components should be determined

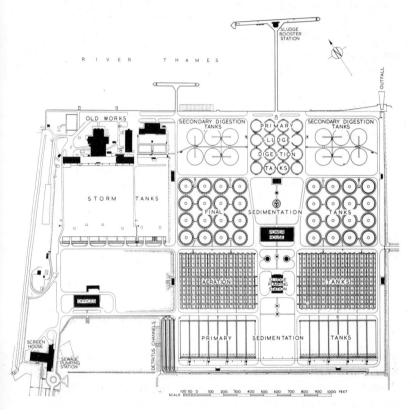

FIG. 64—General plan of Crossness sewage-treatment works

during the design of the lay-out, and should be decided not merely by habit or by copying from previous plans, but by choosing the cheapest and most convenient relative positions as determined by conditions of site.

The excavation plan. The excavations for the works, particularly tank-work, associated manholes and inverted pyramids, can be unduly complex. On the other hand, they may be comparatively simple if they are studied during the course of the design. Take as an example the diagrammatic representation of a settlement tank with two pyramidal bottoms illustrated in Fig. 65. The excavation for this tank, if properly designed, would consist of a simple one-lift "dig" rectangular on plan and with flat bottom and vertical sides, together with two simple pyramids excavated below the general level of the main excavation and of comparatively small cubic capacity. Any manholes adjacent to the tank should, if possible, be dug to the same depth as the main excavation, and would be timbered separately. The cost per cubic yard of the whole of this work would be low.

But it should be observed that the designer in the case illustrated in Fig. 65 has made an oversight, in that he has shown the intersection between the two pyramids at a level above the bottom of the main excavation, meaning that either a small ridge of earth must be left in position (which one could not expect the contractor to do) or else the cavity marked "X" must be filled with concrete. This could have been avoided by designing the tanks so that

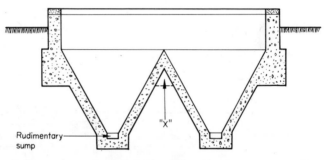

Fig. 65—Faulty design of tank

the intersection of the exterior faces of the pyramids occurred only slightly above the bottom of the main excavation.

This is only one example. There are many instances in the design of sewage-tank work where simplification of the excavation plan assists the work of the contractor and reduces cost. Every attempt should be made to arrange such simplification, provided it is not achieved at the expense of additional excavation or refilling with concrete. Where, however, complex excavations are unavoidable, a separate excavation plan should be included in the set of drawings.

Embankment plan. The formation of embankments has much to do with the general appearance of the completed works and the cost of earthworks. In by far the greater number of sewage-treatment works the excavated earth can be disposed at a minimum cost in the embankments without the necessity of providing spoil-heaps or the expense of carting from the site. Spreading earth other than in the embankments is not good, because it produces "made ground," adding to the cost of future works. The designer should therefore always attempt to obtain a perfect "balance of cut and fill," and proportion his embankments accordingly.

The style in which embankments are treated varies considerably. There are many ways in which slopes may intersect. Some designers prefer to follow closely the outlines of the tanks, others to frame the tanks in a simple rectangle of neatly trimmed earth: some like sharp hipped angles, others circularly swept angles. All methods are legitimate; some are better than others from the aesthetic point of view. But on no account should the designer be content to show outlines that are vague, merely because he is not sure how they will appear on plan or what are the true representations of the intersections. Earthworks will not appear well formed unless properly executed by the contractor, and the contractor cannot be blamed for a bad job if he has to work to an incorrect drawing. The proper intersections of complex embankments are very interesting, well deserving of study, and are very pleasing when well executed.

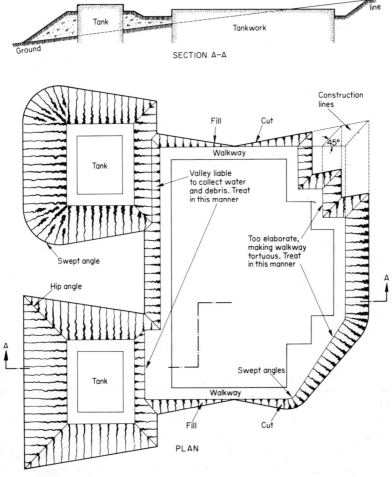

FIG. 66—Diagram illustrating the development of embankments in various circumstances

There is one practical point to be kept in mind when forming embankments, and that is the avoidance of valleys where surface water or debris may collect (see Fig. 66).

Flexibility of operation. As a general rule all arrangements of pipework and valves should be designed to facilitate the utmost flexibility of operation. All operators do not think alike regarding how the works should be used and,

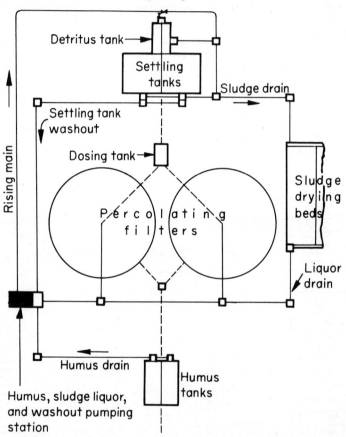

Fig. 67—Example of a good lay-out for small sewage-treatment works but with uneconomical arrangement of pipework

almost invariably, different methods have to be tried experimentally before the best results are obtained. Moreover there is always the possibility of an emergency that might require the discharge of a fluid other than to that part of the plant normally intended to deal with it.

The objection to be raised against design to allow for alternative operation is that this facilitates misuse of the plant. But it is a matter of experience that if the staff intend to use the plant in a manner that was not intended or is likely to cause damage they will do so no matter what obstacles are put in their way and regardless of any warnings given on printed instructions or notice boards. The designer has to hope that they will learn by experience before too much harm has been done.

Thus broadly one should design to permit various means of operation but give warning of the dangers that certain methods of operation may involve.

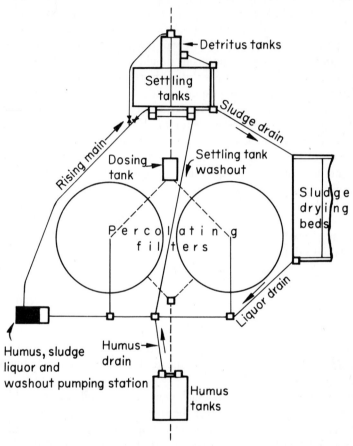

FIG. 68—Example of a well-conceived small sewage-works lay-out with economical arrangement of pipework

There are, however, certain arrangements that should never be made. These include such things as secret outlets for crude sewage to watercourses from sewers, pumping stations and sewage-treatment works.

The following are some of the permissible alternatives of flow. If there are sludge and wash-out pumping stations the rising mains should be so interconnected that either sludge or other liquids may be discharged to any process where they will not be harmful. For example, it should be possible to discharge crude sludge to primary digestion or, by-passing primary digestion, to secondary digestion or, by-passing digestion altogether, to sludge-drying beds. Similarly, humus, after discharge to the general wash-out pumping station, should be capable of delivery to the incoming flow of sewage so as to settle out in the sedimentation tanks, or, alternatively, of being passed to digestion tanks or sludge-drying beds. The sludge pumping station should be able to serve not only for the purpose of pumping crude sludge to the digestion tanks, but also as circulating pump to stir up the digestion tanks as may be found desirable, or as a stand-by to relieve the general wash-out pumping station in case of breakdown. In view of the possibility of pumps having to serve different purposes, rising mains should be short because, on the commencement of pumping, there is always the contents of the rising main to be discharged, and this may be a liquid not suitable for discharge to the section of the works in question, if in large quantity. If the rising main must be long, arrangement should be made for it to be emptied back to either suction well, and the wells should be made of ample capacity accordingly.

Lay-out of feed-pipes and drains. All drains should be laid to suitable gradients according to the liquid that will pass through them. Those of the wash-out drains liable to discharge crude sewage should be laid to gradients that will give a velocity of $2\frac{1}{2}$ feet per second; those that will only have to discharge liquids comparatively free from solids may be laid to slacker gradients, but preferably to gradients sufficient to give velocities of not less than 2 feet per second. In English practice it is usual not to lay any sludge pipes to a gradient of less than 1 in 100 (see section on flow of sludge in pipes in Chapter XXVIII).

The general lay-out of feeds and drains is a simple matter, and yet drawings are frequently prepared which show most extravagant arrangement of both rising mains and drains. A common fault is an attempt on the part of the engineer to make the drainage plan symmetrical or too orderly, at the expense of additional length of pipe or the sterilisation of land that might otherwise be useful in the future.

There is little justification for such practices, for while an ordered lay-out of tanks and filters is desirable for the sake of economising land, improving the appearance of the works and facilitating future extension, exaggerated tendencies towards rectilinear pipe lay-out work in the opposite direction. Take, for example, the two lay-outs shown in Figs. 67 and 68. Both are alike in lay-out of plant, but in Fig. 67, the designer has been at pains to prepare an attractive pattern on paper and, in so doing, has greatly increased the total length of drains and mains required, increased the number of manholes, reduced the available falls or added to the depth of the suction-wells at the pumping station. Figure 68 may be much less attractive on paper, but in this case the designer has economised in his lengths of pipe, his number of manholes and his falls so as to achieve an efficient scheme, and he has not actually detracted from the appearance of his works, which on the ground will be negligibly different from the scheme shown in Fig. 67.

CHAPTER XXIV
PRELIMINARY TREATMENT: SCREENING AND GRIT AND GREASE REMOVAL

The whole of the sewage arriving at the sewage works, including storm water, whether the sewerage scheme is separate, partially separate or combined, should pass through the screens, if they are provided, and through the detritus tanks or channels.

Practice varies as to the position of screens relative to the detritus tanks: sometimes the screens come first, sometimes the tanks first, but this should be decided according to the type of detritus-removal plant.

Screens. Screens are provided at sewage works mainly and, in this country, only for the purpose of protecting pumps and small pipe-lines from damage or chokage by the larger solids that can be delivered to the works. For although theoretically nothing should enter the sewers that is larger than can pass through a 4-inch drain, it is found in practice that occasionally very large solids are illicitly introduced by the lifting of manhole covers. Also, timber used during reconstruction or temporary damming of sewers may find its way to the works. Overseas, fine screens have been used as a substitute for primary sedimentation. They do not, however, effect solids removal so completely or simply as sedimentation tanks, and for this reason their use is limited. In Great Britain very few fine-screen installations exist.

Types of screen. Screens are classified as:

1. Coarse bar screens.
2. Medium bar screens.
3. Fine screens.

Coarse bar screens are screens consisting of vertical, inclined or horizontal bars having clear openings of more than $1\frac{1}{2}$ inches and in some instances as much as 6 inches. These screens, although they collect enough rag, paper and other fibrous matter to warrant frequent cleansing and mechanical cleansing at large works, do not give sufficient protection to the works against smaller suspended particles which can choke pumps, be scum-forming and form objectionable accumulations on tank weirs. While it is admitted that such coarse screens are largely used, this policy is questionable, for the use of medium bar screens, which are far more effective, does not add to complexity of design or greatly add to the cost of a mechanical installation and maintenance.

Medium bar screens have clear openings of $\frac{5}{16}$ to $1\frac{1}{2}$ inches. The disadvantage of too narrow an opening is that mechanised screens are difficult to design to work satisfactorily if the spaces between the bars are too small, for the tines of the raking mechanism, being of necessity narrow, are not easily made adequately strong. For this reason spaces of less than $\frac{3}{8}$ inch are not common. On the other hand, if the openings are too wide, protection of pumps becomes inadequate. Experience has shown that, as a general rule, clear spaces should not exceed $\frac{3}{4}$ inch or be less than $\frac{3}{8}$ inch. (It might be mentioned that while one well-known manufacturer considers $\frac{3}{4}$-inch spaces normal and provides screens of this type whenever the clear spaces are not specified, another considers $\frac{1}{2}$ inch spaces normal for all except wide screens.) Bars are usually $\frac{1}{2}$-inch thick on the upstream side and taper slightly to the downstream side. These bars are specially rolled and, when inviting tenders, the purchaser should make it clear that no tender will be accepted unless the manufacturer can guarantee that he has, or will be able to obtain, taper bars and not ordinary flat-section bars.

Fine screens include rotating horizontal discs, rotating vertical discs and rotating drums constructed of perforated or slotted plates having clear openings of $\frac{1}{4}$ inch down to $\frac{1}{32}$ inch. The screens are rotated continuously and are cleansed by brushes or reverse sparges. These screens can make an appreciable reduction to the solids content of the sewage.

In addition to the foregoing are Comminutors, which combine the operation of screening and disintegration of screenings. These will be dealt with separately.

In English practice crude sewage is screened almost invariably by:

1. Hand-raked bar screens.
2. Mechanically raked bar screens of various kinds.
3. Comminutors.

Accordingly, only these types will be discussed here.

Hand-raked screens. Hand-raked screens are applicable only to small sewage works where the quantity of screenings can be removed manually without difficulty, and also to those works where the incoming sewer is not very deep below ground-level. It is usually assumed that such screens are raked not more frequently than three times a day, and on this basis it is recommended that the submerged area of the screen should not be less than $1\frac{1}{2}$ superficial feet per 1000 head of population. Such screens are often arranged in small hopper-bottomed detritus tanks and set at an angle of about 45° to the horizontal so that screenings can be raked up on to a platform from which they can be shovelled to barrows for disposal.

Although the following type of screen does not require much hand-raking and does involve machinery, it is perhaps most correctly classed as hand-raked, being similar to the truly hand-raked type in that it is applicable to comparatively small works. In this arrangement bar screens are set horizontally and the flow is brought from below upwards to pass through the screen then over a weir and thence on to the works. Periodically a disintegrator is set in action, drawing from the upstream side of the screen and discharging on to the screen on its downstream side, so as to cleanse the screen of screenings by back-washing. The disintegrator breaks up the particles to small size, rendering them unobjectionable from the point of view of treatment. The screen area should be more or less similar to that allowed for hand-raked screens, unless disintegrators are made to work at frequent intervals by clock or differential-float control.

Mechanically raked screens. Mechanically raked bar screens are of several types, the type used depending on local conditions.

When sewers enter the treatment works at a considerable depth below ground-level it becomes almost essential that the bars should be nearly vertical in order that large chambers do not have to be excavated to great depth. The raking gear then has to be one of the types that drags the screenings vertically to the surface. Of the methods involved are the arrangement of a number of rakes carried on endless chains;* and a single rake or grab, which travels vertically downwards, moves into the screens and then goes vertically upwards, taking the screenings with it. When screens of this type are installed on deep sewers, it often means that they are followed by sewage-pumping plant that lifts the screened sewage to the treatment works and the storm flow to the storm tanks. The screens thus serve to protect both the works and the pumps. If the incoming sewer arrives at the works not too deep below ground-level, vertical screens may still be used; but it is also possible to use various types of inclined screens, including the Dorr-Oliver pattern (see Fig. 69). Comminutors are mainly applicable when incoming sewers are not deep below ground-level and are not applicable when water-level of incoming sewer varies by more than about 30 inches.

Mechanically raked screens are operated:

1. *By manual starting and stopping.* This is applicable to comparatively small works and only those where there is no risk of the screens becoming choked with screenings in a short time and holding up the flow. Accordingly, where screens are hand-started they should be of fairly large area so as to allow some accumulation of the screenings before they choke.

2. *By clock control plus auxiliary hand operation.* In this arrangement the screens are made to operate for a short period at regular intervals—e.g. a few minutes once or twice an hour as may be found desirable. The period between starts should be adjustable to allow for results of experience. This arrangement must mean that the raking gear sometimes runs when the screens have not become sufficiently foul to justify raking. Alternatively, there may be occasions when the screens become foul before the predetermined time has elapsed. When raking gear is operated by clock, a time-lag should be arranged so that before the machinery stops the entire operation of raking the screens clear, lifting the screenings to the surface and transporting them to disintegrator or other means of disposal is completed.

* Often essential before automatic pumping plant because effect of starting mechanism is almost immediate. Speed of 6 inches per second is usual.

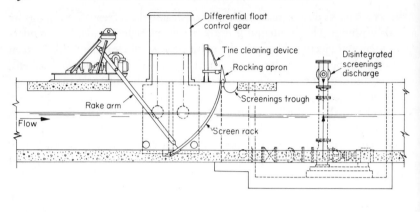

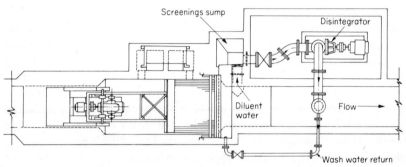

Fig. 69—Dorr type "T" screen and disintegrator
By courtesy of Dorr-Oliver Co., Ltd.

3. *By differential float control.* This is perhaps the best arrangement for general application, and essential in all cases where hold-up of flow could cause an emergency—e.g. at an automatic pumping station. It is not only the safest but also the most economical in that the raking mechanism is brought into operation automatically as soon as the screens have become partly foul, but the rakes do not operate when the screens are clean. Two floats are arranged indicating sewage levels above and below the screens respectively and connected to a single drum which is suspended in such a manner that if the two floats do not rise by equal amounts, it moves laterally, eventually operating a starting switch (see Fig. 69). Usually such gear is arranged to start the raking mechanism when the differential head has reached 6 inches, but there are occasions when closer adjustment is desirable. Such automatically operated screens are, of course, provided with auxiliary manual starting gear.

Proportions of mechanical screens. While it must be admitted that mechanical screens have often been installed on the basis of a crude rule of thumb as to the submerged area required, there are occasions when fairly close calculations are desirable. Unlike the chambers for hand-raked screens at small works, mechanical screen chambers cannot be permitted to settle out detritus and, with it, offensive organic material. Nevertheless, it requires care to design such chambers so as to avoid settlement, and, in fact, many screen chambers are unduly foul with detritus and sludge.

To overcome this objection, the designer has so to proportion the screen chamber both upstream and downstream of the screen that the velocity of flow at some time during the day is upwards of $2\frac{1}{2}$ feet a second. This is rendered difficult by the fact that the width of the chamber near the screen must be at least that of the screen itself, and with some designs of machine is 1 foot more. Thus, if the velocity in the chamber is $2\frac{1}{2}$ feet a second and the chamber is of the same depth as the screen, the velocity through the bars of the screen must be considerably higher. Furthermore, if the screens have to deal with storm water and the number that can be brought into operation is not proportionally greater than the number applicable to dry-weather flow, the condition of a self-cleansing velocity during dry weather may call for excessive velocity through the screens at storm time.

It should be mentioned here that the author does not recommend certain rules for screen size suggested by manufacturers and has used, with satisfaction, much higher velocities through bars than manufacturers would suggest.

A further complication arises if the screens are followed by a pumping station which relies on the storage capacity of the sewer itself for preventing too frequent starting and stopping of automatically operated electric pumps. In this case it is of the utmost importance that the screens shall not so interfere with the flow from the sewer to the pumps as to influence the sequence of operation of starting and stopping the pumps. This virtually necessitates that the loss of head through the screens at all times must be less than the distance between cut-in levels of the various pumps in the station and that, with the aid of differential float-gear, the screens are automatically brought into operation before they have become choked sufficiently for the loss of head to reach the maximum permissible figure.

Thus, when deciding the submerged area of screen, it is necessary to bear in mind what loss of head can be permitted through the screens at any time; how great an area of free space between screens is required to pass the various possible rates of flow of sewage to the works without exceeding this loss of head; how often must the raking gear operate for the screens to be kept sufficiently clear of screenings; and how can the chamber be proportioned so that, without militating against other requirements, the velocity of flow reaches $2\frac{1}{2}$ feet a second at least once a day.

A natural reply to this problem is that a number of screens can be provided and each in turn brought into operation as rate of flow increases. This is, of course, possible but not altogether desirable; for to lay off a screen it is necessary to close the large penstock upstream and downstream of its individual chamber, which means, of course, automatically operated electric penstocks. This is perhaps a small added complication but another thing to

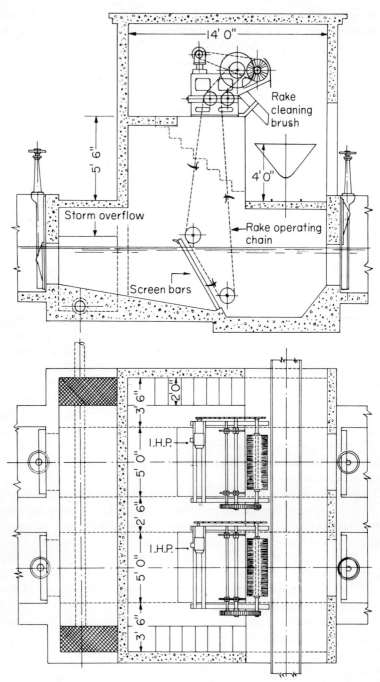

Fig. 70—Mechanical screens in chamber
By courtesy of Hartleys (Stoke-on-Trent), Ltd.

go wrong, and it could involve a serious danger should electric power fail at a time when suddenly increased rate of flow necessitated the opening of penstocks. Thus, as far as practicable, screening chambers should be designed for all, or most, screens to be available and in use at the same time and submerged area to increase with increase of depth of flow.

Screenings. The quantity of screenings also enters the calculation, for this determines how often the screens will have to be raked. On this matter not much general guidance can be given, for the quantity of screenings collected varies considerably from one locality to another. The quantity depends on:

1. The clear spaces between the bars.
2. To some extent on the size of the sewerage system, for sewage that travels long distances has most of its organic content broken up by the time that it reaches the treatment works, in which case the screenings largely consist of cloth and similar fibrous matter, rubber and plastics which are not broken up during transit and some of which may even tend to aggregate together on the way. At small works faeces predominate.
3. Special local conditions—e.g. types of trade waste.

The quantity of screenings is to some extent related to head of population served, but unfortunately most figures given by writers on the subject give quantity of screenings in proportion to million gallons of sewage treated without stating the gallons of sewage per head of population. And as to variation—Babitt states that the amounts removed vary between $\frac{1}{2}$ and 50 cubic feet per million gallons of sewage treated, depending on the character of the sewage and the sizes of the screen openings. Perhaps a fair average for design purposes could be taken as in Table 42.

TABLE 42

QUANTITIES OF SCREENINGS

Clear spaces between screens (inches)	Cubic feet per day per 1000 head of population
$\frac{1}{2}$	0·8
$\frac{3}{4}$	0·5
1	0·3

Fair and Geyer[79] give, for $\frac{1}{2}$-inch and 1-inch screens, 0·547 and 0·274 cubic feet per 1000 head of population per day, and peak collections as five times the average.

Screenings disposal. Screenings have a moisture content of about 85 per cent and weigh about 50 lb per cubic foot: about 85 per cent of the dry matter is volatile, and this material is largely putrescible and offensive. Accordingly, screenings should be removed and disposed without delay.

At small works where screenings are removed manually, they can be dug into the ground or composted to form manure. But this method of disposal is highly undesirable at large works, as without the greatest care and supervision nuisance and danger to health may be involved. Even where screenings are carefully buried there is the possibility of rat infestation.

In some instances screenings have been disposed of at near-by destructor units. Also, in America it is not uncommon to incinerate screenings at the treatment works. These methods, however, are not popular in England for use at large works because the amount of heat required to incinerate screenings can involve excessive expenditure on fuel. Moreover, unless the screenings are incinerated at high temperature, foul odours are produced.

It is for these reasons, and also because screenings are objectionable material to handle, that it has now become general practice at large works in Great Britain to disintegrate screenings with the aid of the Hathorne Davey (Sulzer) disintegrator or, in suitable circumstances, to use Jones & Attwood Comminutors. In all cases the solids in the sewage are broken up into fine particles and returned to the flow of sewage. In the case of disintegrators this return should be upstream of the screens.

It has been suggested in some quarters that organics, once removed from the sewage, should not be returned. In the case of screenings, however, the quantity is generally so small as not to have any significant effect on the treatment plant, and it is, in fact, doubtful if the total removal of screenings, instead of their return, would make any recordable difference in the B.O.D. of the final effluent. On the other hand, dealing with screenings, whether disintegrated or not, by any other means than by returning to the flow of sewage involves many problems. In the first place, disintegration came into use because previous methods of disposal of screenings were not satisfactory, and disintegration has remained in use for as long as most public health engineers can remember, despite such drawbacks as it has.

If, *after disintegration*, the screenings are not returned to the flow of sewage, a difficult problem of their disposal arises, for in the process of disintegration they are diluted by large quantities of water which would make it impracticable to treat them with the sludge. It has been suggested that this water could be removed by settlement and the liquor returned to the flow of sewage, but the effect of this on the treatment of the sewage would be exactly the same as the return of the screenings to the flow of sewage for separation in the sedimentation tanks.

Design of bar screens. Calculations relating to mechanically raked screens are not always simple, and there is no general rule applicable to all cases. The majority of installations have been put in according to rules of thumb which are various but generally make certain that screen area is sufficient to pass the maximum flow at moderate loss of head, so that differential float gear may be installed. These rules of thumb are of limited application and, if applied to exceptional cases, are liable to lead to unsatisfactory results.

SCREENING AND GRIT AND GREASE REMOVAL

To take an example—one arbitrary rule for sizing screens is to allow a velocity of $2\frac{1}{2}$ feet a second between the bars at the maximum rate of flow: another (a manufacturer's rule which ensures low pressure on the screens) is to allow $1\frac{1}{2}$ feet a second in the same conditions. It is obvious that either of these rules must mean a velocity considerably below the self-cleansing velocity of $2\frac{1}{2}$ feet a second upstream of the screens if the width of the chamber upstream is equal to the total width of the screens, particularly having regard to the fact that some makes of screens have side frames occupying an additional foot of width more than the width occupied by the screen bars. In designing the screens for the Crossness works, the author allowed a velocity of 4 feet per second through the screens.

As an alternative to the foregoing rules, the designer could commence with the assumption of a self-cleansing velocity upstream of the screens then find the loss of head across the screens necessary to pass the flow. This approach could be quite satisfactory in some circumstances, but it could possibly lead to a very varied difference of head across the screens at different rates of flow, complicating the installation of differential float-gear. And it might result in a heavy water pressure on the screens, calling for strong construction.

There are other matters which also have to be kept in mind. For example, the head loss through the screens must not cause backing-up of the incoming sewer to such an extent as to result in silting: also allowance must be made that, at all rates of flow, the head downstream of the screens is satisfactory, having regard to the works that follow.

Calculating screen size. The spaces through a bar screen (or the total number of bar screens to be used at any time, allowing one or more screens as stand-by) must be sufficient to pass the maximum rate of flow under the maximum head permissible and assuming that the surface of the screen is partly fouled by screenings. But also, the arrangement should be such that the velocity through the *screen chamber* upstream of the screens is, at peak daily rate of flow, sufficient to prevent settlement, say, $2\frac{1}{2}$ feet per second.

The loss of head across the screen can be calculated by the orifice formula:

$$Q = mA\sqrt{(2gH)}$$

where: Q = cubic metres per second
H = head in metres from water-level upstream to water-level downstream
A = submerged area of openings in screen calculated by multiplying the sum of the width of the openings in metres by the vertical depth less one-third of the head in metres (assuming that the upstream water-level is not above the top of the screen)
g = acceleration under gravity, 9·80665 metres per second per second
m = a constant for the type of orifice which in the case of clean bar screens is about 0·6.

This is usually maintained at as moderate a figure as possible, say, a few inches, but Fair and Geyer give 2 feet 6 inches as the maximum differential head across clogged screens and the author has seen a mechanically raked screen working quite satisfactorily with a differential head of 3 feet 6 inches. Whenever the head across the screen is liable to be more than a few inches *the fact must be stated* in the specification in order that the manufacturer shall make the screen strong enough to withstand the water pressure.

By the use of the orifice formula the maximum permissible water-level downstream of the screens can be found for any water-level upstream and for any rate of flow, or alternatively, where head losses are limited, area of spaces between bars determined to suit. Of course, the actual water-level downstream may be less than the maximum permissible if the hydraulic conditions downstream permit this, and it may be almost constant if weir-controlled. The purpose of the calculation is to relate upstream and downstream structures and water-levels mathematically for the best working conditions, and to determine number and size of screens.

In making this calculation, it must also be borne in mind that the lower the level of the bottom of the screens relative to invert of incoming sewer, the greater will be the area of screen submerged. In large installations it is usually desirable for the bottom of the screen to be only slightly lower than the invert of the incoming sewer or the invert of the channel downstream, as an almost level invert through the screen chamber simplifies self-cleansing design. In smaller installations, however, the bottom of the screen may be well below the invert of the sewer and outgoing channel, thereby obtaining a larger submergence of screens in proportion to width. But when this arrangement is adopted, the screen chamber must be hopper-bottomed and provided with means of sludging and washing out, otherwise it will become foul.

The calculation of loss of head through the screens may be a single calculation in the simple case where water-level does not vary, or may have to be repeated for several different rates of flow between maximum and minimum rates and, by trial and error, a suitable size of screen found that will give satisfactory flow under a reasonable head in all circumstances.

Having arrived at a total width of spaces between bars and a screen invert which satisfies the requirements of the case, it is necessary next to see if the chamber upstream of the screen can be so arranged that the velocity in it is not less than $2\frac{1}{2}$ feet a second at the peak daily rate of flow. At the same time the velocity in it can be tabulated for all other rates of flow. After this is done, it may be found advisable to adjust the width and depth of the screen, making the calculations all over again.

Number of screens in normal operation. A matter on which confusion has been known to occur is whether the number of screens in operation should be varied according to the rate of flow at the time, or whether all screens should be normally in operation without regard to rate of flow. There is a rule which decides this matter *but which depends on circumstances*, the alternative conditions being:

1. that water-level downstream of screens remains virtually constant at all times;
2. that water-level increases more or less as flow increases.

Dealing with the former condition—this applies when the outflow from the screen chamber is controlled by a weir such as a local weir belonging to the chamber itself, a storm overflow weir, or the weir of detritus or settlement tanks downstream. If the downstream water-level is constant, the submergence of the screens remains virtually constant at all rates of flow. Therefore, if a reasonably constant velocity and constant loss of head through the screens are to be maintained, the area of screens must be increased as flow is increased. This means that more screens must be brought into operation on increase of flow, which may be done by automatic or manual control of penstocks, as circumstances justify. The arrangement makes possible reasonably constant velocities through the screen chambers, which, if properly designed, should be self-cleansing.

The second condition applies in circumstances such as the following: where screens discharge direct to the suction-well of a pumping station, the downstream level is determined by the cut-in and cut-out levels of the pumps, and consequently submergence of screens varies more or less in direct proportion to rate of flow. Thus, constant velocity and constant loss of head through screens can be automatically achieved and, accordingly, any variation of number of screens in operation is undesirable because it upsets the maintenance of desired conditions. In this circumstance it is usual to design screens to work under ideal conditions as to velocity and loss of head when one screen, considered as a stand-by, is laid off and the remainder are in operation.

An intermediate condition occurs when screens discharge direct to constant velocity detritus channels controlled by rectangular standing-wave flume. Then, if invert of screen is level with invert of flume, the rate of flow varies as $H^{1.5}$, where H is the depth through the standing-wave flume.

Disintegration of screenings. After screenings have been carried to the surface by the raking mechanism, they are brushed or washed by sparge off the tines either on to a travelling belt or, preferably, into a flow of water—sewage-effluent or screened sewage—to be discharged into a *small* well or sump (not more than 4 feet wide) from which the disintegrators draw. An ample flow of water is required, generally sufficient to give a dilution of 100 volumes of water for every one volume of screenings in the case of Hathorne Davey (Sulzer) disintegrators.* Thus, if the quantity of screenings per day is

---

* The word "macerator" is often used to denote disintegrator. To macerate can mean to soften by soaking or to digest and as the term has been applied to digestion tanks it is best avoided in other senses.

determined according to the table given earlier and the maximum discharged to the works taken as being in the region of nine times this figure or such other quantity as has been found by records, the maximum rate of flow of wash-water can be calculated and then a sufficient capacity of disintegrator installed to take this rate of flow.

Discharge rates for disintegrators are given by the manufacturers. For example, a 12-inch Hathorne Davey disintegrator discharges 1100 gallons a minute against a head of 2 feet, or 900 gallons a minute against its maximum working head of 8 feet. It is usual to return part of the discharge from the disintegrators to the flow of wash-water and bleed off from the circuit the surplus through a valve or controlled orifice and discharge it upstream of the screens (see Fig. 69). If disintegrated screenings are not discharged upstream of screens, they must be discharged into a rapid flow of water, otherwise their hair content causes them to felt together into mats.

To obtain satisfactory working of disintegrators, it is desirable that screenings should be discharged to them almost immediately after they are brought to the surface: they must not be permitted to accumulate and settle out in the sump from which the disintegrators draw, or such accumulation is liable to lead to chokage. It is a good arrangement to have the screens, disintegrators and the pumps or other means by which wash-water is supplied connected to the same float control, so that when a differential head of, say, 6 inches has built up across the screens the raking mechanism starts to operate, the wash-water flow commences and the disintegrators start to rotate simultaneously. A time-switch should be installed in order that the complete operation of bringing screenings to the surface, delivering to the disintegrators and discharging the disintegrated screenings to the sewer takes place before raking mechanism, flow of wash-water and disintegrators are switched off by the differential float gear.

Comminutors. Comminutors (Jones and Attwood) are installed in lieu of screens and disintegrators, for they combine the dual purposes of intercepting and breaking up the screenings. They are applicable to all circumstances where the head and rate of flow do not vary excessively, and contrary to some opinion they can be used economically for large rates of flow. The Comminutor consists of the three main castings of column, housing and base, and a vertically rotating drum of high-grade cast-iron provided with slots $\frac{1}{4}$ inch in width in the smaller sizes and $\frac{3}{8}$-inch width in the larger sizes. The Comminutor is housed in a concrete chamber into which the sewage is brought. The sewage then flows through the slots into the drum and passes downwards and out of the lower end of the drum to be carried by pipe or culvert to the works downstream (see Fig. 71). The screenings which are intercepted by the slots are carried by the slow rotation of the drum until they are brought up against a steel comb, against which they are cut up by teeth arranged on the trailing edges of the drum slots.

The size of Comminutor can be determined from charts which the manufacturers have prepared from practical tests. Broadly, it depends on the flow through the orifices under the permissible loss of head. Loss of head is necessary, not only to pass the flow through the orifices but also to hold the screenings in position while the cutting operation is taking place. The minimum head loss varies from 2 inches for the smallest machine to 4 inches for the largest. The maximum permissible head loss is 7 inches for the smallest size and 15 inches for the largest size.

The proper locating of the Comminutor relative to level of sewage is important. The level should be such that the head loss at maximum rate of

FIG. 71—"Comminutor"
By courtesy of Jones and Attwood, Ltd.

flow is not excessive, and at $1\frac{1}{2}$ times average daily rate of flow is sufficient to hold the particles in position during the cutting operation. A standing-wave flume can be installed downstream so as to adjust the depth of flow to suit the characteristics of the Comminutor.

From the foregoing figures it will be seen that when depth of submergence of Comminutor is constant, the maximum rate of flow through the Comminutor must not exceed three times the average dry-weather flow. If a slightly greater maximum rate is required, more Comminutor capacity must be used and standing-wave flume control applied. If the maximum rate of flow is greatly in excess of the average dry-weather flow, arrangements must be made for bringing more Comminutors into operation as the rate of flow increases or, alternatively, for using Comminutors to deal with the peak rate of flow of sewage to be treated and bar or other screens for dealing with storm flow.

Whereas bar screens are usually arranged upstream of detritus channels, Comminutors are better arranged downstream, because the wear on the cutting parts is roughly proportional to the amount of grit. However, these parts are faced with Stellite and the teeth tipped with tungsten carbide, so that wear is reduced to a minimum: accordingly, the Comminutors may be arranged upstream of the detritus channels if this is considered desirable from the point of view of the functioning of the latter. There is no need to protect Comminutors with coarse or other bar screens, as has sometimes been done: in fact such an arrangement virtually does away with the advantage of using Comminutors.

The designer who has not had a wide experience of Comminutors is well advised to discuss his proposals in detail with the manufacturers.

Detritus tanks. Detritus tanks are not essential at all sewage works. Their purpose is to remove heavy grit and large particles, which, if settled in the sedimentation tanks, might render sludging difficult: efficient detritus removal is particularly desirable where sludge digestion tanks are installed. It is usual for municipal sewage works to be provided with detritus tanks or channels of some form, but they may be dispensed with to advantage at small private works treating domestic sewage only, where the only heavy grit to be expected is that produced in small quantity by vegetable washing. Detritus tanks or channels are necessary at works treating sewage from combined systems because of the road-grit discharged thereto.

At one time it was regular practice to install no fewer than two tanks, the total capacity of which was not less than one-fiftieth dry-weather flow. As a result, most English grit-chambers were too large, and a great deal of organic matter settled-out with the grit and, being infrequently removed, became septic and highly offensive.

The simple form of detritus tank which has been used for many years, and which is still installed at small works, is a small tank usually serving as screen chamber also, having a sloping bottom with channel or sump and outlet controlled by valve or penstock discharging to a manhole. The method of operating such a tank is to divert the flow from the tank to be sludged, decant the top-water by means of a decanting valve (merely a penstock in the wall of the tank) set above the level of the sloping floor. When the greater part of the sewage has been decanted, the remaining contents of the tank are run out to the sludge manhole and swept, if necessary, or dug out should the outlet become choked.

Detritus, being heavy material, does not flow easily, and is inclined to pack round the outlets of hopper-bottomed tanks. For this reason it is advantageous if the outlet can be arranged to discharge directly with little or no pipe-line. Detritus tanks can seldom be sludged under "hydrostatic head" in the same manner as are pyramidal-bottomed sedimentation tanks.

It used to be the practice at large works to install long detritus tanks which were cleansed by mechanical dredger or moving grit-pump. This practice has given place to special mechanised tanks, such as the Dorr "Detritor" or to constant-velocity detritus channels.

The Dorr "Detritor," manufactured by Dorr–Oliver Co., Ltd., is designed to remove detritus and to cleanse it of organic content so that the detritus may be used for various purposes or dumped without any risk of nuisance. The installation consists of two separate parts, the collecting tank with its sweeping mechanism and the grit-washer. The collecting tank is circular in shape and is arranged for cross-flow, having adjustable deflectors on the inlet side and a submerged weir on the outlet side. The grit settles on the flat floor of the tank and is swept by a rotating mechanism to a pit at the periphery, whence it is discharged to the grit-washer.

The detritus-settling tank of the Dorr "Detritor" is designed on Hazen's principle of settlement (see Chapter XXVI), the surface area being proportioned according to the rate of flow as described under the head "Principles of design" in the present chapter. Often the installations are designed to settle the siliceous particle that falls 1 foot in 16 seconds (see Table 43). For this purpose the tanks have superficial areas of 1 square foot for every 33,750 Imperial gallons per day maximum rate of flow. Cross velocity should not exceed 1 foot per second at any rate of flow.

The grit-washing mechanism has an inclined "reciprocating ladder" arranged in a ramp and moving in such a manner as to work the grit up the ramp against a downward flow of wash-water. Eventually the comparatively clean grit is discharged from the end of the ramp into a tip-wagon (see Fig. 72).

Aerated detritus channels. Reasonably clean grit can be collected in detritus channels in which a controlled degree of turbulence is produced by aeration. A perforated pipe is passed along one side of a spiral-flow aerating channel near the bottom but in a position where it will not foul any detritus-removal mechanism. The spiral flow can be adjusted so as to give just the right degree of cross-velocity, to permit settlement of detritus but not of organic solids, by regulating the flow of air.

The channels should have a surface area similar to that of a Dorr "Detritor" as described above: otherwise the proportions of the channel are not important. Various means of detritus removal are possible, a travelling dredger (similar to that in Fig. 75) probably being best for obtaining clean grit. Arrangements should also be made for intercepting and decanting grease at the outlet ends of the channel.

Constant-velocity detritus channels. It was some years after the publication of the Final Report of the Royal Commission that much interest was shown in this country in the principle of constant-velocity detritus settlement. It was found that if the flow of sewage through detritus channels could be maintained at a velocity of 1 foot per second at *all rates* of flow, the settled grit would be reasonably free from organic content and not liable to become offensive. To achieve this condition, fairly strict control of velocity is necessary, for if the velocity falls to ⅔ foot per second a large amount of

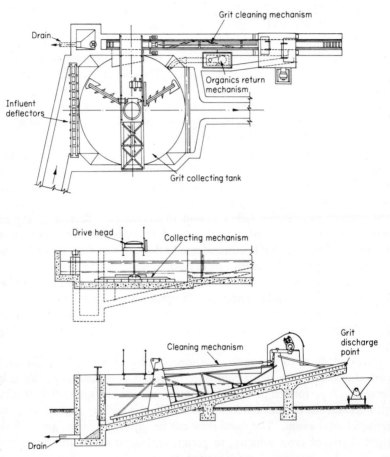

FIG. 72—Dorr "Detritor" type "A" with grit-washing mechanism
By courtesy of Dorr-Oliver Co., Ltd.

organic material is deposited; and if it exceeds 1 foot per second by more than about 25 per cent too much grit is swept out of the channel and is deposited in the sedimentation tanks. One foot per second has been found the most satisfactory velocity for channels of moderate or shallow depth; theory suggests that, if the channels are very deep, a slightly higher velocity is desirable, but on this matter there is insufficient information.

The detritus settled in a well-designed constant-velocity detritus channel has an inorganic mainly siliceous content, about 25 per cent of which will not pass a 16-to-the-inch mesh, and 50 per cent will not pass a 100-to-the-inch mesh. The detritus, as taken fresh from the channel, looks black with dirt, yet the organic content of the dry matter can be as low as $4\frac{1}{2}$ per cent, although it is more often higher than this; a high organic content can result from poor design. Detritus taken out by dredger or pumped into a vessel from which water is rapidly decanted has a moisture content of about 30 per cent. When detritus is dried it weighs 85 lb per cubic foot. Nearly all the material will pass a $\frac{3}{8}$-inch sieve.

The quantity of detritus is very variable, being the least in domestic soil-sewage from separate systems and the most from combined systems taking the flow from heavily gritted roads. Also it varies from time to time, for where sewers are old and badly graded much detritus that has settled in the sewers comes down in storm times. In one instance of a typical combined system, the recorded quantity of detritus averaged 1 ton per annum per hundred head of population. A design figure given by William E. Farrer, Ltd., is 0·17 cubic foot per head of population per annum.

Principles of design. Hazen's theory that settlement tanks should have the surface area proportioned according to the rate of flow to be treated applies more strictly to detritus than to any other sewage solids, for the greater part of detritus is sandy material the particles of which settle according to their size, and there is no flocculation. Sand grains of diameter smaller than 0·013 cm obey Stokes' Law, and the speed of subsidence varies as the square of the diameter. Above this size there is transition between two laws, and the falling speed of siliceous grains in water at normal sewage temperature are as given in Table 43.

TABLE 43
FALLING SPEEDS OF SILICEOUS PARTICLES IN WATER

Diameter (cm)	Falling speed (cm per sec)	Seconds per foot of fall
0·013	1·0	30·5
0·02	1·8	16·9
0·05	5·0	6·0
0·1	10·0	3·0
0·4	25·0	1·2
1·0	43·0	0·7

Present-day practice is usually to design for the settlement of the particle that falls at a rate of 1 foot in 16 seconds, which means that a detritus channel or tank must have a surface area of 1 superficial foot for every 33,750 Imperial gallons per day maximum rate of flow to be treated by that channel. This is the principle adopted in the design of Dorr "Detritor" installations. Practice in the design of constant-velocity channels is to allow from one to

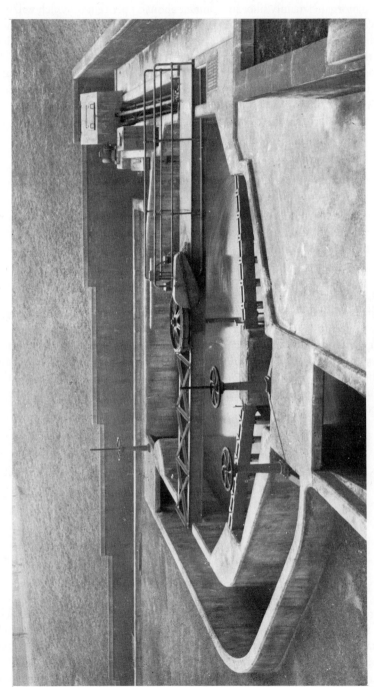

Fig. 73.—Dorr "Detritor"
By courtesy of Dorr-Oliver Co., Ltd.

two times this surface area. Having decided the surface area of tank or channel, the figure is recorded for determining, in the case of a constant-velocity channel, the length of channel after the other dimensions have been calculated.

Before calculating the other dimensions of the channels, the depth has to be determined. This is arbitrary, but should be kept within reasonable limits. If the channels are made very deep the number of individual units is reduced (in the case of the parabolic type), but depth occasions a loss of head which

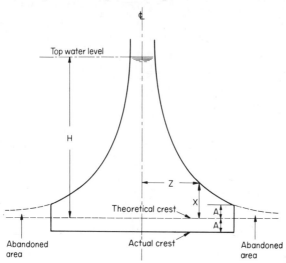

FIG. 74—Proportional-flow plate weir

may not be practicable. As a general rule for a gravity system, it appears that the most satisfactory arrangement is to have channels of overall depth equal to the overall depth of the incoming sewer. If sewage is pumped to the works it may be found advisable to make the channel depth equal to the depth between the highest cut-in and the lowest cut-out level in the suction-well of the pumping station.

Constant velocity can be maintained by several methods. In earlier installations several channels were arranged side by side, the number in use being varied in accordance with rate of flow by hand- or automatic-electric control of penstocks. This method is not too satisfactory, particularly when the human element is involved. But it is still used where there is not sufficient head available for the installation of weir- or flume-controlled channels.

One of the earlier type of weir-controlled channels, much favoured in America but little used in Great Britain, is a rectangular channel having, at the downstream end, a specially shaped weir plate (see Fig. 74). This plate is shaped according to the formula:

$$Q = \tfrac{1}{2} C \sqrt{2g}\ b\pi h$$

where: Q = discharge in cusecs
h = depth of flow in feet over theoretical crest
b = weir constant
C = coefficient for thin plate weir, usually taken as 0·62.

From this formula the value of b is calculated. Then the curves of the sides of the plate are plotted by the formula:

$$z = \frac{b}{2\sqrt{x}}$$

where: x = the vertical distance in feet measured from the theoretical crest of the weir to the curve of the plate
z = the distance in feet from the centre line of the opening to the curve of the plate.

It will be noted that the two arms of the inverted T run-off to infinity, and therefore in practice they are shortened, and an additional area added to the orifice by lowering the actual crest below the theoretical crest sufficiently to compensate for the lost area. (A reasonable approximation to constant

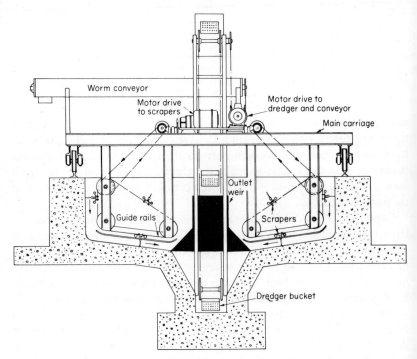

FIG. 75—Cross-section of constant-velocity detritus channel controlled by specially shaped weir plate
By courtesy of Adams-Hydraulics, Ltd.

velocity in a rectangular channel can be obtained by arranging a standing-wave flume and, at the invert level of the flume, an orifice in the form of a horizontal slot which has a discharge similar to that of a flume running full.) A weir of this type needs to have a free fall at no time backed up on its

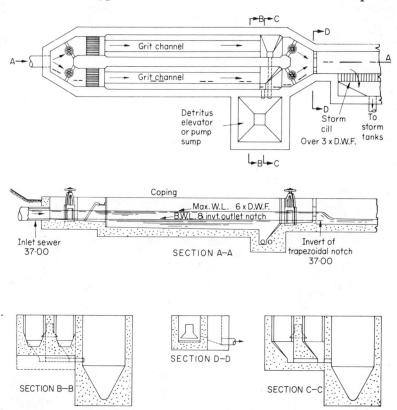

FIG. 76—Constant-velocity detritus channel installation

This diagram, like that illustrated in Fig. 75, involves a specially shaped weir plate. The design varies from the flume-controlled constant-velocity channel, as described in the text and shown in Figs. 77, 79 and 82 in the following respects:

1. There is only one controlling weir for the two channels.
2. The inverts of the channels are below the crest of the weir and therefore velocity falls at slack rates of flow.
3. There is a sump for grit removal at the outlet end of each channel. This last arrangement is sometimes incorporated in flume-controlled constant-velocity detritus channels: it has the fault of collecting organic solids with the detritus.

By courtesy of Adams-Hydraulics, Ltd.

outgo. Thus, considerable loss of head is involved, particularly where deep channels are used. Moreover, the channel of rectangular cross-section, while simple to construct, requires a fairly complicated dredger mechanism for the removal of its grit content. For this reason, channels approximating to parabolic shape are far more popular in this country.

Parabolic flume-controlled channel. The most credit is due to C. B. Townend for expounding the theory of the parabolic channel controlled by rectangular standing-wave flume, although the same principles were individually discovered by Professor Camp in America, somewhat later by the writer in this country and had also been known in Holland as a means of controlling constant velocity in irrigation channels and afterwards applied to detritus channels. The firm of Ames Crosta Mills & Co., Ltd., also arrived at the same principle independently.

Fig. 77—Ames Crosta Mills grit extractor in parabolic detritus channel, at Troqueer sewage-purification works, Dumfries

By courtesy of J. H. Shennan, Esq., B.A., M.I.C.E., M.Cons.E., Consulting Engineer, and Ames Crosta Mills & Co., Ltd.

The design principles are as follows. The discharge of a rectangular standing-wave flume is approximately according to the formula:

$$Q = 3 \cdot 09 \, B.H.^{\frac{3}{2}}$$

where: Q = cubic feet per second
B = width of flume in feet
H = depth in feet from invert of flume to upstream top water-level.

The cross-section of a channel, which increases in area directly as the rate of flow through a rectangular flume increases, is parabolic, the sharpest part of the curve forming the invert. The width of the channel at any point above the invert is as given in the formula:

SCREENING AND GRIT AND GREASE REMOVAL

$$X = \frac{3Q}{2H}$$

where: Q = rate of flow at that depth in cubic feet per second
X = width of channel at water-level in feet
H = depth of flow in feet.

Thus, if a flume is designed to pass the maximum rate of flow of one channel according to the former formula, and a channel arranged with the invert level to that of the flume and according to the latter formula, at all rates of flow the velocity through the channel will be 1 foot per second.

Channels designed to this principle may be almost truly parabolic but with the invert curved to the arc of a circle, so that the Ames Crosta swinging-arm type of detritus-removal mechanism may be used (see Figs. 77 and 78). A more popular alternative is to provide a channel of W cross-section so designed that at any rate of flow the velocity through the channel is approximately 1 foot per second (see Fig. 79). This has the advantage of steep sides

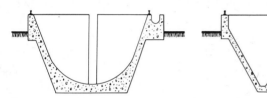

FIG. 78—Cross-section of parabolic constant-velocity detritus channel

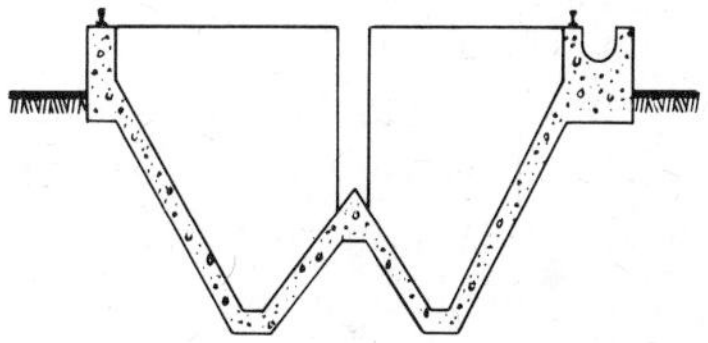

FIG. 79—Cross-section of W-shaped constant-velocity detritus channel

on which detritus will not hang up, the possibility of using a simple sludging mechanism and the avoidance of curved surfaces. It is sludged by a moving bridge carrying two grit-pumps which simultaneously suck from the two channels of the W section (see Fig. 80).

In an installation of several channels there should be sufficient channels to take the maximum rate of flow, plus one or more stand-by. Each parabolic or W-section channel should discharge to its own standing-wave flume. Then it does not matter whether the required number for the rate of flow or a greater number of channels are brought into operation: the velocity in all the channels will always be approximately 1 foot per second, giving satisfactory settlement of reasonably clean grit.

Having found the cross-section and width of channel, the width can be divided into the surface area already determined and the length found. It should be noted, however, that the minimum length required by this rule to settle a particle falling 1 foot in 16 seconds is 16 feet of length for every foot of maximum depth of water. In practice, the length can be increased up to twice the theoretical minimum.

Fig. 80—Constant-velocity detritus channels, Crossness sewage-treatment works

Detritus channels should be arranged with their inverts level with, or only slightly lower than, the inverts of the flumes that control them, and they should *not* contain any deep sump to collect grit. Each channel should have isolating penstocks at inlet and outlet ends and a wash-out penstock for emptying any channel that is to be laid off.

Detritus removal. Shallow channels for very small works can be cleansed by decanting the sewage and manually digging out the grit. The use of such methods is to be recommended at very small works; but first, a calculation should be made of the quantity of detritus liable to be accumulated and the amount of man-power needed. In the smallest installations it is advantageous to make the channels extra long so as to store reasonable quantities of detritus between digging-out operations.

For larger channels it is usual to provide either dredger mechanisms or

Fig. 81—East elevation of Crossness detritus channels

travelling grit-pumps. Grit-pumps are now far more popular than dredgers. They are arranged to suck from the invert of the channel through suction pipes that are given flexibility so as to be able to trail over any excessive bank of detritus. Rate of upstream motion of the carriage should be about 3 inches a second, although there is no harm in the carriage moving as fast as 12 inches

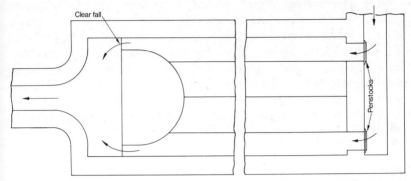

FIG. 82—Constant-velocity detritus channels for small sewage works: plan

This figure and Fig. 83 illustrate the extreme simplicity of the flume-controlled constant-velocity detritus channel. It will be noted that a simple cross-section varies little from the true parabola: only one penstock is required to control the inlet of each channel, back-flow through the outlets being prevented by a clear fall. The small amount of detritus settled in a small installation can easily be removed manually; in a larger installation mechanical dredgers or travelling grit-pumps can be installed.

a second when going in the direction of the flow of sewage. The grit is drawn off with sewage at a dilution of about 98 per cent moisture content. This liquor can be discharged into a settling vessel on the travelling bridge, in which the grit is deposited and from which the supernatant water overflows back into the channel. At the end of each journey to the distant end of the channel and back again the grit, with a remaining moisture content of about 30 per cent, can be tipped to wagons or lorries for disposal.

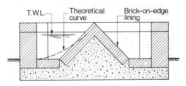

FIG. 83—Constant-velocity detritus channels for small sewage works: cross-section

An alternative method of removal of grit from channels is by water carriage. This method is advisable when the installation is so large that settling vessels are impracticable. The deliveries of the pumps discharge into a small open grit channel arranged by the side of each detritus channel, care being taken in the design that the discharge into channels does not cause splash (see Figs. 78 and 79). The grit channel should be of sufficient capacity to take the maximum rate of flow from both or all pumps discharging into it, and should be laid at a gradient sufficient to produce a velocity of 4 or 5 feet a second so as to reduce the possibility of settlement or stoppage to a minimum. Should a stoppage occur in the water-carriage system, the channel or channels can be flushed by running the pumps with the detritus-dredging mechanism standing still.

When water carriage is used, detritus can be delivered to long, narrow lagoons, where it will settle and whence the supernatant water can be drawn off over a board formed to a weir of proportions calculated to give a velocity of flow through the lagoon of a few inches a second, thereby ensuring adequate deposition of grit. Supernatant water is then drained back to a sump and pumped to the flow of sewage upstream of the detritus channels. Alternatively, the water-borne grit may be discharged to a grit-washing mechanism for removal of a proportion of its remaining organic content and separation of part of its moisture content.

TABLE 44

DETRITUS AS TAKEN FROM CHANNEL

	At inlet per cent	At middle per cent	At outlet per cent
Moisture	30	30	32
Organic content	5	6	5
Remainder	65	64	63
	100	100	100

A sample taken from truck had a moisture content of 30 per cent and 3 per cent organic matter.

MECHANICAL ANALYSIS OF NON-ORGANIC CONTENT

	At inlet per cent	At middle per cent	At outlet per cent
Above 16 mesh	25	23	10
Above 30 mesh	13	13	16
Above 100 mesh	50	50	57
Silt	12	14	17
	100	100	100

Limits of application. Since the publication of the previous edition of this book several channels of the parabolic type have been constructed and a large range of sizes has been covered. The channels installed at Mogden were intended to deal with a flow of 575,000,000 gallons per day. There are six channels, which would, theoretically, have been 30 feet wide, but they were, for convenience, distorted from the true parabola and constructed only 24 feet wide. The channels are 90 feet long and 9 feet deep. The two inverts to each channel have a sump or grip, extending for the full length and below the theoretical curve, to store grit—a questionable feature—and a special pump is arranged so that its suction can move along the sumps to remove detritus.

This installation is capable of dealing with the full storm flow, but at times of reduced flow the number of channels in operation can be reduced accordingly with the aid of electrically operated penstocks, for each channel has its own controlling flume.

A smaller installation is that at the Wandle Valley Sewage Works. This consists of two channels of small dimensions but proved efficiency. As at Mogden, electrically operated penstocks are installed, but in this case they do not appear to be necessary. The grit removed from these channels is coarse and clean, although intermingled with matchsticks and similar objects. Like those at the Mogden works, the channels are distorted from the true parabola to avoid curved surfaces. Both channels are used together during high rates of flow.

The writer designed for the Crossness sewage-treatment works constant-velocity channels of W section capable of dealing with a maximum flow of 324,000,000 gallons a day, each channel having a width of 24 feet and an overall depth of 13 feet from invert to coping. The invert of the channel is 1 foot below the invert of the controlling flume. These channels differ from those at Mogden in that the effective length is 322 feet (see Figs. 80 and 81).

The constant-velocity detritus channels installed at Kingston, Jamaica, are of very different design, not having been influenced by any previous pattern. These are almost truly parabolic, and have no sumps or grips in the invert. There are two channels intended to work alternately, each being capable of taking the total flow liable to be discharged to the works. The channels are 3 feet deep, including freeboard of about 1 foot, and 4 feet 3 inches wide. The length of each channel is 70 feet 6 inches. The installation is capable of dealing with a maximum rate of flow of about 3,000,000 gallons per day.

An example of a constant-velocity detritus channel installation at a comparatively small works is that which was constructed for the Luxborough Main Outfall works of the Urban District of Chigwell. This plant was intended for a dry-weather flow of 1,000,000 gallons per day from a population of 33,000. But to begin with the population was no more than 26,000. Mr. B. G. Kirk[149] described the channels in the following words:

> ... the flow passes into two detritus channels. These channels each have a total length of 52 ft 0 in, and in section they may be likened to a "V" with sides set at an angle of 30° to the horizontal. Sluice gates are provided on both channels so that either may be shut down for cleaning while the other handles the entire flow. The grit is shovelled out by manual labour and the operation is carried out daily. Excellent results are obtained with these channels, even in times of heavy flow.

The Chigwell channels were probably descended from the Kingston, Jamaica, design, although they are very different in design and do not maintain constant velocities at all rates of flow. Mr. Kirk, after several years' experience with these channels, maintains that the manual cleansing is easy and efficient. More recently a hand-operated machine has been made for the removal of detritus from such small channels.

From the above descriptions of a few installations it will be seen that the parabolic channel is applicable not only to plants of the largest size, but also to those of very moderate capacities. The questions that remain are: to what extent can the size of channel be reduced and the method practically applied to still smaller works, and how should such channels be designed and operated?

It will be appreciated that the settlement of grit is not a very complex phenomenon, and that it is similar over a very wide range of scale. The sandbanks that form in a river of the greatest magnitude owe their origin to the laws that also determine the formation of the small banks of sand that one frequently sees in a roadside channel. It would therefore appear that detritus channels could be constructed to almost any degree of smallness, the limits being only those of a practical nature—i.e. the depth of flow should always be sufficient to transport the largest organic solids, and the channels should not be so small that detritus cannot be removed efficiently with either a spade or a specially constructed scoop. Just how far we can go in this direction has yet to be determined by practical experiments, but from the above considerations it would appear that it should be possible to apply the method to all small municipal works, except the very smallest, such as are used for minor villages, hospitals, institutions, etc., where detritus removal is not necessary.

Grease removal. In America it is usual to remove grease from sewage prior to sedimentation, but in Great Britain this has generally been found unnecessary.

The reason for the difference would appear to be different culinary habits. During the Second World War the sewage-treatment works provided to British design for American army camps in Great Britain were found to be unsatisfactory and in one instance the author made an inspection because of a complaint that it was impossible to dry the sludge. It was found that, in a small typical sewage-treatment works with primary sedimentation tanks, percolating filters, humus tanks and sludge-drying beds, there was an excessive amount of light-brown scum floating on the primary tanks and a heavy light-brown sludge which did not flow easily, tending to block the sludge pipe, a 6-inch pipe laid at the usual gradient of 1 in 100. The sludge would not dry on the drying beds and, even when covered over to keep off the rain, would not dry on an experimental bed where ordinary sludge would have dried very quickly. Instead of the usual sewage-works odour there was an odour reminiscent of a piggery.

Inspection of the kitchens showed that there were typical open-topped boarded-over grease-traps heavily laden with floating scum and bubbling with gas from digesting sludge at the bottom. The kitchens were inspected during meal preparation and washing-up time and it was noticed that large quantities of grease cut from the outside of various canned meats or washed from butter dishes that were far from empty were being washed with hot water down the sinks. It was evident that this excessive load of grease was the cause of trouble at the sewage works.

Normally at British sewage works the grease content comes away with the sludge and scum that are separated in the primary sedimentation tanks. It is very seldom that arrangements for grease removal are provided at municipal works, but there are occasions in the treatment of trade wastes when grease separation is desirable.

The use of simple grease-traps has been described in Vol. I, Chapter XX. Grease separation should normally come after screening, before primary sedimentation, but can be combined with detritus settlement. One of the favoured methods is aeration by the diffused-air method as described in Chapter XXX, allowing a detention period of about three minutes and an air supply of about $\frac{1}{30}$ cubic foot per gallon of sewage.

This method was used at the Greater London Council Beckton (Northern) sewage-treatment works where it was claimed that some reduction of B.O.D. resulted. Grease was removed at the lower ends of the aerated channels, usually in the form of grease balls.

In designing the Crossness (Southern) sewage-treatment works the writer decided to omit this pre-aeration on the grounds that it was unnecessary and that the amount of benefit negligible compared with the benefit that would be obtained by a similar amount of aeration in the activated sludge unit. It was found, when the works were put in operation, that in spite of there being no pre-aeration, grease balls of all sizes up to 6 or more inches diameter accumulated downstream of the standing-wave flumes of the detritus channels and grease could be collected and discharged to the sludge drains from specially constructed inlet manholes of the primary sedimentation tanks.

As the sewage had been screened and passed through centrifugal pumps which, undoubtedly, would have destroyed large grease balls, the grease balls must have been formed in the long rising-main culverts leading to the detritus channels or in the long detritus channels themselves without the aid of any aeration.

There was no evidence of any inferior performance of the works due to the omission of pre-aeration.

Grease removal can be effected by applying a partial vacuum to the sewage or waste after aeration for a period of about fifteen minutes. This is done by passing the flow through a closed tank which is above the hydraulic gradient so that the liquid must be drawn up to the top of the tank by vacuum and maintained at that level by vacuum pump as air or gas is given off. The bubbles emitted by the sewage carry the grease to the surface, where it may be removed by mechanical device. The Dorr-Oliver Co., Ltd., can provide equipment of this kind.

The foregoing should not be confused with pre-aeration of sewage prior to primary sedimentation for the purpose of reducing the load on a treatment plant. For this kind of pre-aeration detention periods of one-half to three-quarters of an hour are desirable if a noticeable B.O.D. reduction is to be effected. The advantage is controversial.

CHAPTER XXV

SEPARATION AND TREATMENT OF STORM WATER

As a result of Royal Commission recommendations it has been, and still is, usual to design works serving combined or partially separate systems to treat three times dry-weather flow to Royal Commission standard and for all flows in excess of this quantity to be passed to storm tanks for partial treatment by sedimentation or for subsequent return to the works for full treatment after the storm is over.

On the other hand, separate soil sewers have been designed to accommodate up to six times dry-weather flow and many treatment works serving them, having no arrangements for separation of storm water, also have had to take this flow.

Storm-water separation. Storm water should usually be separated after screening the sewage and settlement of detritus, for it is not unduly expensive to provide screens and detritus channels of adequate capacity to deal with the peak storm flow, and storm water should not be discharged without being screened as well as being freed from sludge and scum by settlement.

Storm water can be separated by any one of three methods, according to circumstances. At works of moderate size and where the sewage does not have to be pumped at the treatment works, the flow of three times dry-weather flow to full treatment is gauged through an orifice or special module, and the storm water spilled over weirs of adequate length. At larger works the orifice or module can be replaced by electrically operated float- or electrode-controlled penstocks, the storm water again passing over weirs. Where the whole of the sewage is pumped from the sewers to the works, the best arrangement is usually to provide a battery of pumps which deliver up to three times dry-weather flow to the treatment plant and another battery which automatically cuts in to deliver the remainder of the flow to the storm tanks as soon as a flow of three times dry-weather flow is exceeded.

Separating weirs. A common arrangement for separating storm water is a submerged orifice to control the rate of flow to the treatment works and long storm weirs discharging to the stand-by tanks. A good type of orifice is that which has its invert flush with the invert of the channel upstream, in order that no solids shall be retained. The orifice should be adjustable so that the rate of flow may be controlled accurately.

The required length of storm overflow weir depends on the depth of orifice below weir level and on the degree of accuracy required. If it is

280   PUBLIC HEALTH ENGINEERING PRACTICE

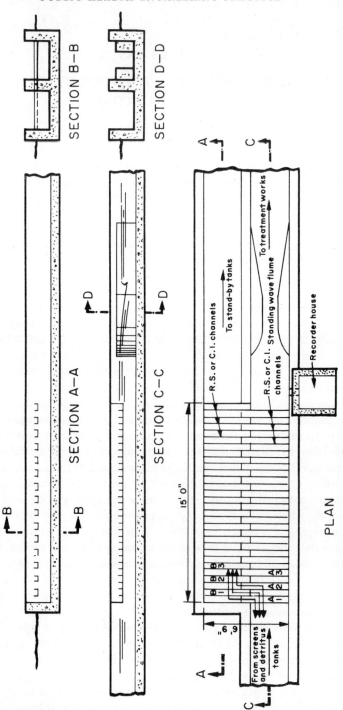

possible for the orifice to be set very deep and to have free discharge on the downstream side, the storm weirs need not be long; but if, as is usually the case, available head is limited and the orifice cannot be placed very deep, and if, at the same time, accuracy of storm discharge is desired, the storm weirs must be long. When in place of a submerged orifice a standing-wave flume is used to control the rate of flow to the treatment works (and at the same time serves as gauging and recording flume), as might be the case when available head is very restricted, still longer storm weirs are required.

A number of methods has been applied to make long storm weirs economical in construction. In one arrangement (see Fig. 84) two chambers are placed side by side, one from which overflow takes place, and another into which overflow discharges. Arranged across the surface of the first chamber is a number of rolled-steel channels, the flanges of which serve as weirs. Storm water overflows into the channels, the ends of which discharge into the second chamber. In the latter is arranged a similar series of rolled-steel channels, the flanges of which are set at the same level as those of the former series. Sewage flows into these channels from the first chamber and overflows their flanges into the second chamber. Thus, on the wall between the two chambers rest alternately the ends of both sets of channels. The effective weir-length is the effective length of all the flanges.

Separating weirs, being free from mechanical contrivances, are reliable and comparatively simple, and they may be used for small and large works alike. Self-charging siphons used in place of simple overflow weirs economise in space, but do not gauge so accurately as well-designed weirs, and are economical only if there is plenty of head available for deep downstream legs.

Storm tanks. When sewers are strictly on the separate system there should be very little increase of infiltration in wet weather, and no special arrangement need be made for the treatment of storm water at the disposal works. Works should be fully able to treat up to three times dry-weather flow according to the recommendations of the Royal Commission on Sewage Disposal, but it is considered desirable for the pipework, etc., to be made capable of taking somewhat greater flows so as to permit moderate overloading of the components of the works. In this connection it should be remembered that efficiency of sedimentation tanks falls only slightly on considerable increase of flow, and percolating filters of typical present-day design often increase in efficiency on overload: and these facts, together with the reduction of strength of sewage on dilution, can mean that overloading the works is not accompanied by the production of an effluent of reduced quality.

When the sewerage is on the combined system, or on a partially separate system in which more than a very small proportion of roofed or drained areas drains to the sewers, the position is very different. The peak storm flow from a small catchment can be very large indeed, say forty or more times the dry-weather flow, making it impracticable to put the total flow arriving at the works through either sedimentation tanks or aeration units, both from the point of view of hydraulic design of the pipework, etc., and the mechanical and biological functioning of the plant. It therefore becomes necessary to arrange for full treatment of a maximum of three times dry-weather flow, as recommended by the Royal Commission, and to separate by weirs or other means any flow in excess of this figure for storage until such time as it can be treated, or when the quantity exceeds the storage capacity of the storm tanks, or partial treatment by settlement.

There is a not uncommon misconception that storm tanks should treat three times dry-weather flow only and any flow in excess of this figure be discharged to natural watercourses without any treatment whatsoever; but this is contrary to the specific statement in the Commission's Report that there should be no storm overflows at or near the treatment works, and such practice is technically unsound in that it does not take anything like full advantage of the storm tanks' ability to reduce the amount of polluting matter discharged. The Royal Commission recommended that the storm water to be given partial treatment should be passed to storm tanks or stand-by tanks having a total capacity of six hours' dry-weather flow and that the whole of the storm water—no matter what its quantity—should be dealt with in this manner.

Purpose and capacity of storm tanks. Storm tanks serve a dual purpose. In the case of the short, intense storms that occur mainly during the summer months and south-west winds, peak run-offs from small and medium-size catchments are very high; but as these storms do not last long, the amount of rainfall precipitated at any one time is not great and can be stored in tanks of moderate capacity. Thus, the run-off from these storms usually fails to fill the storm tanks, which then act as storage or balancing tanks. Then, when the flow coming down the sewers has fallen to less than three times dry-weather flow, the contents of the storm tanks can be passed to the works for full treatment. In these circumstances no sewage or storm water whatsoever escapes to the watercourse without full treatment.

At times of prolonged rainfall, intensity of precipitation is, on the average, low, according to rainfall statistics, but the total amount of water precipitated during a period of rainfall is considerable. It is then that the storm tanks will become completely filled, after which the excess of flow will spill over the outlet weirs and the tanks will function as sedimentation tanks, giving partial treatment to the diluted storm water. While this means that some storm water is discharged after settlement only, the degree of pollution of the watercourse is less severe, particularly as the first flush of storm water is not allowed to escape. Again, on the tailing-off of rainfall, the remaining contents of the storm tanks is passed to the works for full treatment.

It is probably true to say that the full value of storm tanks is not generally appreciated, for too often they are considered from the point of view only of settlement tanks for storm water. This is far from being the case for, in a normally designed and operated installation on an average combined-sewerage system, by far the greater part of the storm water will be returned to the works for full treatment and on only one of about every three wet days will a storm cause spill-over of highly diluted and adequately settled storm water with very little risk of harm to the watercourse to which discharge is made. This can be illustrated by the following calculation.

In a representative area of Greater London there are about 150 persons per impervious acre and the dry-weather flow is about 57 gallons per head per day. If then we take the storage formula given towards the end of Chapter IV:

$$C = \frac{3354 \cdot 5 \, Ap^{\frac{3}{2}} N^{\frac{1}{2}}}{P^{\frac{1}{2}}}$$

and apply it to the data above, we can find the frequency of occurrence of the storm that will just about fill a storm tank having a capacity of six hours dry-weather flow.

Then:
$$C = \frac{57 \times 150 \times 6}{6 \cdot 23 \times 24}$$
$$= 342 \text{ cubic feet per acre of impervious area.}$$

And the value of P will be twice dry-weather flow because that is the difference between the average dry-weather flow and the maximum flow to be treated. Then:

$$P = \frac{57 \times 2 \times 150}{24 \times 60 \times 6 \cdot 23}$$
$$= 1 \cdot 9 \text{ cubic feet per minute}$$

Then:
$$342 = \frac{3354 \cdot 5 \, 1^{\frac{3}{2}} N^{\frac{1}{2}}}{1 \cdot 9^{\frac{1}{2}}}$$

$$N = \frac{1}{50 \cdot 6}$$

In other words, the tank would be just filled to the point of spilling-over about once every week (provided that the formula can be extrapolated to include such frequent storms).

Next we can consider what capacity of storm tanks would be required if the tanks were to be just filled to the point of spilling-over once a year. This would mean multiplying C by $\sqrt{50 \cdot 6}$ giving a detention period of about 42 hours.

From the foregoing calculations it will be seen that the usual practice of allowing a storm tank detention period of six hours dry-weather flow must make a very considerable difference to the amount of pollution caused by spill-over of storm water in that it is one storm only per week that will cause spill-over, that the spill-over will be a small part of the total run-off due to that storm, that it will occur when there is considerable dilution of sewage by rainwater and that it will be of settled, not crude sewage. The calculations show also that virtually to do away with spill-over of storm water would require the provision of tanks of much greater capacity than usual.

It is, however, usually possible to pass up to six times dry-weather flow through the works for full treatment. A calculation shows that, if this is done and the storm tanks have a capacity of 48 hours dry-weather flow, there will be no spill-over of sewage that has been treated by settlement only except at those works where the ratio of peak storm water flow to dry-weather flow is greater than the average.

In those cases where the river authority has insisted on a final effluent of better quality than Royal Commission standard it would be irrational to permit a storm water spill-over that could cause in one wet day more pollution of the river than the treated effluent could cause in a week or more.

Storm tank details. The traditional form of storm tank is a rectangular tank almost identical with the quiescent sedimentation tank which was not uncommonly used some sixty years ago. These tanks were about five times as long as they were wide; their bottoms were almost flat, sloping towards the inlet end to simplify sludging manually. Inlets were usually simple and penstock-controlled. The outlets at the other ends of the tanks were flat-crested weirs protected by free-floating scum-boards which rose up and down with the water-level in grooves in the side walls of the tanks. After use the tanks were decanted to the level of the top of the settled sludge by floating-arms placed near the outlet weirs and surrounded by floating scum-boards in the form of square boxes open at top and bottom. After decanting top water, which was pumped or gravitated to the flow of sewage arriving at the works, the sludge was squeegeed towards the inlet end to a channel in the floor, whence it was drawn off through a penstock.

The only marked difference between storm tanks and quiescent sedimentation tanks was that storm tanks have a series of weirs arranged at the inlets so that each tank came into operation in turn, in order that a small flow of storm water would not necessitate the sludging of all the tanks. A further modification that can be made is to arrange for the first tank that comes into operation by virtue of its having the lowest inlet-weir level to have no outlet weir; this prevents the first flush of very foul storm water from escaping with no treatment other than settlement.

Following tradition, it became common, even at modern large works, for storm tanks to be of the rectangular form described; but with the adoption of mechanisation such tanks were often made capable of mechanical sludging by machines of the types used for rectangular primary sedimentation tanks. As storms do not occur every day and, even in wet weather, storm tanks need not be sludged every day and may not have to be sludged until after storm water spill-over has ceased, storm tanks do not need to be sludged as often as the sedimentation tanks at the same works. This can mean the provision of a different type of sludging mechanism from that used in the sedimentation tanks. But another obvious solution of the problem is to have storm tanks and sedimentation tanks of similar proportions and general design, to provide the sedimentation tanks with a sufficient number of scrapers and transfer carriages and to so arrange the storm tanks and sedimentation tanks that the scrapers can be transferred from the sedimentation tanks to the storm tanks whenever the latter need to be sludged.

There are two ways in which this can be done. First, the storm tanks can be in a line parallel with the sedimentation tanks and the scrapers passed over the transporter carriages from one set of tanks to the other. The second method is to have storm tanks and sedimentation tanks in one long line. The method to be adopted depends on the site, and such arrangements as described are practicable only if conditions are such that weir level of storm tanks and sedimentation tanks can be the same.

While rectangular storm tanks are by far the most common, it is not essential that this type should be used, and circular tanks with mechanical sludging gear can serve quite satisfactorily, provided that they are arranged for decanting after use and their scum-boards so located that scum does not pass under the boards when the level of the sewage is rising in the tanks. If the tanks are not at the sewage-treatment works the mechanisms can be automatically started and stopped (see "Storm tank policy" in Chapter V).

Very often, when old works are converted, existing sedimentation tanks, disused contact beds or other structures of suitable proportions are converted to storm tanks. This may be one reason for the perpetuation of the rectangular tank. In a recent design allowance has been made for a battery of rectangular sedimentation tanks of simple type to be converted to storm tanks by minor modifications. In this instance sewage arrives at the works at such a level that gravitation can take place through the old tanks. But the new treatment works are at a higher level. Consequently, up to three times dry-weather flow can be pumped to the treatment works, while any flow above the pumping capacity passes over weirs bringing in one storm tank after another until all are full; then a further increase in water-level causes spill-over at the outlet weirs to a watercourse. When storm flow begins to fall off, the dry-weather pumps pull down the water-level and decantation of the contents of the storm tanks takes place automatically through tide flaps at a low level in the inlet weirs. At a still lower level are weir penstocks by which any remaining top water may be judiciously drawn off. Finally, when the tanks are empty, any one to be sludged is isolated and the sludge removed by mechanical plant.

Balancing pumping flow. Balancing the rate of flow that is due to pumping is frequently essential at small sewage-treatment works. There is a minimum size of rising main (3 inches or 4 inches) suitable for delivering crude sewage, and as the velocity in the rising main should not be less than 2 feet per second, the pumps should be sized accordingly. At a small sewage works this minimum pumping rate may be greater than three times dry-weather flow, and therefore, unless special precautions are taken, sewage will spill over the storm weirs every time the pumps function. If there are no storm weirs the performance of the works can be impaired by overload of sedimentation tanks, siphons and percolating-filter distributors.

In such cases it is usual to attempt to balance the flow to the extent only necessary to prevent overloading of sedimentation tanks, dosing siphons, filter arms, etc., or to prevent discharge over storm weirs when the *average* rate of flow is not greater than the maximum rate of flow that the works are designed to treat.

The usual procedure is to install, at the top of the rising main, a balancing tank, the working capacity of which is equal to the working capacity of the suction-well of the pumping station. Sewage is drawn off from this balancing tank by means of a constant draw-off floating arm, floating weir of module, adjusted so as to discharge at all times at the rate of three times dry-weather flow. With this arrangement the rate of flow to the sewage works will exceed the maximum rate of flow to be treated only when the flow to the pumping station itself also exceeds the maximum rate. Thus, although the pumps may be of any size, their discharge will neither interfere with the normal functioning of the storm weirs nor influence unduly the quality of the final effluent.

Balancing tanks of this kind serve as detritus tanks, and are best constructed in such a manner that they may be easily sludged and at regular intervals. The upper portion of the tank serves as the working portion from top water-level at which overflow weirs are provided (these should be the storm weirs themselves) to the bottom water-level, which is the lowest level at which the floating arm will function properly. Below this level is the storage for detritus and sludge. Hopper bottoms may be provided and sludge removed under hydrostatic head; or bottoms may be flat and provision made for sweeping. The floating arms and overflow weirs should be provided with floating scum-boards.

In large works the balancing tank should preferably be a separate tank serving no purpose other than that of balancing pumping flow and separating storm water. At small works it may accommodate the screens. Generally the balancing capacity should not be in part of the sedimentation tanks, as this arrangement would impair their efficiency.

The disadvantage of balancing tanks of this kind is that floating weirs or arms are not 100 per cent efficient mechanisms. They require careful attention if they are to function regularly, and they have to be specially designed if they are to give truly even flow.

It has sometimes been suggested that to avoid the use of floating arms of complicated floating weirs an orifice placed at a low level or a well-designed siphon will serve. This is a fallacy, for any fixed orifice or siphon not controlled by a module must give varied discharge according to head, although it will smooth out the flow to a small extent: analysis of any particular case will show that the degree of error involved in using an orifice or siphon is much greater than would at first appear.

Generally, balancing tanks should not be required other than at small works for once the flow is of such magnitude that the rising main required to accommodate three times dry-weather flow is greater than 4 inches diameter, the pumps should be of such sizes that, when the rate of flow of sewage is three times dry-weather flow, the pumping rate is three times dry-weather flow. Only when this rate of flow is exceeded should an additional pump cut in. Then the additional pump may deliver via a separate rising main direct to the storm tanks, doing away with the necessity for separating weirs or, if this arrangement is not practicable, the storm weirs should commence to function only when the additional pump cuts in.

Smoothing out the midday peak. Attempts at balancing the effects of the maximum rate of flow that occurs daily round about midday are comparatively recent and few in number. One method was described by H. Lorain Folkes,[83] and a further brief description was given by H. C. Whitehead.[254] The advantages, however, of balancing out the midday peak are of questionable value, for this balancing is possible only during dry weather, when the works should not be liable to overload. Moreover, the maximum load with which the sewage works have to contend during the day does not necessarily occur at midday, for, unless the sedimentation tanks are unduly small, the maximum rate of flow entering these tanks displaces the weak sewage that fills them during the night. Thus, to a great extent, *the rate of flow of organics to be treated* is automatically balanced out, and smoothing out the rate of flow of liquid could possibly have an adverse effect.

Should, however, it be desired to arrange for balancing flow, the ideal arrangement is to provide balancing tanks before the sedimentation tanks, thereby increasing the efficiency of sedimentation.

Flow-regulating devices. Every balancing tank depends for its proper operation on some mechanical device which gives a truly constant rate of discharge at all heads in the tank. When the balancing tank is intended to be of some service as a settlement or detritus tank, either a constant draw-off floating arm or a floating weir must be used. Most of these devices are imperfect in some respect. Floating arms vary in the depth at which they float, according to the amount of liquid passing down the arm, and they tend to stick at the joint about which they rotate. But they are probably most frequently used. Floating weirs need to be frictionless, and not liable to leak in the telescopic or jointed section, which means that, to be efficient, they must be very well made.

When balancing tanks are not intended to serve as sedimentation tanks, sewage may be drawn off at a level below the lowest working water-level, and a simple module* can take the place of the floating arm or weir. A module usually consists of a controlling orifice and some kind of float-operated gate which maintains a constant head above the orifice and, therefore, a constant rate of flow. Modules for regulating the flow of clean water are comparatively easy to design, but when the flow of sewage, particularly of crude sewage, has to be controlled, the module must be so designed that any stoppage due to solids will cause further movement of the gate and a tendency for the gate to cleanse itself without any marked variation of rate of flow occurring.

Figure 85 illustrates a module designed to function in this manner. The height of the float above the orifice in the gauging chamber determines the height of the module tube above the outlet cone of the balancing tank. Should the narrow slit between the tube and outlet become choked by solids, the outflow will immediately reduce slightly, causing the level of the float to fall, lifting the module tube until the offending solids are released. Once cleared, the module falls back to its normal position. Should the flow exceed average dry-weather flow for a sufficient period of time to fill the balancing tank, sewage will spill over the top edge of the module tube, and the rate of flow to the float-chamber will exceed the rate of flow of the float-chamber through the orifice. The float will then rise, lowering the module tube on to the cone, closing the inlet at the bottom. Once this has happened, the upper end of the tube will act as a weir discharging up to three times dry-weather flow. Any excess over this rate of flow will pass over the storm weirs.

Adjustment of the module is obtained with the aid of either the adjustable orifice in the control chamber or the adjustable stop nut underneath the float. (Incidentally, the float should be able to rise from this nut, but not move below it.) Adjustment of rate of flow over the top end of the module is obtained by altering the level of storm-overflow weir-plate.

A balancing tank arranged with an efficient module or floating-arm, and having a capacity of three hours dry-weather flow, would smooth out all intense temporary peaks of flow and have a very great effect on the main midday peak. The capacity of the balancing tank depends on the average daily flow-curve, and should be designed according to the height and duration of the peak flow. In some instances a longer detention period might be necessary, but seldom, if ever, would a capacity of more than six hours dry-weather flow be needed even were a high degree of balancing required.

Flow and strength regulation. If both flow and strength are to be regulated, and consequently rate of discharge of organics is to be balanced, a

* This is, unfortunately, the only accepted dictionary term for a device for maintaining the flow of water at a constant rate under varying head. The word is used to mean also any small measure, a small-scale plan or model, an empty representation, a model of exemplar, 100 litres per second, the size of an architectural part used as a measure of the whole, the diameter of a coin, modulus, pitch of a gear wheel and, in modern electronics, any component part of a complex whole!

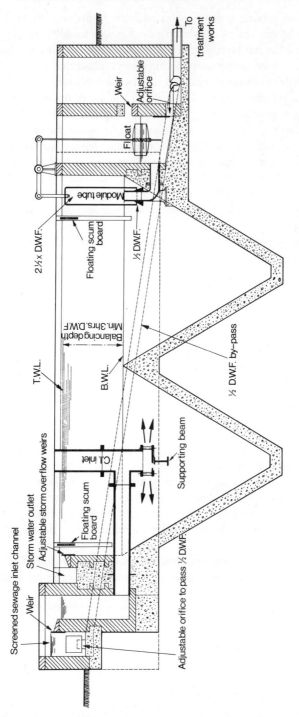

FIG. 85—Balancing tank
By courtesy of The Institute of Sewage Purification

fair degree of mixing is required in addition to an even flow. Sufficient mixing can be ensured by combining the flows of freshly incoming sewage with sewage that has been detained for several hours. This arrangement is included in the balancing and mixing device illustrated in Fig. 85. The incoming flow of sewage is separated into two streams by means of an orifice plate and an overflow weir. The orifice plate passes one-half dry-weather flow (or other quantity as may be found by experiment to be the optimum quantity) direct to the sewage-treatment works, while the remaining flow, whatever its amount, is passed over the overflow weirs to the balancing tanks. The module of the balancing tanks is adjusted so as to pass one-half dry-weather flow during dry weather, while in storm time it will pass two-and-a-half times dry-weather flow when functioning as a weir (see Fig. 85). In the morning when the flow is at about, say, one-half dry-weather flow and the strength is at a minimum, weak sewage arriving at the works will combine with strong sewage which has been stored in the balancing tank and is being discharged in equal quantity to the incoming sewage (or as otherwise determined). This will continue until the rate of flow rapidly increases towards midday and the strength correspondingly becomes higher. Then, while sewage of increased strength passes via the by-pass direct to the works, a larger quantity spills over the weirs to fill the balancing tank.

A tank of adequate capacity should by these means give almost 100 per cent efficiency of balancing rate of flow and a good degree of balancing of strength. Further, unavoidable, mixing would take place in the sedimentation tanks.

The size of balancing tank required for balancing strength would again have to be determined by examining the average daily curves of flow and strength. While a tank of three hours dry-weather flow capacity would most probably suffice for evening out the rate of flow and with the above arrangement would give an average time-lag of six hours, a larger capacity might be desirable so as to ensure high-efficiency mixing. But again, a capacity of six hours dry-weather flow giving an average time-lag of twelve hours should, in nearly all cases, be ample. The optimum proportions of by-passed and balanced flow could be determined by experiment after the tanks had been installed.

The foregoing describes arrangements that are not required at ordinary sewage-treatment works but could well be needed to control discharge of trade wastes to sewers.

CHAPTER XXVI

THEORIES OF CONTINUOUS-FLOW SEDIMENTATION

QUIESCENT, or fill-and-draw sedimentation, which is described in Vol. I, Chapter VII, is obsolescent for the treatment of sewage. Sedimentation tanks are now designed as continuous-flow tanks through which the sewage flows steadily from inlet to outlet, according to the rate at which it is delivered to the treatment works. At one time continuous-flow sedimentation tanks were designed entirely by crude rule of thumb, with more or less satisfactory results. But, by care in design, cost of tanks can be reduced without loss of efficiency.

Settlement of particles. The velocity of settlement of particles in quiescent conditions depends on the size, shape and specific gravity of the particle concerned. Small particles behave in accordance with Stokes' Law—i.e. the falling rate is controlled by the viscosity of the fluid, in which case it varies as the square of the diameter of the particle and also increases as the temperature of the fluid is increased. Large particles are not affected by the fluid viscosity, and their velocity of settlement varies as the square root of the diameter of the particle. Intermediate between the large and small particles are those particles the falling rates of which are dependent on both density and viscosity (see Table 43).

Before the publication of the Fifth Report of the Royal Commission, A. Hazen[104] had devised a theory of settlement which was made possible only by the assumption of simplification of conditions. Hazen assumed that particles subsided at regular rates according to their size, the small particles falling in accordance with Stokes' Law, and that once a particle had reached the bottom of the tank it remained undisturbed. On this basis it would be necessary for a particle to fall from the point of entry to the bottom of the tank during the detention period if it was not to be swept from the tank with the effluent. As, in fact, conditions of turbulence interfere with flow in tanks, Hazen assumed a state of complete turbulence, giving an equal distribution of suspended matter throughout the tank. He then produced mathematical formulae intended to predict the amount of sediment remaining in an effluent according to the amount in the influent, the type of settlement—i.e. quiescent, continuous flow, etc.—and the ratio of detention period to time taken for a particle to fall from the surface to the bottom of the tank.

According to Hazen's theory, depth of tank had little effect on sedimentation, but the size of particle that can be settled depended on the surface area of the tank or, more correctly, the capacity divided by average depth.

In support of his theory Hazen gave a series of experimental figures. These are plotted on Figs. 86 and 87. It will be noted that in Fig. 86 the smallest size of particle to be settled is equated against gallons per day per square foot of surface area, and that plotted to logarithmic ordinates and abscessae the curve produced is a straight line, the equation of which is:

$$\text{Gallons per day} \times C = (\text{Diameter of particle})^2$$

As the particles are small ones that settle in accordance with Stokes' Law, it would appear that the flow per unit surface area is in direct proportion to the falling velocity of the smallest particle settled. In Fig. 87 are plotted the same experimental results, but in this case hours' detention period is equated against size of particle. In this case detention period is related to the falling velocity of the smallest particle settled, and cursory inspection of Figs. 86 and

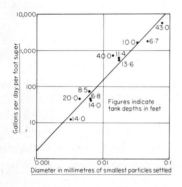

FIG. 86—Relation of sedimentation to surface area of tank, according to Hazen's figures

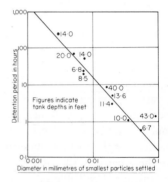

FIG. 87—Relation of sedimentation to capacity of tank, according to Hazen's figures

87 gives the impression that capacity and surface area are similar in their influence on efficiency of settlement. But it will be noted that in Fig. 87 the figures for deep tanks are generally on the right of the average line; and this suggests that surface area has more effect than capacity. The reason that capacity also appears to be important is that all sedimentation tanks vary in depth between narrow limits and that therefore the capacity is, in practice, more or less related to the surface area. The depths in this instance were more varied than those found in English practice, where generally the depth of tanks varies between 5 and 10 feet. Imhoff[124] followed Hazen in believing that depth, and therefore capacity, was not of importance, and C. H. Capen[37] also believed that depth had little effect.

Hazen's theory is interesting, and a useful indicator of the importance of surface area in settlement-tank design, but contains assumptions that are not valid except in the simplest conditions. The theory can be applied with limitations, to the settlement of detritus, and it is now usual to consider sur-

THEORIES OF CONTINUOUS-FLOW SEDIMENTATION

face area, along with other factors, in all sedimentation problems.

The theory that sedimentation depends on the surface area of the tank may be explained as follows:

Efficiency of sedimentation cannot occur unless, in the case of a cross-flow tank, the smallest size particle to be settled entering at top water-level can fall to the bottom of the tank during time of detention.

Therefore: time of detention $= \dfrac{\text{depth} \times \text{area}}{\text{rate of flow}} = \dfrac{\text{depth}}{\text{subsiding velocity}}$

which equation, simplified by cancelling out depth, gives us:

area × a constant for any subsiding value = rate of flow.

In the case of an upward-flow tank, velocity of upward flow = subsiding rate of smallest particle to be settled, and rate of flow = velocity × area.

Thus, it can be proved that it is not possible to obtain efficiency of settlement of the smallest size of particle concerned without the requisite surface area, and it may be said that the same surface area is required for similar conditions in both types of tanks. It is reasonable to assume, without further calculation, that in tanks where the flow is diagonally upwards the condition is still true.

Having proved that a sufficient surface area is necessary for efficient settlement, a further assumption may be made—i.e. that, allowing for a degree of inefficiency and working on the results of practical tests, a theoretical maximum rate of upward flow may be assumed for design purposes, *even if flow is not, in fact, upwards*, between the rates of $4\frac{1}{2}$ and 6 feet per hour upward flow at the maximum rate of flow. Allowing for a maximum rate of flow to be three times the average dry-weather flow, as is usual in English practice, the surface area of sedimentation tanks should not be less than 1 foot super for every 225 or 300 gallons per day dry-weather flow. For good results the figure of 225 is probably necessary, although many tanks have been designed to the figure of 300 for primary sedimentation, and 450 for humus tanks. Allowing for the difference of maximum rate of flow, this is compatible with American practice. Metcalf and Eddy[177] give 300 (U.S.) gallons per day per foot super for granular solids; 600 gallons per day per foot super for sewage solids for tanks 5 feet or more in depth; 800/1000 gallons per day per foot super for final tanks. Imhoff and Fair[125] give 900 (U.S.) gallons per day per foot super for preliminary sedimentation and 900/1800 gallons per day per foot super for secondary sedimentation or humus tanks.

Flocculation. Flocculation is the clinging together of small particles to form larger particles, thereby increasing the falling velocity. The rate at which small particles are brought together is influenced by the pH value of the containing fluid and the velocity of motion of the eddies in the fluid. Even inlet eddies have some effect, although it is believed that this is negligible.

The best results with mechanical flocculation appear to depend on more or less critical velocities, both of the sewage and the agitators. J. Hurley[117] quotes A. J. Fisher, who, writing in the American *Sewage Works Journal*, expresses the opinion that "sudden changes in direction and velocities in excess of 1·5 feet per second will at least partially undo some of the work accomplished," and it is well known that violent agitation will break up previously formed flocs.

Long periods of mechanical flocculation appear to be unnecessary. H. W. Gehm (*Proc. N. J. Sewage Works Assoc.*, p. 108, 1938), quoted by Keefer,[139] showed that, according to laboratory experiments, half an hour's mechanical flocculation produced a result 50 per cent as good as five hours' mechanical flocculation. It would appear from the curves shown by Keefer that after forty minutes' flocculation the curve becomes a straight line when plotted to a normal scale, and further increases are slight.

Flow in sedimentation tanks. Flow in sedimentation tanks is always complex. In the ideal case a sedimentation tank would be of even cross-section throughout its length, and the cross-sectional area of inlet and outlet would be the same as that of the tank. If this were, in practice, the case, and tanks were of great length, in the absence of disturbing factors, flow through the tank would be affected only by friction on the sides, and therefore the velocity at the centre near the surface would be greater than the velocity near the wetted perimeter of the channel. By experiment it has been found that at moderate velocities, and with a good form of inlet, flow behaves as if influenced chiefly by momentum.

The energy liberated by reduction of velocity and as a result of loss of head in the inlet has to be dissipated. The momentum of the incoming sewage may set up a swirling motion that revolves the whole of the sewage in the tank in either a horizontal or a vertical direction, or breaks up the mass of the tank into two or more large eddies. These large eddies cannot be prevented except in extraordinarily long tanks and necessitate study of inlet design.

Effect of turbulence. Turbulence *must* reduce settlement, because local upward flow carries with it fluid containing a more than average concentration of particles, while a local downward flow carries with it fluid containing a less than average concentration of particles.

Inlet effect. If water enters the tank in the form of a jet, even at a comparatively low velocity, the jet persists and, with little spread, continues across the tank until it meets with a surface which deflects it in another direction. Thus, if the entrance is in the form of a weir, the water falls over the weir and, in the absence of other disturbing factors, continues downwards until it reaches the invert, travels along the floor until it reaches the far end wall and turns upwards, striking the surface. Here a portion is deflected at right angles over the outlet weir and a portion turns towards the inlet. Thus, a large vertical eddy is set up.

If the inlet is in the form of a submerged weir without scum-board, the forward velocity continues over the sill and along the surface of the water of the tank. When the water reaches the outlet weir, a portion passes over and a portion is deflected downwards by the end wall and again deflected by the floor back towards the inlet end of the tank. A vertical eddy is thus set up in the opposite direction from that previously described. In both cases there is the mixing of the water in the tank with the water entering the tank, and the water which passes over the outlet weir contains only a portion of that which has passed at a high velocity from inlet to outlet.

Reduction of velocity-head causes the production of numerous small eddies. The converse is true, and increase of velocity due to convergence of flow towards an outlet reduces turbulence that would be present otherwise— a point worth considering in design. The salt solution experiments of Clifford and Windridge[42] showed that the degree of mixing depended on inlet velocity and on rate of flow through the tank. This is confirmed by A. Holroyd.[111]

A high-velocity jet discharging into a tank sets up violent eddies, which interfere with settlement. Eddies are formed by the expansion of the cross-sectional area of a stream, and as it is always necessary to expand from a pipe or other form of inlet to the cross-sectional area of the tank, eddies must be produced. If they are not to interfere with settlement they must be localised and caused to die out before the water containing them enters the main body of the tank. The Clifford inlet is one of the devices designed to produce this effect.

The energy in the eddy currents is dissipated by friction against the walls and floor of the tank and in the water itself. It is also absorbed in the lifting of settled particles. *Thus, the greater the proportion of sediment in the liquid, the greater the amount of energy required to prevent settlement.*

Large eddies have at their centres dead water—i.e. water which remains unchanged for long periods—and thereby the effective surface area in the case of horizontal eddies, and the effective capacity of the tank in all cases, are reduced. Small eddies add to the turbulence of flow in the tank and reduce sedimentation, as well as tending to lift previously settled sludge. This can be done by both horizontal and vertical eddies. Eddies revolving in a vertical direction produce downward and upward flows, and the upward flow may lift particles. Eddies which revolve in a horizontal direction, owing to friction with the bottom of the tank, revolve at a greater speed at the top than at the bottom. As there is greater centrifugal force at the top, there is outward flow at the top, inward flow at the bottom and upward flow in the centre, which may raise particles. This behaviour may easily be observed by placing a few grains of sand in a tumbler and stirring at increasing speed until the particles move towards the centre and finally rise.

Very slight differences of specific gravity, due to concentration of suspended solids, materials in solution or slight differences of temperature, are sufficient to cause marked movement in the body of sewage in a tank. In particular, the difference of temperature between sewage in tank and sewage entering tank influences direction of flow. Warm sewage will flow over the surface,[42] and cold sewage or sewage heavy with suspended solids will flow to the bottom of the tank.

Velocity of horizontal flow, and scour. Sediment is scoured as a result of the turbulence which is determined by the velocity and the hydraulic mean depth of the channel in which the flow takes place, together with the effects of varying cross-sectional area of flow. In investigations on conditions of stability in alluvial channels and rivers, Kennedy[140] and Lacy[155] found that scour was dependent, in the case of a river of great width, on depth, and generally, on hydraulic mean depth. Friction in channels, turbulence and scour can always be related to hydraulic mean depth when the channels are of unvarying cross-section and of great length. Sedimentation tanks are not of great length, and are greatly influenced by turbulence. Therefore one can only assume approximate velocities for the scour of either detritus or sludge, regardless of the form of the channel or its size or depth. Conditions are too complicated for calculations to be made. As previously mentioned, a velocity of 1 foot per second can be taken as being satisfactory for all detritus channels. There is some variation of opinion as to what horizontal velocities should not be exceeded in sedimentation tanks. This lack of agreement is due to the fact that the velocities mentioned by the different authorities are theoretical velocities, the actual local velocity in contact with the sludge being unknown. Metcalf and Eddy[177] refer to velocities of 45–150 feet per hour, and 30–40 feet per hour for the final sedimentation tanks for activated sludge works. Babbitt and Schlenz, and Imhoff, say that cross velocity should not exceed 600 feet per hour. Folwell gives 150 feet per hour, Fuller and McClintock 50–150 feet per hour, Smith 60 feet per hour for preliminary sedimentation and 30 feet per hour for final activated sludge.

Horizontal velocities produce vertical velocities in the order of one-tenth of the horizontal velocity. Thus, if one assumes that an upward velocity should not exceed 4–6 feet per hour, it would appear to be required that the horizontal velocity should not exceed 40–60 feet per hour. These theoretical figures agree with the opinions of some of the writers quoted.

TABLE 45
SLUDGE PRODUCED PER 1000 GALLONS OF SEWAGE, AT DIFFERENT VELOCITIES

Velocity (inches per second)	Sludge (gallons)	Analysis of sludge		
		Moisture (per cent)	Dry matter (per cent)	Dry matter (lb)
$\frac{1}{6}$	4·040	95·57	4·43	1·79
$\frac{5}{6}$	2·474	92·87	7·13	1·76
$1\frac{2}{3}$	1·838	91·34	8·66	1·59

Variation of horizontal velocity within limits influences the density of sludge to a greater extent than the efficiency of sedimentation, as illustrated by the figures given in Table 45, which are quoted by Kershaw[143] from Dunbar, who quotes them from Steuernagel. (The last column is worked out by the writer.)

Also quoted from Dunbar by Kershaw are the figures of Bock and Schwartz, who showed that "settling tanks respectively 162 feet long and 243 feet long, and worked with a velocity of $\frac{1}{6}-\frac{1}{3}$ inch per second, deposited 55·7 per cent of the suspended matter from the day sewage in the 162-foot tank, and 61·5 per cent in the longer tank. On increasing the velocity to $\frac{4}{5}$ inch per second, the figures for the longer tank were only reduced to 57 per cent."

Detention period. There is now no question that detention period is a most important factor in sedimentation-tank design, for a long detention period ensures (apart from any tank of absurd design) adequate surface area, low velocity of flow, a large body of water to absorb inlet energy and sufficient time for flocculation and biochemical changes to take place. Thus, present-day practice in Great Britain is to consider such details as adequate surface area, inlet velocity and direction, outlet-weir length, etc., and then to allow such detention period as appears to be convenient and adequate. Broadly, the tendency of British practice has been towards allowing a detention period of 6 hours dry-weather flow for primary sedimentation tanks in percolating filter and activated sludge works, and between $4\frac{1}{2}$ and 6 hours (generally the latter figure) for final sedimentation tanks of activated sludge works. In the last case, as will be described in Chapter XXX, the capacity of the final sedimentation tanks has to be considered not only from the point of view of sedimentation but also having regard to the fact that oxidation with the aid of activated sludge is continuing in these tanks, which may therefore be made larger at the expense of reduction of capacity of the aeration tanks, provided the latter are still large enough to house the aerating mechanism.

When speaking of detention period it is necessary to make clear what is meant. A primary sedimentation tank does most of its work during that part of the day when the rate of flow and the density of suspended solids load are highest: thus, sedimentation tanks have to be designed on peak, not average conditions. In Great Britain it is usual to speak of a tank having a detention period of "6 hours' dry-weather flow," but at the peak rate of three times dry-weather flow the theoretical detention period will be 2 hours. Figures in overseas publications probably refer to detention periods at the average daily rate of flow: just what is meant is rarely specified.

Occasionally writers on the subject have referred to the actual as against the nominal detention period on the grounds that, as a result of short-circuiting, the fluid is not retained in the tank as long as the calculated period. In this connection it should be pointed out that it is very rare indeed that there is not complete circulation of flow in sedimentation tanks resulting in part of the sewage passing over the outlet weirs long before the end of the detention period but other parts long after. It is not possible to state what is the actual average detention period without making exhaustive experiments.

Formerly in England capacities were largely based on the Royal Commission recommendation of 10 to 15 hours' dry-weather flow. In America capacities of $1\frac{1}{2}$ to 3 hours' dry-weather flow have been recommended by Keefer[139] for percolating filter works, 20 minutes to $1\frac{1}{2}$ hours for activated sludge preliminary tanks, 1 to $1\frac{1}{2}$ hours for humus tanks, and $1\frac{1}{2}$ to 2 hours for final activated sludge tanks. Metcalf and Eddy[177] refer to periods of $1\frac{1}{2}$ to 6 hours for percolating filter works, but 2 to $2\frac{1}{2}$ hours in very recent plants. Capen[37] considers that the detention period at the average rate of flow should not be less than 2·4 hours.

The difference between British and American sedimentation tank capacity appears to be largely a matter of traditional practice, although the fact that British works are designed to treat up to three times dry-weather flow may have been partly responsible. Various calculations made to find the economic optimum capacity for primary tanks for either percolating filter or activated sludge works have all shown that the American capacities are more strictly economical. The conditions, however, are not at all critical and a very large increase in tank capacity makes but a small over-all cost on the works as a whole, taking both capital and running charges into account. And extra capacity has several advantages from an operational point of view.

Curves of settlement. A considerable proportion of sludge is settled during the first hour: thereafter less and less is settled until there appears to be no advantage in further increasing tank capacity. What is particularly noticeable is that on the average it is necessary roughly to increase the capacity by eight times to double the quantity settled when working within the normal limits of practice. *There is no critical point.*

One of the most complete sets of experimental curves which give percentage solids remaining in the effluent after different periods of detention is that which was published by Holmes and Gyatt[110] (see Figs. 88 and 89). These results were obtained by experiments with a tank 100 feet long, 30 inches wide and at depths of 2, 4 and 5 feet. When replotted to logarithmic scales these curves became straight lines, which is true of most similar data (see Fig. 89).

There is very little information of the type obtained by Holmes and Gyatt that is accompanied by unquestionable evidence.* On Fig. 89 are plotted a

---

* Since this was written the writer has plotted on logarithmic paper results of study of some eighteen military camp sewage works which were published in "Sewage Treatment at Military Installations—Summary and Conclusions" by the N.R.C. Subcommittee on Sewage Treatment, *Sewage Works Journal*, January 1948. The curve so plotted completely confirmed those based on Holmes and Gyatt's work, the only noticeable difference being the very low efficiency of the military installation sedimentation tanks (presumably due to some detail of design) as compared with Holmes and Gyatt's long, narrow, experimental tanks.

THEORIES OF CONTINUOUS-FLOW SEDIMENTATION

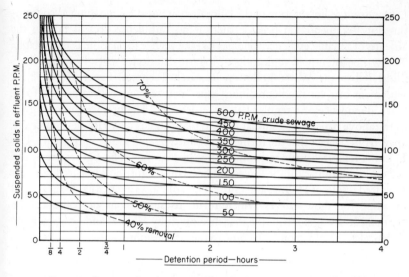

FIG. 88—Degree of settlement produced at various detention periods

curve quoted by Professor Slade which gives the experimental results of another writer, and a curve which is attributed to Johnson. These two curves reasonably agree with those of Holmes and Gyatt in slope. Curves given by Metcalf and Eddy,[177] together with the information from which they are plotted, agree with those of Holmes and Gyatt as regards the proportional effect of initial concentration.

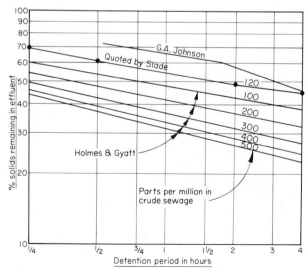

FIG. 89—Degree of settlement produced at various detention periods
(plotted to logarithmic scales)

Quiescent sedimentation in the laboratory is not the same thing as continuous-flow sedimentation because there is no turbulence present but, in the absence of reliable evidence, one has to take such information as is available. The writer attempted replotting to logarithmic scales several of the curves illustrated in the Water Pollution Research Technical Paper No. 7[229] for settlement of river mud containing 5 per cent sewage. There proved to be marked evidence of a regular law of settlement. The first section of Fig. 43 of the above paper is of a similar type to that produced by Holmes and Gyatt, apart from the fact that the former is obtained by quiescent experiment. This figure is replotted to logarithmic scales in Fig. 90. (The figures are scaled off, and the position of the lowest is uncertain.) Figure 90 compared with the replotted results of Holmes and Gyatt (Fig. 89) shows a remarkable agreement of slope for those curves which relate to suspensions of concentration normal for sewage. From the evidence given by Holmes and Gyatt and the technical paper No. 7 one can come to the conclusions:

1. The slope of the curve increases as percentage solids in crude sewage increases, and that therefore a long detention period has more effect in proportion to a short retention period when the initial concentration is strong than when the initial concentration is weak. This means that there is more rapid settlement of highly concentrated sediments.

2. In the case of sewage of normal range of concentration of suspended solids, the percentage of solids remaining in the effluent is reduced inversely as the fourth to fifth root of the time of detention.

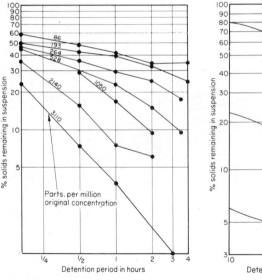

FIG. 90—The effect of detention period on settlement of river mud

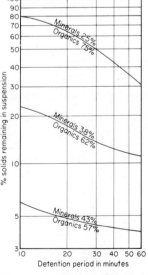

FIG. 91—The effect of detention period on settlement of activated sludge

Experiments with materials other than normal sewage show that the slope of the curve, and therefore the power (mathematical) of detention period, are influenced by the type of the material to be settled. For example, on Fig. 91 are plotted three curves from information obtained from Activated Sludge, Ltd., each of the curves being different, as the result of the different proportions of organic to mineral matter. The figures obtained by Clifford and Windridge[42] are not very regular, but appear to be more or less similar.

Egolf and McCabe[64] found that in controlled conditions rate of settlement could be predicted.

The writer, after examining several curves for settlement, is of the same opinion and has suggested the following formula as being representative of normal conditions of primary settlement:

$$S_2 = \frac{S_1}{C_1 \times \log S_1 \times D^n}$$

where: S_2 = suspended solids in tank effluent in parts per million
S_1 = suspended solids in crude sewage in parts per million
C_1 = a constant, which in the main experimental data was approximately 1·1
D = detention period in hours
$n = \dfrac{\log S_1}{C_2}$
C_2 = a constant, which in the main experimental data was approximately 10
(Logarithms are to the base 10).

This formula is not an exact interpretation of data and cannot be extrapolated beyond the normal detention periods or intensities of suspension other than those normal in sewage, but it gives a fair relation of the various factors and can be used for determining the effects of varying detention periods in any particular case, once the values of the constants have been found by experiment.

Recently the writer came across a print of a graph giving a series of curves of percentage reduction in suspended solids plotted against detention period for rectangular tanks 200 feet long, 42 feet wide and 11 feet deep. The print was indistinctly marked "from Journal of American Society of S.W. May 1940." The data of these curves replotted to logarithmic ordinates and abscissae showed that between the limits of one and ten hours and crude sewage suspended solids content of 50 to 300 parts per million, the value of C_1 was 0·725 and the value of n was 0·4 throughout and not varying according to the logarithm of the initial suspended solids content.

Making use of the principal eddies. As it is impossible to prevent main swirls from taking place in tanks, except in tanks of great length in proportion to width and depth, the designer can with advantage so design his tank that the direction of flow remains constant as far as practicable, and so arrange his weirs that they collect top water in the position where the concentration of suspended solids is minimum. Norval E. Anderson[6] showed that, in the case of final sedimentation tanks for activated sludge works, the weight of the sludge carried the incoming flow to the bottom of the tank, and that consequently the tendency was for sludge to rise at the outlet weirs, particularly in the case of rectangular tanks. By setting the outlet weirs away from the far end of the tank a distance of not less than 0·7 of the depth at that point, the sludge which was raised by the impinging of the current on

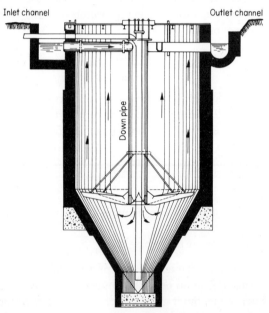

FIG. 92—Dortmund upward-flow sedimentation tank

the end wall had time to fall back, and a very clear effluent could be drawn off by the weirs which formed a bridge over the tank. The same principle was applied successfully to central inlet tanks, in which case, in place of a peripheral weir built on the outer walls, a channel was suspended at the surface as an annular ring intermediately between the central inlet and peripheral walls. The author included this arrangement in the final sedimentation tanks for Crossness Works, using pre-cast concrete channels supported on reinforced-concrete piers.

The principle of upward flow. In the original upward-flow tank installed at Dortmund, Germany, with the aid of an elaborate inlet distributor, a grid of surface weirs and a considerable depth in proportion to width, true upward flow was obtained (see Fig. 92). The intention of the upward flow was that particles of relatively small size should be carried up with the sewage, falling at a velocity lower than the upward flow of the sewage. Flocculation would cause these particles to increase in size before they reached the surface and to fall at a greater rate than the upward flow. In falling they would filter out small rising particles, thereby increasing the degree of flocculation. This theoretical principle of the upward-flow tank is still employed, but most so-called Dortmund or upward-flow tanks are merely pyramidal- or conical-bottomed tanks in which flow is usually radial and only partially upward and therefore they do not behave truly in accordance with the upward-flow principle, and should not be called Dortmund tanks. They are an extremely useful form of tank when properly designed, but their design is full of pitfalls, and when badly designed they are naturally unpopular with sewage-works operators.

Two-stage sedimentation. Most sedimentation tanks are single stage—that is, all tanks are placed in parallel. Two-stage sedimentation was known at the time of the Royal Commission, but has not been greatly favoured. Some early tanks were in the form of two-stage units, primary and secondary portions being both rectangular. The use of deep baffling in tanks, at one time common, may be considered similar in effect to two-stage sedimentation, and both forms are alike in that a large quantity of heavy sludge is settled in the first stage and a small quantity of light sludge in the second stage. This in itself is a disadvantage, in that it makes the handling of sludge more difficult. Between the World Wars there was a revival of the two-stage method, in which generally the first stage was a small-capacity tank continually raked and the second stage a large-capacity rectangular tank infrequently sludged. Whitehead considered that, as rectangular tanks were economical to construct, but economy of labour and efficiency were provided by mechanically cleansed tanks of radial-flow form, in large installations it may be found advantageous to combine these forms of construction by dividing sedimentation into two stages, the first stage being the radial-flow tanks and the second stage rectangular tanks of double the capacity. The method, however, is one on which there can be considerable variation of opinion, particularly as there is little information available on comparative results.

The efficiency of two-stage or multi-stage settlement is comparable with that of deep baffling, on which subject there is some variation of opinion. Metcalf and Eddy[177] express the view that too much baffling is likely to agitate sludge and mix it with sewage. Capen[37] considers that baffles, although sometimes useful, are mostly harmful. Clifford, replying to the discussion on his paper said:[42]

> Baffles in tanks were used many years ago, the idea being that they would prevent stream flow through the tank, ensure mixing and equalise the temperature. Model tank experiments showed that this idea was entirely fallacious at anything like the rates of flow obtaining in sewage works tanks. (He) had put baffles in a model horizontal flow tank and by allowing tap water to flow into coloured water

had obtained after the first baffle a ribbon of clean water passing through the tank; more than 50 per cent of the tank was "dead" water.

There would never appear to be any advantage in installing a second-stage tank after a first tank of very high efficiency. J. Hirst[108] found that the amount of purification in terms of oxygen absorbed brought about by secondary sedimentation in tanks with a capacity of four hours dry-weather flow after primary sedimentation in tanks of radial upward-flow type and eight hours dry-weather flow capacity was almost negligible. He quoted figures taken over a period of three years. He also expressed the view that in summer-time the advantage derived from secondary tanks was more than offset by the fact that their detention of the sewage tended to cause septicity.

Thus it would appear that the extra complication of two-stage settlement is generally not justified.

Experimental work. Even the simplest experiments on sedimentation are laborious, because the results of settlement are erratic, being influenced by causes sometimes so slight as not to be detectable. Very slight differences of chemical composition or temperature can produce considerable differences in results—a fact that has been observed by a number of writers. A good deal of experimental work has been carried out with the aid of coloured fluids and saline solutions, but it has been noted that the substances used, even when in diluted form, by their specific gravity influenced the flow. Camp[36] found that experiments with salt solution were invariably influenced by the specific gravity of the solution.

Full-scale experiment carried out in the actual tank of a sewage works is complicated by the variation of the rate of flow and of the strength of the sewage, the variation of temperature, the irregularity of pH value, etc., so that only by the averaging of readings taken over a very long period of time can data of value be obtained.

There are so many factors in play that the investigator may overlook a possible cause of variation, and, without justification, assign a particular cause to any observed effect. This is further complicated by attempts by experiment to prove preconceived theories, a not uncommon form of approach which invites error.

Although, under the guidance of the Ministry of Housing and Local Government[187] methods of testing sewage effluents have been standardised, there are all manner of methods applied to testing samples for determining the efficiency of sedimentation tanks. Some experimenters use the Gooch crucible, some centrifuge their samples, while some prefer to determine "settleable solids." By "settleable solids" is usually meant the amount of solids that settles during the theoretical detention period of a tank, settlement being made in quiescent conditions in a laboratory. The results of such a test depend on the differences of temperature in laboratory and sewage tank, and the depth of the settling vessel, which may be a shallow one (although, according to Imhoff and Fair[125] and Camp,[36] the settling vessel should be of equal depth to the tank). The results are also influenced by the quantity of sample taken from the settling vessel, as observed by Clifford and Windridge.[42]

To find the efficiency of a sedimentation tank under working conditions it is necessary to take samples of influent and effluent at intervals over at least twenty-four hours, and preferably seven days. The not uncommon practice of taking one sample of the influent and one sample of the effluent after a period of time equal to the theoretical detention period must lead to error, except when applied to flows of unvarying strength—for example, river-water required for water supply.

An example of how such work should be carried out is found in the Water Pollution Research Technical Paper No. 7, which is an investigation into the effect of discharge of crude sewage into the Estuary of the River Mersey on the amount and hardness of the deposit in the Estuary. In this description of work carried out by the Department of Scientific and Industrial Research[229] full details of the methods employed and lengthy tables of data are given, so that any investigator can use the material obtained for the checking of a theory of his own.

Model experiments. Model experiments make possible control of rate of flow, temperature and composition of sewage, but at the same time they involve fresh difficulties due to scale effect.

According to the theory given at the end of Chapter XVIII, experiments show that in an open channel the critical velocity occurs at $\dfrac{2000v}{m}$ cm per sec, where v is the kinematic viscosity which may be taken at 0·01 for average temperatures, and m is the hydraulic mean depth in cm. In the case of sedimentation tanks it will be almost certain that turbulent flow will be present throughout the tank, and therefore the ratio $\dfrac{v}{V} = \sqrt{\dfrac{l}{L}}$ applies. In this case the scale of the tank will be as follows:

If the diameter of the tank is reduced by dividing by the factor x, the surface area is divided by x^2, and if the scale is maintained completely, the capacity by x^3. Then all lineal dimensions are divided by x, including depth of flow over weirs, and therefore the flow in gallons per hour is divided by $x^{2·5}$. As velocity of upward flow or falling rate of the smallest particle to be settled is rate of flow divided by surface area, the velocity is divided by $\dfrac{x^{2·5}}{x^2} = \sqrt{x}$. Thus, a true scale-model tank reduced lineally by the factor x is capable of settling a particle which falls at a velocity reduced by $\sqrt{x}$, and this particle would be, in the case of large particles of more than 1 cm in diameter, reduced lineally in the same scale as the tank. However, the particles concerned in sewage sedimentation are not those the falling rates of which are due to friction, but are those which subside according to Stokes' Law, and therefore the particles are lineally out of scale with the model.

If, however, as is often the case, the particles to be settled are those found in the actual crude sewage as settled in the prototype, the subsiding rate in the model must be the same as that in the full-scale tank. Then it is necessary to make adjustment, and this can be done in some experiments by reducing the vertical scale—i.e. the depth of the tank, the depth over weirs, etc.—by dividing by $x^{\frac{1}{3}}$ instead of by x. Then the scale in the tank is as follows:

All horizontal lineal dimensions are reduced by dividing by the factor x. The surface area is reduced by the factor x^2. All vertical dimensions are reduced by the factor $x^{\frac{1}{3}}$. The capacity is reduced by the factor $x^{2\frac{1}{3}}$. The rate of flow is reduced by x^2. Ratio of flow to surface area then remains constant, as surface area is reduced by x^2 and rate of flow by x^2, and therefore upward velocity or sinking rate of smallest particle to be settled is the same in prototype and model.

It will be obvious that variation of the vertical scale will not be legitimate in all models, particularly those in which the prototype is of considerable depth and the ratio of size extreme.

All the above is based on the assumption that no flocculation takes place. Flocculation, which cannot be avoided in the sewage, will, in model experiments, give added scale effect for which adjustment can again be made within limits.

But it is the falling or terminal velocity of the particle which determines the scale reduction in settlement-tank model experiments, and if this is not kept in mind, the experiments will be out of scale and of reduced, if of any, value.

CHAPTER XXVII

SEDIMENTATION-TANK DETAILS

Having dealt with theoretical considerations, current practice should be discussed. The following are probably the average figures of present-day English practice. Primary sedimentation tanks have capacities of two hours detention at the peak rate of flow through the works (which means, of course, in most cases a detention period of six hours dry-weather flow); and a surface area of 1 superficial foot for every $4\frac{1}{2}$ cubic feet per hour or 675 gallons per day at the maximum or peak rate of flow. The inlets of the tanks should be so designed that the velocity at the point of inlet to the main body of water does not exceed $\frac{2}{3}$ foot per second at the peak rate of flow. Outlets should be in the form of weirs over which it is generally held that depth of flow should be shallow. According to some American opinion the peak rate of flow over outlet weirs should not exceed 166,666 Imperial gallons per day per foot of weir; some other American authorities give a maximum figure of 8340 Imperial gallons per day per foot of weir.

Types of tank. There are three main types of primary tank:

1. Hopper-bottomed or conical-bottomed tanks with peripheral weirs and central inlets. These include the (now unusual) true Dortmund or upward-flow tank and the more usual semi-upward- or outward-flow tank.

2. Circular tanks with bottoms sloping slightly towards the centre and sludged by rotating mechanism.

3. Rectangular longitudinal-flow tanks formerly hand-sludged by decanting and squeegeeing to the outlet, but now usually mechanically sludged without emptying.

These tanks should not be used indiscriminately, but selected according to circumstances.

At small works hopper- or conical-bottomed tanks are in every respect the best to use, for they do away with the need for the cost of installing and maintaining mechanical raking gear. The decision to use other than hopper-bottomed tanks arises when the works are so large that the number of hopper bottoms becomes too great for it to be easy for one man to sludge all the tanks in a day, giving proper attention to the "bleeding-off" of the sludge.

Bleeding-off sludge. The recognised method of drawing off sludge under hydrostatic head is known as "bleeding-off" the sludge. First the operative opens the sludge valve or penstock, if necessary rodding the outlet pipe, which should be straight and brought up to above sewage level to facilitate this. Then, as the sludge flow speeds up, he reduces the flow to the rate which he has found by experience to be optimum. In time sludge hangs up on the sides of the hopper and sewage is drawn down in a central cone. The operative, noticing the thinning of the sludge, then reduces or stops the flow until the sludge has subsided. This process, which requires careful attention by the operative, ensures the withdrawal of a sludge of a moderate water content: its neglect results in thin sludge making digestion tanks inadequate. While thin sludge can sometimes be dewatered in special tanks this should not be made necessary by bad operation of the sedimentation tanks.

Pyramidal-bottomed tanks. Pyramidal-bottomed tanks should be designed on capacity and surface area according to the figures already given. In English practice it is usual to slope the sides of the pyramids at not less than 60° to the horizontal so as to reduce the hanging up of sludge. These sloping sides may be brought right up to weir-level, or the upper part of the tank may have vertical walls, according to the relation of surface area to capacity, of ground-level to water-level, or any peculiarities of the site.

The inlets of the tanks should be so arranged that sewage enters the tank above the highest level of the amount of sludge likely to be collected just before sludging. At small works sludging may be as infrequent as once a week but for design purposes it is best to allow for sludging at intervals of four days. The amount of sludge that should then be allowed for should be based on the suspended-solids content of the sewage and a sludge moisture content of $97\frac{1}{2}$ per cent. It should be borne in mind that whereas the settlement efficiency of the tank may be only 70 per cent the remaining 30 per cent of suspended solids that escapes over the weirs will be intercepted in the humus tanks or final sedimentation tanks of a percolating filter scheme or activated sludge scheme respectively, and returned to the primary tanks for settlement. It will be noted that the sludge settled in primary tanks occupies a very considerable depth of a pyramidal bottom; if the inlet is not placed above the level of the sludge, efficiency of settlement will be greatly reduced. In one case, a hopper-bottomed settlement tank was giving no settlement at all, because the inlet was too low, and in the form of a downward jet, so that sludge was kept in continual suspension until it became septic, gassed and rose to the surface to form scum.

The inlet should be central, and may be in the form of a bend turned upwards towards the surface surrounded by a baffle box, or a cast-iron or concrete T-pipe, the flow coming in through the junction, which should be about one-third of the diameter of the main pipe. Or again, a Clifford inlet with eddy bucket may be used.

Unless the tanks are true upward-flow Dortmund tanks, the weirs should be peripheral and be protected by scum-boards projecting about 12 inches below water-level and 2 inches above, and arranged at least 9 inches from the weirs.

As pyramidal tanks are usually of moderate size, it is safe to say in all cases that the depth over the outlet weirs will be moderate.

Design method for pyramidal tanks. A number of the worst faults possible in the design of pyramidal tanks can be avoided by adhering to the following principles of design:

A. If a sedimentation tank is to be efficient, the surface area should not be less than 1 square foot per 225 gallons per day dry-weather flow.

B. The weirs should be completely peripheral: the inlet should be central, and so designed that it does not produce a high-velocity jet.

C. The peripheral weir should be as long as possible, but not so long that it would flow in patches instead of continuously over its length. (With certain of the materials used in weir construction, surface tension prevents flow when the depth over the weir is less than $\frac{1}{10}$ inch.)

D. The inlet should not be so low in the tank that it would at any time be below the level of settled sludge if the tank should be left unsludged for some time.

The following is an example calculation for a tank that is part of a percolating filter scheme:

Population: 27,000 persons.
Dry-weather flow: 30 gallons per head per day = 810,000 gallons per day.
Maximum rate to be treated: three times dry-weather flow.
Quantity of sludge: $\frac{1}{2}$ gallon per head per day at $97\frac{1}{2}$ per cent moisture content.
Tanks to be complete pyramids with sides sloping at 60°.

A rule-of-thumb figure of 2 gallons sludge to be stored per head of population is based on the assumption that tanks are sludged once every four days and the moisture content is $97\frac{1}{2}$ per cent, or once every eight days (a week and a day's grace) and the moisture content is 95 per cent. These are usually ample allowances.

The calculation would then be, for the above example:

$$4 \text{ days' quantity of sludge} = \frac{27{,}000 \times 4}{2 \times 6\frac{1}{4}} = 8640 \text{ cubic feet.}$$

If this quantity is to be stored in the lower two-thirds of the depth of the pyramidal tanks (a satisfactory rule of thumb), the capacity of the tanks becomes:

$$8640 \times \left(\frac{3}{2}\right)^3 = 29{,}160 \text{ cubic feet} = 5 \cdot 4 \text{ hours' dry-weather flow.}$$

$$\text{The required surface area} = \frac{810{,}000}{225} = 3600 \text{ square feet.}$$

It has been pointed out by the writer[69] that if a pyramidal tank which is a complete pyramid with 60° slopes is less than 27 feet square, the weirs will flow in patches for most of the day unless they are castellated, and to avoid castellation one should assume that tanks are approximately 27 feet square. Then the number of tanks required is obtained as follows:

$$\frac{3600}{27^2} = 5 \text{ tanks to the nearest whole number}$$

$$\text{Capacity of one tank} = \frac{29{,}160}{5} = 5832 \text{ cubic feet} = 0 \cdot 289.x^3$$

x being length of side of tank.*

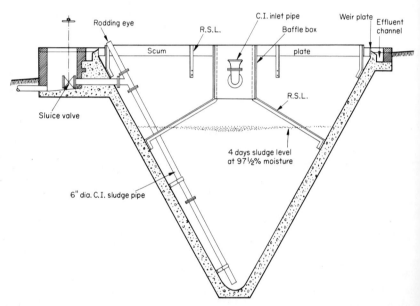

FIG. 93—Cross-section of pyramidal sedimentation tank

Circular mechanically-raked tanks. Where works are too large for the use of simple hopper-bottomed tanks, circular mechanically-raked tanks are usually indicated. Circular tanks have the advantage over rectangular mechanised tanks in that a much simpler mechanism can be used for raking the sludge and operation is simple, for the gear merely has to be switched on and off again as required, or may even be left running indefinitely. There is no need for automatic stops, lifting and lowering scraper blades, as in the case of rectangular tanks, or transfer of mechanism from tank to tank.

The sludge-raking gear for primary circular sedimentation tanks is almost invariably in the form of rotating arms, two or four in number, with a blade or blades pushing the sludge towards a central hopper having at least one

* The capacity of a 60° pyramid is approximately $0 \cdot 289.x^3$, when x is the length of base.

day's capacity and from which the sludge is withdrawn under hydrostatic head. One of the earliest forms of blade and, theoretically, the most perfect, was a single spiral plate. This proved difficult to erect and maintain in rotation without jamming, and multiple plates are more common, either as two parts of a spiral arranged on a moving bridge extending from the centre to the periphery of the tank or several short echelon blades arranged on two or more arms.

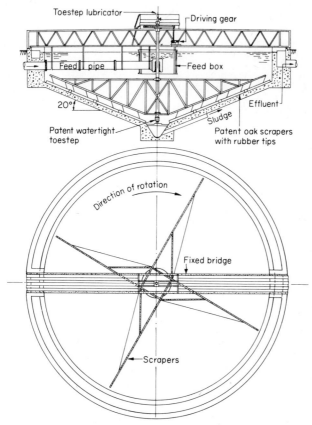

FIG. 94—Bridge-type sedimentation tank for activated sludge
By courtesy of Hartleys (Stoke-on-Trent), Ltd.

To overcome jamming of the blade of the spiral single- or two-piece blade against the bottom of the tank, a modern arrangement is to have the blade well clear of the bottom of the tank and to have it followed by numerous trailing short blades that pick up the sludge escaping through the gap.

There is a second reason for having trailing blades which is particularly important in the case of final tanks of activated sludge works. It is not uncommon practice when constructing circular tanks to screed the concrete with the aid of boards attached to the rotating mechanism so that the mechanism will rotate just clear of the face of the concrete. This, of course, is done when the tank is empty with the result that when the tank is filled with sewage or effluent the arms tend to float up and leave an appreciable gap between any fixed blades and the concrete: in one instance this was measured as $1\frac{1}{2}$ inches at the periphery. To overcome this trouble trailing blades can be used or, as illustrated in Fig. 97, lengths of chain.

An objection that has been raised to multi-arm echelon blade scrapers is that the sludge is very slowly moved from the periphery to the centre,

FIG. 95—Sludging mechanism for circular tank
By courtesy of William E. Farrer, Ltd.

being pushed a few feet by each short blade and left on the floor of the tank until the next section of blade comes round and pushes it another few feet. Thus (in theory) the sludge may take hours to be removed, which, incidentally, does not matter in the case of primary tanks. In practice, it has been found by experiment that echelon blades do not work in this way; actually they cause a general streaming of the sludge towards the centre of the tank.

Scum is usually removed by a blade which pushes the scum over the ramp into a chamber at some point on the periphery. This, of course, applies only to primary tanks or humus tanks after percolating filters—not the final tanks for activated sludge works, where no allowance is made for scum removal. Primary tanks must have sludge storage hoppers of sufficient size to store all the sludge collected between desludging times, otherwise satisfactory "bleeding-off" will not be possible.

Rotating scrapers are driven in two ways. In the fixed-bridge type of mechanism a bridge and gangway lead from the periphery to the centre, at which point is an electric motor and gear-head that drives the rotating arms below the bridge. The weight of the central end of the bridge and the rotating gear is carried on an up-take tube which serves as inlet to the tank. Scrapers not involving a fixed bridge are usually arranged to pivot only at the centre or merely be supported there, and are driven by a motor at the periphery with wheels that run on a rail at the outer wall or on a rubber tyre on the wall itself. Peripheral speed should not exceed 4 feet per minute, particularly in the case of activated sludge tanks.

Rectangular tanks. Mechanisms for scraping rectangular tanks are largely used and particularly were in the inter-war years for the mechanisation on conversion of old quiescent and other hand-swept flat-bottomed tanks. For this purpose they are ideal, but they are also used for new rectangular tanks in favourable circumstances. Rectangular tanks occupy less space than circular tanks of equal surface area, and they usually involve less complex feed and effluent channels. On the other hand, they have the distinct disadvantage of more elaborate and less easily operated raking gear, and if a machine is provided for each tank, they generally prove more expensive than circular tanks. The last disadvantage, however, can be offset if the machine can be transferred from one tank to another, as will be described. In most instances the mechanism of the circular tank is easier to maintain and operate.

Rectangular tanks are normally about five times as long as they are wide, and vary in depth from 5 to 10 or more feet. The bottoms should slope at about 1 in 100 towards the inlet end where the sludge hoppers should be provided. These should be large enough to hold all the sludge collected in the sludging period, which may be one day, two days or four days, as the designer may consider desirable.

The inlet should be arranged not only to ensure a low velocity of inflow but also so to direct the incoming sewage that it does not disturb the sludge settled in the hoppers. The weirs should be at the outlet end of the tank and protected by scum-boards.

Scrapers for rectangular tanks are, in Great Britain, generally of the moving-bridge type running on rails and carrying scraper blades for removing sludge and scum. Sometimes one machine is provided per tank, but more often each machine is supplied with a transfer carriage in order that it may be used to sweep one tank, then be transferred laterally to sweep another. To arrive at the most economical arrangement, the amount of sludge likely to be collected should be calculated and a decision made whether the tanks are to be swept once a day,* once every two days or twice

---

* In some instances tanks have been swept twice a day, but this produces a thin, watery sludge.

a week. Then, bearing in mind how long it takes to sweep each tank and what work can be done in a day shift, the minimum number of machines required can be estimated.

Scum should be swept towards the inlet end, not to the outlet end as is sometimes done, and it should be pushed over a ramp into a channel from which it can be washed away by the sludge, which can also discharge into the same channel. This is better than the arrangement of drawing off scum and, unavoidably, vast quantities of sewage over weir-penstocks at the sides of the tanks; for the latter method dilutes the sludge to which the scum is eventually discharged.

The blades of scrapers for rectangular tanks should lift right out of the sewage when the scrapers are moving towards the outlet end: they must, of course, lift out if a transfer carriage is used. The speed of motion should not exceed 4 feet per minute while the blades are submerged.

There is some difference of opinion on the speed at which scraper mechanisms should travel. A recent English experiment made on final sedimentation tanks for activated sludge showed that increasing the peripheral velocity of the scrapers of a 95-foot diameter tank from 3·25 to 11·6 feet per minute had the effect of increasing the suspended-solids content of the final effluent from 16·75 parts per million to 24·50. It should be appreciated that this means that the increase in velocity did as much harm as would have been done by reducing the detention period of the tank by an appreciable amount.

To prevent cross-winding causing jamming, jerking or even derailing of the bridge, the wheel-base of the bridge should not be less than one-fifth of the span between rails. The bridge should be rigid and the drive should be by two wheels on one axle, the wheels turned with great accuracy to the same diameter. These wheels are usually double-flanged; but one maker has flanged wheels at only one side of the tank together with rollers bearing on the sides of the bull-head rail, while at the other side of the tank the bridge is supported by rollers bearing on the top of the rail only with no flanges or other lateral control. An additional advantage of this arrangement is that it makes ample allowance for temperature expansion where the tanks are wide.

Electricity is supplied to transfer carriages and travelling bridges either by bare wires and contacts or by insulated cable. The latter arrangement reduces the possibility of accidents to workmen. The cable should be wound on drums of the "catherine-wheel" type not of the "bobbin" type, as the former pick up and lay the cables in straight lines. The cables of the transfer carriages should plug in at various convenient points along the transfer track; those of the bridges should plug into sockets on the transfer carriages. In America, Link-Belt mechanisms are used, which consist of continuously running chains carrying a series of blades which sweep the sludge to one end of the tank and, on their return journey along the surface, sweep the scum to the other end. Similar mechanisms have been installed in Great Britain but have been found to involve excessive maintenance.

SEDIMENTATION-TANK DETAILS

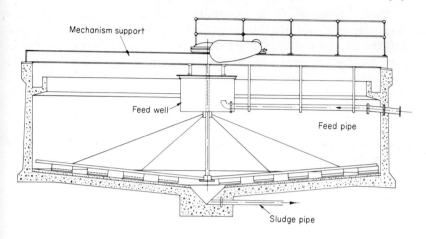

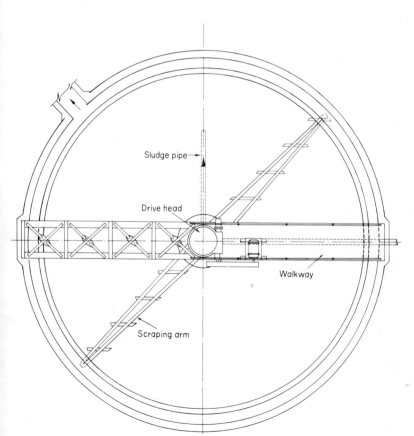

Fig. 96—Dorr "Clarifier" type "A"
By courtesy of Dorr-Oliver Co., Ltd.

Sludge removal from pyramidal tanks. Some early pyramidal-bottomed tanks had bottoms sloping at 45°. These were found to be unsatisfactory, and consequently the slopes of the bottoms of pyramidal tanks have increased progressively. At the present time in England, a slope of 60° to the horizontal is almost universal, this angle, like the previous angle of 45°, presumably being chosen because it was easiest drawn with a standard set-square. In America somewhat steeper slopes—for example, slopes of 1·75 in 1, or steeper—are common. Steep slopes reduce the hanging up of sludge, but unless they are very steep they do not entirely prevent it. They add considerably to the cost of construction.

The sludge pipes from pyramidal-bottomed tanks usually consist of 6-inch-diameter pipes in England, regardless of size of tank, but are generally not less than 8 inches diameter in America. They may be either cast-iron pipes erected inside the tanks and controlled by sluice valves or clay pipes set in concrete outside the tanks and controlled by penstocks. The disadvantage of the first arrangement is that pipes add to the possibility of sludge lodging on the sides of the pyramids; the advantage claimed is that they may easily be repaired. External sludge pipes do not add to the internal ironwork of the tank with the above-mentioned disadvantage, and although it has been suggested that they are inaccessible for repair it might be pointed out that such repair is seldom required, and simply accomplished by disuse of the external pipes and replacement with internal pipes of cast-iron. Bellmouths to sludge pipes are not considered good practice, in that they can lead to choking: simple spigot ends are preferred. In too many installations there are one or more bends, often of sharp radius, which make rodding difficult. Every sludge pipe should have a rodding eye and, as is often quite easy to arrange, the pipes should, where possible, form a straight line between this eye and the bottom of the tank. It should hardly be necessary to state that the rodding eye should be above top water-level, but there have been occasions when this has been overlooked. The hydrostatic head required for starting the sludging of a pyramidal tank is considered to be at least 3 feet. This may be increased with advantage, but too great a head adds to the difficulty of operation.

Deep sumps at the bottoms of pyramids are preferred by operators because they render easy the complete removal of sludge. They have the disadvantage of being costly to construct. The nominal sump often seen in design, which consists of a cubical box at the bottom of a pyramid (see Fig. 65), is of no value, and can be omitted.

Wash-outs. Sewage-works managers appreciate the installation of means for emptying sedimentation tanks, and are inclined to consider pyramidal tanks ill-designed if they have not means for complete emptying. On the other hand, to lay pipes so as to gravitate from the lowest point of an inverted pyramid to a general wash-out pumping station usually means expensive excavation at a considerable depth and complete refilling of trenches with

SEDIMENTATION-TANK DETAILS

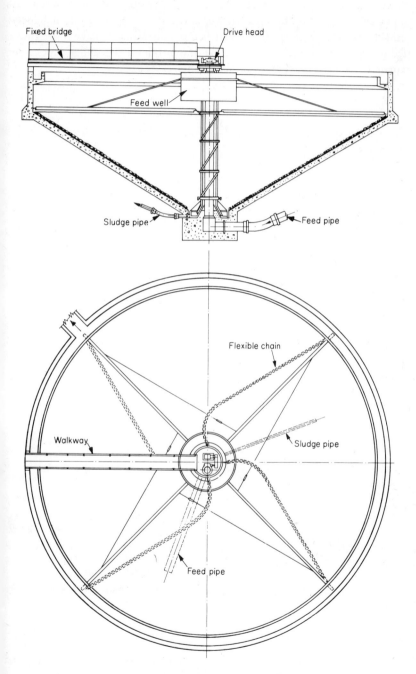

FIG. 97—Dorr "Clarifier" type "SC"
By courtesy of Dorr-Oliver Co., Ltd.

concrete where they are under or near to tank work, etc. Such wash-out drains can be astonishingly expensive, and for this reason some designers omit them entirely, suggesting that when pyramidal tanks have to be emptied a large part of the sewage can be drawn off via the sludge pipes, the remainder being removed by portable pump.

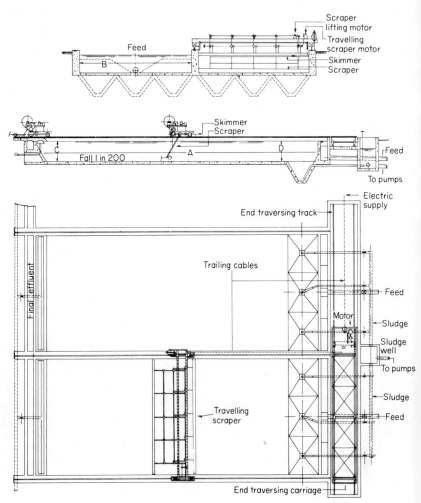

FIG. 98—Rectangular sedimentation tanks with travelling scraper and transporter carriage
By courtesy of Hartleys (Stoke-on-Trent), Ltd.

When, however, wash-out pipes are omitted, the designer should make sure that the lowest part of the sedimentation tank is not beyond practicable suction lift of a pump placed at coping level. He should also provide for the inclusion, in the contract, of a suitable petrol or electric pump, a place for it to be housed satisfactorily and, if an electric pump is chosen, outdoor plug-in power points, for where there is electricity at the works reliable electric motors are to be preferred to prime movers.

When tanks are partly above ground-level, wash-out pipes can be arranged about 3 feet below ground-level, so as to discharge the greater part of the tank contents by gravity, then, with proper arrangement of valves on the wash-outs, the latter may serve as suctions for a portable pump, or, alternatively, a permanent pump housed at a low level close to the tanks. The latter, however, is a provision suitable for larger works only.

FIG. 99—Hartley sludging mechanism, Crossness sewage-treatment works

Scum removal. Sedimentation tanks serve an additional purpose in intercepting floating matter or scum. To prevent the scum from passing over the outlet weirs, scum-boards are provided near the outlets and, formerly, at intermediate positions throughout the length of horizontal-flow tanks. The last arrangement is not favoured now, for it is appreciated that scum-boards or baffles additional to those at the outlet weirs generally tend to reduce efficiency of sedimentation.

Scum-boards should not be placed too near the outlet weirs, for they induce high upward velocities and reduce efficiency of sedimentation in the same manner as if the weirs were submerged to the level of the bottoms of the boards.

Scum-boards are usually constructed of well-creosoted soft-wood or suitable hard-wood (not elm, because this material rots particularly rapidly when between "wind and water"). Steel scum-plates are also used, but these rust. Reinforced concrete is more durable, and slate, although expensive, has been employed.

Inlet and outlet channels. Quite a large proportion of the earlier sedimentation tanks still in use are faulty in the design of inlet and outlet channels. In works of recent construction conditions are, on the whole, much better in these respects.

Fig. 100—"Mieder" scraper
By courtesy of William E. Farrer, Ltd.

As a legacy from the quiescent method of sedimentation it was at one time the usual practice to feed rectangular tanks by means of one long, diminishing channel. The tanks nearest the source of supply received a greater quantity, owing to the hydraulic gradient of the channel producing a greater head above top water-level in tank or inlet weir than in the case of the tanks fed farther down the channel. The effect of this was that some tanks having high-velocity flow coupled with large rate of flow settled out large quantities of heavy sludge: other tanks settled out small quantities of light sludge of high moisture content. As would be expected, this added to sludging difficulties.

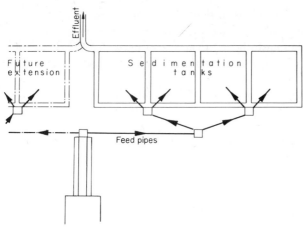

FIG. 101—Diagram illustrating the system of progressive bifurcation of feed-pipes of equal length

It is now becoming accepted that the best practice is to feed individual sedimentation tanks by means of separate leads of equal length from a common distribution chamber, or, alternatively, to split up the flow to the tanks by a series of bifurcations so as to distribute the flow equally (see Fig. 101). In arranging the splitting of flow, the effect of velocity-head should not be overlooked. This can be very large.

Inlet pipes and channels, in spite of previous screening and settlement of detritus, are liable to contain some heavy solids, and therefore the pipes and channels should be designed to secure high velocities of flow.

The assumption that after sedimentation there is no silt that can cause fouling of outlet channels has been proved by experience to be erroneous. Even where sedimentation is quite up to the usual standard there is a proportion of settleable material carried over the outlet weirs due either to excessive velocity of flow in tanks or to flotation caused by gassing of sludge. Over a period of time this sediment can require the manual cleansing of any outlet channel if the velocity of flow is too low. There can, however, be no sedimentation in the outlet channels if they are designed to give a velocity of flow equal to that in the inlet channels. Perhaps the practice of making channels "amply large" is the cause of low velocities, and although inadequate channels have been known, those of excessive size are more common. *Sizes should always be calculated.*

A quick and easy method of designing channels is as follows. The diameter and gradient of the circular pipe required to pass twice the maximum rate of flow at a velocity of 3 feet per second are found (by reference to Vol. I, Fig. 68). Then rectangular channels are constructed of equal width to the diameter of the pipe and greater depth than the radius of the pipe to give freeboard. In calculating the falls on the invert, loss of head at right-angle bends is assumed to be equal to the friction in a length of channel of forty diameters.

The difference between the area of a semicircle and the area of a half of square gives a factor of safety of about 6 per cent allowing for the discharge of channels tending to be less than that of pipes (see Vol. I, Chapter X). For freeboard a depth of more than the maximum possible velocity-head may be necessary.

Construction of pyramidal sedimentation tanks. Experience has shown that the sloping sides of pyramidal bottoms of tanks are very strong, and normally do not require to be of more than a few inches in thickness. This sometimes surprises structural engineers who have not had experience of this particular line of design and who, if left to themselves, would be inclined to make the inverted pyramids of moderate-size tanks as much as 18 inches thickness of mass concrete, or else to use reinforcement. Such thicknesses or reinforcement are quite unnecessary in the majority of instances.

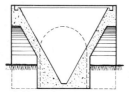
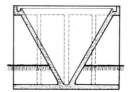
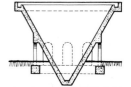

FIG. 102—Pyramidal sedimentation tank constructed above ground-level in mass concrete

FIG. 103—Pyramidal sedimentation tank constructed above ground-level in reinforced concrete and resting on reinforced-concrete raft

FIG. 104—Pyramidal sedimentation tank constructed above ground-level in reinforced concrete to an economical design

When pyramidal bottoms are constructed below ground-level in good ground, a thickness of concrete of about $\frac{1}{2}$ inch per foot length of side of tank at top of pyramid will often suffice, but this figure should be increased slightly in the case of other than the best subsoils.

When tanks are constructed in bad ground where, owing to the nature of the subsoil or the presence of subsoil water, it is necessary to make the excavation in a series of lifts, employing stages of vertical timbering, the space left between the face of the excavation and the outside of the concrete of the pyramid is filled with 10:1 concrete. This concrete fill forms a series of foundations to the tank which makes it possible for the good-quality, watertight concrete of the pyramid to be maintained at the minimum thickness.

When pyramidal tanks are constructed wholly, or mostly, above ground-level, three methods of construction can be considered. The first of these, which is applicable to those circumstances where reinforcement is undesirable —e.g. in out-of-the-way parts of the world—is to construct the tanks of mass concrete so designed that the structure is amply strong to withstand the internal water pressure without the aid of earth support (see Fig. 102). This construction, however, is too costly for general application.

Reinforcement is particularly useful for the construction of tanks above ground-level, and the form of construction which usually occurs to the designer is that illustrated in Fig. 103, where a reinforced-concrete pyramid is supported from a reinforced-concrete raft by means of reinforced-concrete wing walls. This construction is applicable to tanks supported above ground-level, but resting on a subsoil which will not take a heavy load. It should be considered too extravagant for use on good ground.

The form of construction illustrated in Fig. 104 is very economical for tanks which are erected above ground-level on sound ground; or are supported on piles, the tank being either above the ground or in the ground where the subsoil is very unstable—e.g. peat or alluvium. The pyramid is of reinforced concrete supported on its apex, and at a point near the top, by foundations carried to sound ground, or else by piles.

The maximum bending moment in a reinforced pyramid occurs at one-third the depth when the pyramid is filled with water. Unless beams are provided, the main reinforcement should be horizontal.

CHAPTER XXVIII

TREATMENT AND DISPOSAL OF SEWAGE SLUDGE

THE amount of sludge produced depends on the nature of the sewage and on the method of treatment adopted.

The average quantity of sludge dry solids at British sewage-treatment works is about 0·173 lb (0·0785 kg) per head of population per day which, at an average moisture content of about 95·52 per cent gives 0·387 gallon (1·76 litres) per head per day. Of the dry solids 72·5 per cent are volatile matter or 0·125 lb per head per day. These figures include both the solids that come down the sewer and those produced by the aeration process, percolating filters or activated sludge treatment.

When sewage is settled only and not given any secondary treatment that will produce additional solids and returned unsettled solids for settlement, the quantity of sludge is much less, being about 0·12 lb (0·0545 kg) per head per day dry solids which, at an average moisture content of about 92·5 per cent, gives 0·16 gallon (0·727 litre) of sludge per head per day. Of the dry solids 72·2 per cent are volatile matter.

The average quantity of humus settled in humus tanks after percolating filters is 0·059 lb (0·0268 kg) per head per day or 0·766 gallon (3·48 litres) per head per day at a moisture content of 99·23 per cent. Of the dry solids 64·2 per cent are volatile matter.

At activated sludge works (diffused air) the daily weight of dry solids in the surplus activated sludge is 0·058 lb (0·0253 kg) per head which at a moisture content of 99·33 per cent is 0·868 gallon (3·94 litres) per head per day of which 77·1 per cent is volatile matter. The quantity from surface-aeration plants is said to average about twice as much, but as there is some possibility of error in the data this opinion should be viewed with caution.

The above figures are average, based on a "Report on the Water Pollution Research Laboratory Survey of the Performance of Sewage Sludge Digesters throughout England, Scotland and Wales" by Swanwick, Shurben and Jackson, May 1968.

As moisture contents can vary considerably higher figures should be allowed in design calculations than given above, for a small change of moisture content makes a very big difference in the amount of sludge. It should be noted that a change of moisture content from 90 to 95 per cent or from 95 to 97½ per cent *doubles* the quantity of sludge, this being because it is the quantity of dry matter which remains constant per head of population at any particular sewage works and addition of water which increases the amount of sludge to be disposed. Suppose you have a sludge of

90 per cent moisture content, the dry-matter content will be 10 per cent. If this 10 per cent is 1 ton, there will be 9 tons of water. The addition of a further 10 tons of water will alter the moisture content to 95 per cent and the dry matter content to 5 per cent, at the same time doubling the quantity of sludge (see Fig. 105).

One interesting fact that the foregoing figures illustrate is that the amount of sludge solids returned from percolating-filter humus tanks is not less than the surplus activated sludge solids produced in a diffused-air plant. This is contrary to the bulk of opinion held in the past which was that activated sludge works produced far more sludge than percolating-filter schemes. It is true that sludge has a high moisture content but so has humus and, as will be seen, the quantity of surplus activated sludge is not much greater than the quantity of humus in gallons per head per day.

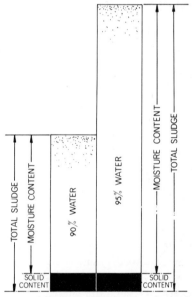

FIG. 105—Diagram illustrating how the moisture content of sludge affects the volume

Calculating quantities of primary sludge. The quantity of sludge to be dealt with in any part of a sewage-treatment plant depends on the weight of suspended solids and the percentage of water in which the solids are suspended. Thus, the easiest way of making sludge calculations is to determine the weight of solids received into the process, deduct any losses and then add the appropriate moisture content.

The sources of sludge at sewage works are the suspended solids in the incoming crude sewage and the suspended solids produced by biological action from the dissolved solids in the sewage. The losses which have to be accounted for in the various calculations are the loss of suspended solids that escape with the final effluent and the solids that are converted to gas in the process of digestion.

The quantity of primary sludge, excluding any sludges returned from secondary processes, can be estimated from the analysis of crude sewage. It is not usual to make any allowance for loss due to detritus or screenings removal because, on the one hand, sewage samples are often taken after these processes and, on the other hand, the quantities of detritus and screenings that are kept out of the primary sludge are comparatively small.

Suppose we are doing the calculations for sewage works to treat 81,000,000 gallons per day average flow of sewage that has a suspended-solids content of 400 parts per million and a B.O.D. value of 350 parts per million. The weight of dry suspended solids will be:

$$400 \times 81 \times 10 \text{ lb per day} = 324,000 \text{ lb per day}.$$

If this amount of solids forms a sludge of 96 per cent moisture content (or 4 per cent solids content) the quantity of sludge will be:

$$\frac{324,000}{4} \times \frac{100}{10} = 810,000 \text{ gallons of sludge per day}.$$

Efficiency of primary sedimentation can vary considerably, but in the majority of instances it is somewhere in the region of 70 per cent. Thus, at sewage works where sedimentation of crude sewage is the only process or where humus or other secondary sludges are not returned to the flow of sewage for settlement with the primary sludge, the quantity of sludge settled can be found by taking the appropriate percentage of the quantity of sludge as calculated above. For example, if settlement efficiency is 70 per cent the quantity settled would, in the case of the example calculation, be 567,000 gallons per day.

However, wherever a Royal Commission effluent is required, the amount of suspended solids escaping with the effluent is very small, say 15 parts per million in a good scheme. This means that the overall settlement efficiency of the sewage works must be high. In effect, what happens in any modern percolating filter or activated sludge plant is that the solids that will not settle in the primary tanks are altered and made settleable in the aeration units, where additional solids are added to them. These altered and additional solids settle with great efficiency in the humus tanks or final sedimentation tanks.

In percolating filter schemes the humus is often returned to the flow of sewage to be resettled with the primary sludge. The effect of returning humus to the flow is very nearly the same as that of ensuring 95 per cent efficiency of settlement in the primary tanks, for while humus is an altered product consisting largely of the debris of the vegetable and animal growth in the percolating filters, it contains also the sludge which had escaped settlement in the primary tanks in the first instance. And as, where a Royal Commission effluent is secured, the suspended solids in the final effluent do not exceed 30 parts per million, the amount of primary sludge, including humus, cannot be less than the difference between incoming and outgoing solids. Thus, returning to our original calculation and allowing on the average a loss of suspended solids of 15 parts per million with the final effluent, the quantity of *settled* primary sludge including returned humus:

$$= \frac{81 \times (400 - 15) \times 100}{100 - 96} = 780,000 \text{ gallons per day.}$$

(It should be noted that the foregoing calculation has been simplified by assuming that a gallon is not a measure of volume but an equivalent weight of 10 lb.)

In practice the amount of humus is more than that of the solids which have failed to settle in the primary sedimentation tanks, for the humus contains suspended solids produced from dissolved solids by the percolating-filter aeration process. On the average during the year the humus may amount to about one-third of the total sludge produced at the treatment works, but the quantity varies considerably throughout the year.

From the calculation of sludge produced in a percolating filter scheme, it will be seen that where humus is returned to the primary tanks the efficiency of primary sedimentation has practically no effect on the quantity of sludge to be passed to sludge treatment. Settlement efficiency is, however, necessary, because without it an excessive load is put on the percolating filter. This applies to a much lesser extent in activated-sludge works, where, in nearly all instances, surplus activated sludge is returned to the primary tanks. But the efficiency of the primary tanks enters the sludge calculations, for it affects the quantity of surplus activated sludge to be pumped. On the other hand, the quantity of returned sludge—i.e. sludge drawn off from the final sedimentation tanks and returned to form mixed liquor in the aeration tanks—has no effect whatsoever on the quantity of surplus sludge or, in fact, any other calculation of sludge quantities: the returned sludge may be considered as a closed circuit going round and round on its own, as far as sludge calculations are concerned.

The primary sludge settled in an activated sludge plant consists of the whole of the suspended solids brought in by the crude sewage, plus the organic solids produced from dissolved solids by the activated sludge process, minus the solids lost with the final effluent.

Unfortunately the aforementioned report does not include any information on the weight of suspended solids delivered from the sewers to the primary sedimentation tanks, the weight of suspended solids lost with the final effluent or even the efficiency of primary sedimentation. It is therefore not possible to derive from the figures in the report the quantity of suspended solids produced as a result of growth of organisms in the aeration process.

This figure is not easy to derive for it requires very full and accurate information on all the gains and losses of solids in the works, the final figure being a difference between the gains and the losses as measured in the appropriate samples. The best the writer has been able to do has been to suggest that the solids produced in a diffused-air plant (and, perhaps, a percolating-filter scheme) are in the region of 0·2 lb per lb B.O.D. delivered to the works.* (From the figures in the report it would appear that a somewhat similar quantity of solids is produced in percolating filters although it is known that anaerobic digestion takes place below the layer of aerobic organisms on the particles of filter medium.)

Accepting this figure the solids settled in the primary tank, in the example with which we started, are:

$$
\begin{aligned}
\text{solids in crude sewage, } 81 \times 400 \times 10 &= 324{,}000 \text{ lb per day} \\
\text{solids produced by process } 81 \times 350 \times 10 \times 0{\cdot}2 &= \underline{56{,}700} \text{ ,, ,,} \\
\text{Total} &= 380{,}700 \text{ ,, ,,} \\
\text{solids lost with effluent, say 15 p.p.m., } 81 \times 15 \times 10 &= \underline{12{,}150} \text{ ,, ,,} \\
\text{settled solids} &= 368{,}500 \text{ ,, ,,} \\
\text{settled sludge at 96 per cent moisture content} &= \frac{368{,}500}{10} \times \frac{100}{4} \\
&= 921{,}250 \text{ gallons of sludge per day.}
\end{aligned}
$$

Disposal of sludge. Sludge disposal is one of the major problems of sewage treatment because sewage sludge is not a substance of high salvage value, and although to some extent it can be used as manure, the quantity produced exceeds the normal demand. As a consequence, unless efforts are made to popularise the sludge as a manure, or means are found for its destruction or reduction in volume, vast quantities of sludge accumulate at the sewage works.

At one time many towns gravitated their sludge into lagoons or low-lying places, in which it remained wet and putrescent for indefinite periods. This obviously was not a desirable process, but it has not been completely abandoned.

To some extent sludge can be disposed in the wet condition as received from the sedimentation tanks. In the first place, certain towns conveniently situated ship their sludge out to sea and dump it in deep water. This is a cheap method, costing more or less the same as drying on beds, according to circumstances, but it is limited to those sewage-treatment works suitably located and of sufficient size to justify the construction of vessels for this purpose.

Another method of disposing of wet sludge is to deliver it by pipe-line to neighbouring farms, where it can be used as manure. However, as will be pointed out later, sewage sludge is not, as is popularly imagined, an ideal fertiliser, and there are limitations to its utilisation for this purpose.

---

* At Crossness sewage-treatment works they averaged 0·125 lb per lb B.O.D. removed by aeration.

At small sewage works, particularly those serving isolated buildings, sludge is disposed by running it into trenches and covering it over, or ploughing it into land reserved for the purpose.

The approximate relative costs of the various means of sludge disposal are given below in terms of percentage of the cost of drying on beds.

Method of disposal	Cost as a percentage of drying on beds
Disposal in lagoons	25
Drying on land or pumping to farm land	50
Drying on sludge beds	100
Shipping to sea	100
Filter pressing	150
Heat-drying of filter-press cake	300
Vacuum filtration and flash-drying	670

Sludge drying. The reduction of volume of sludge by dewatering is very great. For example, the sludge from pyramidal sedimentation tanks has a moisture content of $97\frac{1}{2}$ per cent, but this can be reduced by drying to about 55 per cent, which means that the original weight of sludge is reduced to one-eighteenth. (A cubic yard of this dried sludge weighs about 14 cwt.) For this reason, in the majority of instances sludge is most easily disposed after the moisture content has been reduced to such an extent that a freely flowing liquid is converted to a dry, friable solid which can be removed from the works by road vehicle.

The generally preferred method of sludge dewatering is that of discharging the liquid sludge on to beds of clinker through which the water content, or "sludge liquor," gravitates, while the solid matter is retained on the surface. When the sludge has been on the beds a week or so—the period depending on the weather and the nature of the sludge—the surface cracks and the sludge becomes dry and spadable. At this stage it is dug out of the beds and dumped near-by for further drying before disposal from the site, in order that the beds may be ready to receive fresh doses of wet sludge.

The survey made by the Water Pollution Research Laboratory showed that, of works operating heated digesters, 75 per cent used sludge-drying beds. In suitable circumstances sludge-drying beds can be merely excavations in which about 9 inches depth of clinker of about 1 to $1\frac{1}{2}$-inch grade is laid to a level surface and topped with $1\frac{1}{2}$ to 2 inches of fine, clean clinker of about $\frac{1}{4}$- to $\frac{1}{2}$-inch grade. The last is required as a filter material to prevent sludge from entering the coarser material: it has to be renewed from time to time because it is not possible to dig out the sludge without removing some of the fine clinker.

The beds are provided with under-drains, usually agricultural tiles, which discharge to manholes, and thence to the sludge-liquor drains The sludge liquor, being very strong, needs to be treated, and this means that it has to be pumped back to the flow of raw sewage, with which it is passed through the works.

The manholes which collect the sludge liquor can serve a secondary purpose, for while the sludge is resting on the beds, supernatant water collects at the surface, and this can be drawn off, thereby accelerating the drying. One method of doing this is to form openings with slotted sides in the walls of the above-mentioned manholes. Pieces of 1-inch by 2-inch timber can then be stacked flat side up in the opening. When supernatant water has collected on the surface, the timber can be removed to such a level that it serves as a weir which draws off the water. Special adjustable weirs also are manufactured for this purpose.

Sludge beds are usually constructed in a more permanent form than simple excavations. Concrete floors are laid and brick or concrete walls built round the beds and false floors of special filter-tiles are provided. This method of construction is more generally preferred to the plain excavation, but it is considerably more expensive. It is necessary where pollution of the subsoil water is not permissible.

Sludge is discharged on to drying beds to a depth of about 9 inches. This determines the depth of the excavation, the height of the wall and the size and number of the beds. It is undesirable for a bed to be partly filled only with sludge at one time and afterwards completely filled, for in the interval between the first and second filling the first dose of sludge will become partly dried, and this will prevent the efficient draining of the second deposit. For this reason, each individual bed should be of such a size that it will not accommodate more than the amount of sludge discharged from the sedimentation tanks on a sludging day.

Another aspect of operation which influences the number of beds is the rotation of operation, filling the bed, leaving it to dry and digging out the dried sludge. While some beds are in the process of being filled, others are draining, and others are being emptied. To make this rotation practicable, there should not be fewer than four beds at the smallest sewage works, and preferably not fewer than six; while at large works it is advantageous to have a considerable number of beds.

At small works sludge is usually removed from the beds by barrow, although portable mechanical loading equipment can be used. For this reason there is some advantage in having a point of easy access to each bed: there should be either a ramp from the level of the clinker to the access road or an opening in the wall between the bed and the access road which is closed by dam-boards except when the sludge is being barrowed out. A properly constructed road must also be brought right up to every bed, in order that the sludge may be carted away. For large works elaborate sludge-removal machines are available, in particular those made by Norstel and Templewood Hawksley, Ltd.

Area of sludge-drying beds. At one time it was usual to allow 1 square

TREATMENT AND DISPOSAL OF SEWAGE SLUDGE

yard of sludge-drying bed for every seven persons. The advent of sludge digestion led to a reduction of area where that method was applied, on the grounds that digestion not only reduced the quantity of sludge but, it was believed, that digestion invariably made the sludge more easy to dry. The general opinion now is that all sludge beds, including those for digested sludge, should be ample in proportions. At entirely new works 1 yard super should be provided for every five or six head of population and one square yard per $3\frac{1}{2}$ persons is not uncommon. Where sludge is digested, the drying area can, perhaps, be reduced, but this is not always the case.

Much attention was at one time given to glass-roofed beds, which were used at many works in the United States. It has been claimed for these that

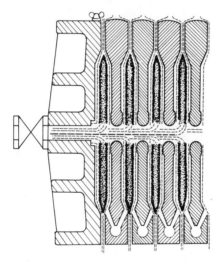

FIG. 106—Cross-section of part of a filter-press
By courtesy of Manlove, Alliott & Co., Ltd.

the sludge takes about half as long to dry on them as on beds in the open; but even if this were so, it is difficult to see where the advantage comes in, for each square yard of bed with a glass roof will usually cost two or three times as much as a square yard of unroofed bed. Another arrangement is the provision of corrugated-iron roofs that can be moved on rails from one bed to another.

Some operators consider that such impediment to air flow can delay the later stages of drying. Recent experience has, in fact, cast some doubt on the advantages which are claimed for covered beds. In drying sludge or any other wet material the access of air is not the least important factor, and if the wind blows freely over the sludge the latter will part with its moisture much more quickly than if it is confined under a roof. Rain is not such a hindrance to the drying of sludge as is generally supposed; and digested sludge which has once begun to dry will not relapse into a sodden condition.

Filter-pressing. Another method of dewatering sludge, which was widely used in this country forty years ago, is filter-pressing.

A filter-press is built up of cast-iron plates, with raised edges, several of which are placed together face to face in a frame, with coarse cloths of jute or hemp between them, and screwed tightly together. The spaces between the plates form a series of narrow cells, into which the liquid sludge, dosed with lime and aluminium sulphate, is pumped or forced by compressed air under a pressure of about 60 lb per square inch. The water escapes through the cloths, and by draining grooves formed in the plates into a channel

FIG. 107—Sludge cake from filter-press
By courtesy of W. K. Porteous, Ltd.

below, leaving the solids in the cells in the form of a stiff cake. The whole operation of filling, pressing, removing the cake and reassembling the press takes about three-quarters of an hour. Filter-presses were very commonly used at works where chemical precipitation was carried out, but they are now rarely installed except in conjunction with special processes. The filter cake has a moisture content of about 75 per cent.

Centrifugal separators. The disadvantages of filter-presses led to experiments with centrifugal separators of various types.

Most experiments did not give satisfactory results although one method is said to have been used with success in Germany and there is some centrifuging of sludge in England.

Vacuum filters. A vacuum filter consists of a cylinder of metal gauze divided internally into several sectors, which rotates slowly, with its undersurface dipping into the liquid sludge. In the sectors which are immersed in the sludge a vacuum is maintained, which draws the water into the cylinder, leaving a thin paste of solid sludge on the outer face of the gauze. As the cylinder rotates, this cake is lifted out of the bath, and air is drawn through it, further reducing the moisture content. When the sectors have risen still higher, the vacuum is shut off from them and air is admitted under a slight pressure, loosening the cake from the gauze. The cake, which contains about 80 per cent of moisture, is then removed by a fixed scraper.

When handling a mixture of undigested primary and humus sludge an average dose of 5 per cent ferrous sulphate and 10 per cent lime of the dry

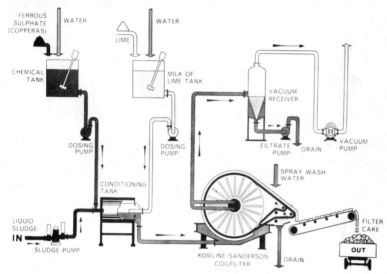

FIG. 108—Flow diagram of a Coilfilter installation
By courtesy of Dorr-Oliver Co., Ltd.

matter in the sludge should be added to the sludge prior to filtration. These proportions will vary with the type of sludge being filtered.

Vacuum filtration can be preceded by elutriation or washing the sludge with several times (at least twice) its volume of clarified effluent in circular upward-flow tanks. The supernatant water is drawn off from the surface and the resettled sludge from the bottom. The chief difficulty that has been experienced with vacuum filters is the choking of the filter fabric with the finer particles in the sludge. It has been found in laboratory tests that this choking can be completely cleared by back-washing, but the *back-blowing* in some types of filters has not proved effective. The difficulty is overcome in the Komline-Sanderson "Coilfilter," which is supplied by the Dorr-Oliver Co., Ltd.

In this filter the medium consists of two layers of small-diameter stainless-steel helical coils. Each coil has its ends connected together so that it forms a continuous loop carried around the filter drum and the discharge and guide rolls of the mechanism. In the lower layer the coils lie side by side but not hard together and in the upper they lie on the lower coils above the spaces between them.

During the cake-formation and cake-drying cycles, these coils are wrapped around the slowly revolving drum, the lower layer being in grooves in the transverse division strips of the drum. During the cake-discharge cycle, the coils leave the drum and are separated from each other by two discharge rolls in such a way that the cake is lifted off the lower layer and discharged from

FIG. 109—Surface of digesting sludge
By courtesy of Birmingham, Tame and Rea District Drainage Board

the upper layer of coils by means of a tine bar. Both layers are then washed by spray nozzles and reapplied to the drum by groove-aligning rolls: final effluent can be used for this washing. As the coils travel over the discharge rolls they bend first in one and then in the other direction, while being washed at the same time. This ensures that the filter medium is clean before the next dewatering cycle.

The filters are made in ten sizes from 56 square feet (5·2 square metres) to 580 square feet (54 square metres) nominal area. The nominal area is the diameter of the drum times the effective width times π. The diameter and width of the smallest size quoted are 6 feet and 3 feet respectively and of the largest size 11 feet 6 inches and 16 feet respectively.

The average output when handling a mixture of undigested primary and humus sludge is in the order of 3 lb of dry solids per square foot of filter area per hour. This figure will vary with the type and character of sludge being filtered, for example, undigested primary sludge alone will give a higher average yield with less chemicals and undigested primary sludge plus surplus activated sludge will give a lower yield with more chemicals.

For determining the size of a filter for works where overtime is not worked, it is usual to base the design on one 8-hour shift only per day for 5 days per week, the actual filter working time per 8-hour day-shift being 7 hours to allow for starting up and and shutting down.

Digestion. Sludge digestion has now become the rule rather than the exception at all except small works for it is a very satisfactory method not only of reducing the total quantity of sludge to be disposed, but at the same time rendering the sludge inoffensive, reducing the concentration of pathogenic bacteria, making the sludge a more suitable material for use on farm land and, sometimes, rendering it more easy to dry. In addition, it results in the production of gas which, at large works, may justify the expenditure of digestion tanks.

Broadly, digestion is a well proved and satisfactory process but troubles have occurred at various works, the majority of these being due to unsatisfactory design or operation. The principal causes of failure under this head are:

1. Sludge escaping via the outlet at the bottom before it has had time to digest because, being cold and heavy, it goes straight to the bottom on entry. This can be overcome in design by making sure that the sludge on entry is mixed with an excess of hot sludge from the tank and by having an alternative outlet that is not at the bottom.

2. Overloading due to the tank being designed to too small a capacity.

3. Inadequate control of temperature.

Trade wastes can give trouble but only to half the extent as poor design or operation. To a very much smaller extent anionic detergents have been blamed.

Types of sludge digestion. The process of sludge digestion is the liquefaction of organic material by anaerobic bacteria which produce an alkaline reaction. When sludge is first placed in a tank, acid digestion sets up, with the production of noxious gases, but eventually alkaline digestion prevails, and, when once established, the alkaline condition remains indefinitely unless upset by improper operation of the works. Alkaline digestion does not produce offensive conditions, the odour of the sludge being tarry and not resembling either sewage or septic sludge. The gas given off by the sludge during digestion is principally methane, with some carbon dioxide and smaller quantities of other gases.

It is stated by Fair and Carlson that the digestion of sludge takes place in three stages:

"(1) a short period of relatively rapid digestion, during which the pH value (or alkalinity) of the sludge decreases, and with it the rate of gasification;

(2) an extended period of slow or inhibited digestion, characterised by low gas production and low, but rising pH value; and

(3) a third stage, when the pH value rises to about 6·8 and the production of methane gas becomes active. During this stage the pH value usually passes the neutral point (7·0), and the production of gas reaches a maximum, dropping off as the digestible material becomes exhausted."

There are three forms of sludge digestion:

1. Cold digestion at the normal day temperature of a temperate climate.
2. Mesophilic digestion for which the optimum temperature is artificially maintained at between 80 °F and 95 °F.
3. Thermophilic digestion for which the sludge is heated to 110°–120 °F.

At low temperatures below 45 °F (which is the lowest winter sewage temperature in London) very little digestion takes place. But at 60 °F the normal temperature of sewage, which is slightly more than average day temperature, digestion is appreciable, and it is practicable to reduce sludge volume by the construction of digestion tanks. If, however, the temperature

FIG. 110—Dorr digester

This is one of the designs of Dorr digester in which the heating pipes are fixed, not moving and the tank is covered by a gasometer. Note scraping and stirring mechanism and outlet for gas (to left). Other Dorr digesters incorporate rotating heating coils, overhead sludge thickening tank, etc.

By courtesy of Dorr-Oliver Co., Ltd.

is artificially raised to about 85 °F, rate of digestion is greatly increased. Moreover, the quantity of gas given off by the sludge during digestion is normally considerably more than sufficient to provide the heat required to maintain this temperature. This mesophilic digestion at about 85 °F is most commonly employed at large works in this country, opinion having been expressed to the effect that it is the most satisfactory from the operator's point of view.

Dr. Willem Rudolfs, of New Jersey, experimenting on thermophilic digestion, found that by maintaining the sludge at a temperature of 110°–120 °F, digestion was greatly accelerated, the time required being reduced to less than that needed for digestion at 85 °F. But it is more difficult to maintain these higher temperatures because of the much greater losses of heat by conduction. Dr. Prüss, of Essen, who also tried thermophilic digestion for nearly a year, found the process so sensitive to changes in the composition of the sludge that the experiments had to be abandoned.

Economic digestion period. If a sufficient period of time is allowed, very nearly complete digestion of organic content is possible, but because the rate of digestion reduces in time, it is not practicable to aim at complete digestion, and normal practice is to digest sludge sufficiently to secure an inoffensive sludge that can be easily disposed.

Two-stage digestion. Two-stage digestion is preferred for all plants, regardless of size, but particularly for those at the larger works. Two-stage digestion simplifies gas collection and assists removal of supernatant water and water-bands which collect below the surface. With single-stage digestion, all the gas is given off in the same tank, and this, together with any mechanical mixing or stirring that may be introduced, militates against the formation of water-bands. But when two stages of digestion are provided, the greater part of the gas is given off in the primary tanks, from which it can be collected for utilisation. Moreover, the primary tanks may be arranged to be heated and provided with stirring or mixing devices, and removal of supernatant water from them can be neglected. The secondary tanks are relatively quiescent, also the amount of gas given off is comparatively small and need not be collected. But from the secondary tanks supernatant water may be removed. It is also usual for primary tanks to be heated but secondary tanks to be unheated, so as to simplify the construction of the latter and because cold sludge is much easier to dewater.

Capacities of digestion tanks. The average daily load on heated primary digestion tanks in Great Britain is about 0·097 lb of volatile solids per cubic foot (1·55 kg per cubic metre) of tank capacity, or 0·134 lb per day of total solids. At the average figure of 4·48 per cent solids content of sludge, this works out at 21 days detention period or 1·29 cubic feet (0·0366 cubic metre) of tank per head of population served.

These are fair design figures, giving a reasonable factor of safety in the majority of instances without extravagance. The average capacity of secondary tanks following a heated primary tank and capable of giving, on the average, 30 per cent dewatering, if the sludge and the operation are suitable, is not less than $1\frac{3}{4}$ times the above capacity of the primary tanks and $2\frac{1}{2}$ times if a good standard (50 per cent or more) of dewatering is desired.

With the above conditions in primary digestion tanks and a temperature of about 90 °F (32 °C) about 40·8 per cent digestion of volatile solids (on the average, 29·6 per cent of total solids) should take place, giving 15·4 cubic feet of gas per lb of solids destroyed, or 0·7854 cubic foot per head of population per day. In practice the quantity of gas varies considerably, being largely dependent on the animal and vegetable oil content of the sludge.

Cold digestion is usually employed at works serving populations of not more than 10,000 persons: apart from rare exceptions the method is not used when the population exceeds 50,000. As a general rule it can be said that if sludge is digested in Great Britain without artificial heat, a sufficient degree of digestion can be effected in about two months, but the favoured period, to be judged by the number of works employing it, is in the region of three months.

If the temperature is raised to the thermophilic range of 110°–120 °F the digestion period can be reduced to about seventeen days or possibly a fortnight at 125 °F. As, however, thermophilic digestion, being unstable, is not popular, mesophilic digestion is adopted at nearly all large and many moderate-size sewage works.

Dewatering. If the quantity of sludge can be reduced by reducing the moisture content, considerable economy can be made in tank capacities. Nevertheless, the desirability of primary sludge thickening remains controversial, and in any case is dependent on circumstances. Very wet sludge as might come from pyramidal-bottomed sedimentation tanks can be improved merely by standing in dewatering tanks for a few hours and decanting the supernatant water: the deeper the tank, the longer the period of quiescent storage required.

Sludges that are already fairly dense, as for example those which come from mechanically raked tanks, do not dewater so easily. In cold weather they can be thickened but in warm weather the dewatering falls off or ceases altogether. For these reasons it is sometimes thought more economical to spend on extra digestion capacity the money that could have been spent on thickening tanks and flocculating mechanisms. Some degree of thickening can be effected by letting the sludge stand and drawing off the supernatant water. In certain circumstances the thickening can be speeded by gently stirring with rods but this method is not used much because of some reports of unsatisfactory results.

But there is an additional reason for thickening which should not be overlooked: wet sludge requires more heat to bring it to the desired temperature, and in some circumstances there may not be sufficient heat available for the process unless the sludge is dewatered to some extent.

Reduction in volume of sludge by dewatering is calculated by Santo Crimp's formula:

$$W_2 = \left(\frac{100 - P}{100 - Q}\right) W_1$$

where: W_1 = original weight of wet sludge
W_2 = weight of pressed sludge
P = percentage moisture of wet sludge
Q = percentage moisture of dewatered sludge.

Dewatering surplus activated sludge. Owing to the very high moisture content of surplus activated sludge there is great advantage in effective dewatering. The procedure that has most commonly been adopted is to discharge surplus sludge into the flow of sewage upstream of the primary sedimentation tank, but this is by no means ideal for it can make primary sludge less dense than it would be if settled alone.

Ames Crosta Mills & Co., Ltd., now manufacture the Komline-Sanderson Dissolved Air Flotation Thickening Process, which can be used for certain industrial purposes and, in particular, for dewatering activated sludge at sewage-treatment works. It is claimed that the process will normally reduce an activated-sludge moisture content from 99·5 per cent to 96 per cent or less and that resultant moisture contents between 95 per cent and 93 per cent are not unusual.

The method consists of the dissolving of air in a liquid under pressure, adding this liquid to the sludge to be dewatered and reducing the pressure so that fine bubbles are produced which carry the suspended solids to the surface forming a sludge blanket, described as the "float," which is then mechanically skimmed off.

The liquid that is aerated is recirculated effluent of the unit. Owing to recirculation certain residual solids tend to build up and therefore provision is made for short-period substitution of effluent from either the primary or the final sedimentation tanks.

The float, which grows to a thickness of 8 to 24 inches, rises above the upper surface of the liquid in the tank and loses moisture content before it is swept into a hopper by a variable-speed skimming mechanism. It has a minimum solids content of 4 per cent and, containing a considerable amount of air, weighs about 7 lb per gallon. If the float is retained in a consolidation tank, the air dissipates and the sludge returns to normal density. This consolidation tank should have a minimum floor slope of $7\frac{1}{2}°$ to the horizontal if provided with scrapers and preferably 60° if without scrapers. A detention period of one or two days is desirable.

The standard loading of the flotation tank is two-thirds gallon per minute per square foot of surface area. As an average for design purposes the solids load can be taken as 2 lb per square foot of surface area per hour although the flotation thickener will accept loading of twice to $2\frac{1}{2}$ times this figure.

It is usual to allow for the possible use of chemicals that will assist flocculation, and chemical-feed equipment is included as a standard part of the plant, although it may be found that chemicals are not necessary.

The manufacturers provide complete prefabricated units in sizes of 15 to 500 square feet surface area: larger units are installed in concrete tanks. Pilot plants are available at nominal hire charge in order that prospective users may assess the effectiveness of the process at their own works.

Shapes of digestion tanks. The shape of a digestion tank depends on the type of digestion. If sludge is to be digested without heating or stirring, the form of the tank should be such that cost of construction is low and short-circuiting is reduced to a minimum. To secure these ends, the tanks should be comparatively shallow, being about twice as wide as they are deep, and comparatively long—say not less than twice as long as they are wide. The inlet should be below the surface at one end and the outlets at the other end, one at the bottom, the floor falling towards it and another at a higher level. Circular tanks with central outlets can be used.

There is no advantage in covering cold digestion tanks unless the gas is required for some purpose. If, however, the gas is to be collected, it is advisable to use floating covers which rise and fall with the level of the sludge as the tanks are filled and emptied. The reason for this is that when tanks are roofed over with a fixed cover, there is a considerable danger of air being drawn in on withdrawal of sludge, producing an explosive mixture.

When tanks are heated and stirred, the best form they can take is circular on plan, and for the sake of economy of cost and to reduce heat losses, they should be deep. A reinforced-concrete circular tank is economical to construct, and, if the depth is not much less than the diameter, the loss of heat through the sides is reduced to a minimum.

If, however, the primary tanks are made too deep, the rate of gas emission at the surface will cause foaming—a trouble that not only calls for cleaning up but is also liable to choke gas mains, putting the plant out of action. Foaming generally commences when gas emission reaches a figure of about 30 or 40 cubic feet per day per foot super of top surface of sludge. Accordingly, quantities of gas have to be considered before surface area, and before depth of tank, can be calculated. The amount of gas obtained varies from about 0·6 to about 1·2 and averages about 0·8 cubic foot per day per head of population served and is about 15·4 cubic feet per pound of solids destroyed. Whenever practicable, the figure should be obtained by either laboratory or large-scale experiments before works are designed. Where this cannot be done it is best to assume that 0·8 cubic foot per day per head of population will be available for power and heating purposes; but for determining surface area of primary digestion tanks and capacities of gas mains, the higher figure of 1·2 cubic feet per person per day should be allowed. Allowance should also be made for fluctuation of gas yield, which is dependent on various factors, such as storm flows in sewers. In one instance it was found that the yield fluctuated between 20 per cent above and below the average, being at these extremes for several days at a time; while production could be as much as 10 per cent above or below the average for as long as a fortnight.

The loss of heat through the bottom of the tank is negligible because of the surrounding earth, but any parts of the walls above ground-level can be

FIG. 111—Heater and circulating chambers of Dorr-Oliver "B.G.H." digesters, Crossness sewage-treatment works

embanked or else heat-insulated with advantage if there is any shortage of heat. In some instances there is ample heat and omission of insulation saves money and may improve the appearance of the tanks (see Figs. 111 and 112). The scum on the surface of the sludge reduces heat loss at the surface and does away with the necessity for heat-insulating the cover. Such tanks can be covered either by floating roofs, with pockets for gas collection, or by gasholders similar in some respects to the gasholders used at gas works.

The bottom of a circular stirred tank is usually sloped to a central outlet at a moderate angle, which may be as much as 40°, but a flat-bottomed tank can be employed particularly if the stirring mechanism includes a raking mechanism by which the sludge is drawn towards a central outlet. Opinion varies as to the desirability of sloping floors: they may reduce the deposition of silt but make manual cleansing difficult if the slope exceeds 1 in 6.

Heat requirements of primary tanks. In summer-time the raw sludge has to be raised from 60° to 95 °F: in the coldest period of the winter it suffices to raise the temperature from 45° to 80 °F. Thus the heat rise can be taken as 35 °F throughout the year, calling for 350 B.Th.U. per gallon of sludge put into the primary tank. By the use of suitable heat exchangers to transfer heat from outgoing sludge to incoming sludge, in the region of a

Fig. 112—View between Dorr-Oliver "B.G.H." digestion tanks, Crossness sewage-treatment works

transfer of 10 °F can be secured without difficulty, thereby reducing the heat demand to 250 B.Th.U. per gallon of sludge treated. This, however, is not a usual feature.

To the heat required for raising the temperature of the sludge in the first instance must be added the heat input necessary to make up that lost by radiation and conduction through roof or gasholder, walls and floor of tank less the heat produced by the digestion process. This is not too easy to calculate accurately, owing to paucity of scientific data, but a fair estimate can be made.

It is certain that some heat is produced by the fermentation of the sludge in the tank but no accurate information is as yet available on this matter and therefore allowance for this is not normally made in calculations. In making some tests on the efficiency of a sludge heating unit in which flows and ingoing and outgoing temperatures were recorded for both heating water and sludge, the author found that there was no recorded loss of heat by radiation, etc., but a slight gain. This could have meant that the heat produced by digestion was greater in amount than that lost by radiation from a tank which, having an annular space outside its walls, was well protected against heat loss. As, however, the tank shared a division wall with another tank that was not withdrawn from service during the test, this possibility should be viewed with caution.

TABLE 46

SPECIFIC HEAT OF SLUDGE AT 60 °F

Per cent total solids	Specific heat
0	1·0
1	0·993
2	0·986
3	0·978
4	0·971
5	0·964
6	0·957
7	0·950
8	0·942
9	0·935
10	0·928

Various tests on existing works have shown that the loss of heat from primary tanks of normal construction is the equivalent of a temperature drop varying from 0·5 to 1·5 °F and averaging about 1·0 °F per day. If one accepts the figure of 1·0 °F temperature drop, the amount of heat required to maintain the temperature in the tank in B.Th.U. per day is equal to the weight of the contents of the tank in lb. This method of calculation, rough as it is, may be more accurate than the following more detailed investigations.

The formula for calculating heat losses through walls, etc., is:

$$Q = A \left(\frac{t_1 - t_0}{\frac{1}{f_1} + \frac{b}{K} + \frac{1}{f_0}} \right)$$

where: Q = B.Th.U. per hour
A = area of wall, roof or floor in square feet
t_1 = °F hot side
t_0 = °F cold side
f_1 = surface coefficient, inside
f_0 = surface coefficient, outside
b = wall thickness in inches
K = thermal conductivity in B.Th.U. (of 1 inch thickness of material) per superficial foot per hour per °F. (See Table 47.)

The loss of heat through each component area is as expressed by the formula, in direct proportion to the area exposed and to the difference of

344   PUBLIC HEALTH ENGINEERING PRACTICE

temperature on each side; and in inverse proportion to the sum of the heat-resisting properties of the various layers of material through which the heat has to pass. The heat-resisting value of any single layer of material is in direct proportion to the thickness of the material divided by its thermal conductivity, or $\frac{b}{K}$.

When a fluid is in contact with a wall surface the problem is complicated by the heat resistance of the film in contact with the surface. For example, the sludge in a digestion tank in immediate contact with the wall loses its heat, but does not rapidly take up heat from the rest of the liquid in the tank,

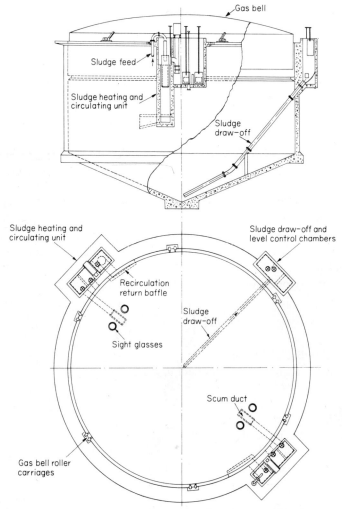

Fig. 113—Dorr primary digester type "B.G.H."
By courtesy of Dorr-Oliver Co., Ltd.

and therefore can be considered as itself acting as a resistance to heat loss. Similarly, the outside wall of the tank exposed to the air is in contact with a thin film of heated air which resists heat transfer. These thin films of fluid determine the surface coefficients f_1 and f_0. They are very greatly influenced by the velocity of motion of sludge, water or air, and much greater heat transfer occurs when the fluids are in rapid motion.

The component parts of a primary digestion tank can be the gasholder roof, the vertical walls exposed to air and the walls and floor below ground-level. Commencing first with the gasholder: the value of $\frac{1}{f}$ for sludge to methane is about 0·6, and from methane to underside of gasholder $\frac{1}{f_1}$ is 0·6.

The value of $\frac{1}{f_0}$ for hot surface of gasholder to outside air in average weather conditions is about 0·25. There is some resistance to heat transfer in the methane that lies between the top surface of the sludge and the under-surface of the gasholder, and as this depth is usually several feet, one can easily make the mistake of allowing for a considerable heat resistance. However, convection reduces this to very little, and it can, perhaps, be neglected. The heat resistance of the metal gasholder itself is negligible, and thus it is probably fair to take the surface coefficients only into account and express the loss of heat through the gasholder by the formula:

$$Q = \frac{A(t_1 - t_0)}{\frac{1}{f} + \frac{1}{f_1} + \frac{1}{f_0}}$$

The loss of heat through vertical walls in contact with sludge on the inside and air on the outside depends on the surface coefficient of external air to wall, for which $\frac{1}{f_0}$ can be taken as 0·4; the values of K for the structural and insulating materials used, as given in Table 47. Surface coefficient for sludge to wall can be neglected.

TABLE 47
THERMAL CONDUCTIVITIES

Material	K
Brickwork (damp)	9·0
Concrete 1 : 2 : 4	10·0
Concrete foamed slag	2·0
Timber	1·0
Cork	0·35
Felt	0·27
Clay (damp)	8·0
Loam	10·0

The loss of heat through underground floors and walls is speculative. In the case of a newly commissioned tank erected in dry ground, considerable losses of heat may take place for a time while the ground is being heated up,

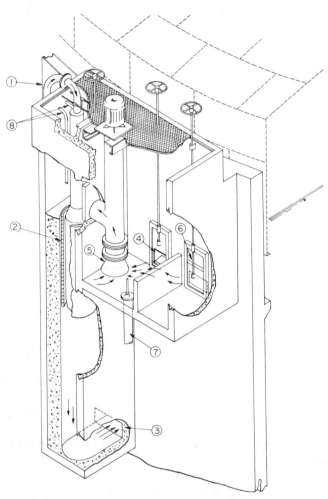

Fig. 114—Dorr type "B" sludge heating and circulating unit

1. Sludge feed-pipe
2. Sludge heater
3. Recirculation return port
4. Sludge penstock opening
5. Circulating pump
6. Scum penstock opening
7. Pump compartment drain
8. Hot water flow and return pipes

By courtesy of Dorr-Oliver Co., Ltd.

but once the ground under the tank is hot it will act as an insulating layer of great thickness. On the other hand, if the subsoil contains water that is percolating through it, and consequently periodically continually changing, the water that is heated by loss of heat from the tank carries heat away with it and cold water takes its place to absorb still more heat. Then losses through underground floors and walls can be appreciable.

Feeding the digestion tanks. There are several hazards involved in the operation of covered primary digestion tanks and the majority of structural failures or other calamities such as fire or explosion at sewage works has occurred in these units. The dangers include the production of an explosive mixture of gas and air inside the tank and the sucking in of the roof by negative pressure. Both of these result from imperfect design together with improper operation.

The early enclosed tanks were often arranged to be operated on a fill-and-draw principle: sludge was delivered into the tank, raising the top sludge level, and later drawn off to the secondary digestion or disposal, lowering the level. With a fixed roof, lowering the level would produce a negative pressure in the tank with the danger of sucking in air or causing the roof to collapse.

This led to the use of floating roofs and gasholder roofs; nevertheless there have been several instances of destruction of floating gasholders due to the sludge level being lowered after the gasholder had come to rest in its lowest position.

Of recent years the author has adopted the practice of always feeding covered digestion tanks on the displacement principle. Sludge is delivered into the tank causing digested sludge to pass over an adjustable weir such as a telescopic weir and so on to secondary digestion or disposal. By this method the sludge level can be maintained within limits close enough to prevent any risk of negative pressure being developed inside the tank during normal operation.

Even this does not altogether do away with the risk, for by improper operation or the accidental opening of a wash-out valve damage could still be done. It is thus becoming usual to have the handwheels of wash-out valves, etc. secured with a padlock and chain, the keys being kept by the superintendent of the works.

Where there are many tanks sludge should be fed to them in equal quantities preferably once a day when the primary sedimentation tanks are sludged. If all the sludge is sent to one tank on one day and to another tank on another day there is almost certain to be temporary overload and foaming.

Seeding and mixing. One reason given for stirring tanks is that bacteria produce toxins, which eventually inhibit their action, and that by mixing the local effect of such toxins is reduced. The main reasons for mixing, however, are the necessity for seeding the whole of the sludge with active bacteria, and thereby immediately setting up the kind of bacterial action required and to prevent sludge from passing out via the bottom outlet before it has been digested. First, raw sludge should be mixed with about 20 per cent of actively digesting sludge before it is passed into the primary tank. This can best be done by arranging pumps of differing capacity in such a manner that both deliver into the same rising main that discharges to the digesting tank, but the suction of the large pump should draw from the crude sludge well, while the suction of the smaller draws from about half-way down the digestion tank. Second, it is good practice to mix incoming cold sludge thoroughly with an excess of heated sludge.

There are various methods of stirring heated tanks. Mechanical stirrers causing rotation of the whole of the liquid in the tank are often used, but it has been objected that these, while they rotate the whole body of the sludge, do not necessarily produce efficient mixing. In the "Simplex" design the sludge is from time to time violently agitated by a pump which, mounted below the floating roof, draws sludge from about two-thirds of the depth and discharges it at the surface (see Fig. 115). Mixing can also be effected in non-proprietary designs by pumps installed outside the tank and used for the double purpose of causing agitation and introducing raw sludge.

A further reason for mixing in a closed digestion tank is that the scum which collects at the surface is broken up by some methods and mixed with the mass of digested sludge. In the case of open tanks, arrangements can be made for taking scum over a ramp. In closed tanks it is usual to attempt to disperse the scum by pushing it below the surface. For example, in the Simplex digestion tank, sludge is drawn from a low level and sprayed over the surface with sufficient violence to submerge the scum. In the Dorr-Oliver "B.G.H." tank (see Fig. 113) scum is withdrawn by a scum horn and passed by the circulating pump into the tank at a lower level. In instances where scum accumulation has given trouble, benefit has been obtained by recirculation of sludge gas which is withdrawn from the gasholder by air pump mounted on the gasholder and delivered back into the tank 10 feet below sludge level by a number of J-shaped inlet pipes. A satisfactory rate of discharge has been found to be 1 cubic foot per minute per 60 feet super of scum surface. Excess pressure valves should be provided on the deliveries of the pumps, but any discharge from these must return to the gasholder so as to avoid the danger of producing a vacuum under the gasholder.

In the "Heatamix" system sludge is recirculated inside the tank with the aid of gas-operated "air-lifts" which are fitted with heaters. In the installation at Mogden four Heatamix steel tubes 27 feet long in 9-foot lengths and 12 inches diameter are fitted in a tank of 70 feet diameter and 875,000 gallons capacity. The detention period during a test was 14 days.

The delivery of each Heatamix is 1300 gallons of sludge per minute for which 40 cubic feet per minute of gas are required per tube. For the whole installation 35 gallons per minute of hot water are required, raising the temperature of the sludge by 26·7 °F. This gives a temperature-drop in the water from 140 °F to 105 °F.

The bottoms of the tubes have bellmouths: the top ends turn over at 90°, 4 feet *below* surface-level. It was found in the tests that, to give effective disturbance of the scum, the direction in which these bends pointed was critical. The tubes are located 12 feet from the circumference of the tank.

The bends of two opposite each other discharge at right-angles to the radii to spin the contents of the tank. One of the other two points at the tube ahead of it and the other to a point half-way between centre of tank and circumference on the radius that passes through the tube ahead of it.

There are some materials in sludge, such as rubber contraceptives, which neither will sink nor digest and, if pushed below the surface, will return once more to form scum. It is advantageous to have some means of drawing these off—e.g. through a Dorr-Oliver scum horn—and discharging them to sludge disposal.

Methods of heating. There are several methods of heating the sludge in a digestion tank, including the following:

1. Fixed coils of hot-water pipes of not less than $1\frac{1}{2}$ inches internal diameter, the pipes usually being located near the walls of the tank. In these pipes clean hot water is circulated (see Fig. 110).

2. Rotating coils or grids of pipes in which hot water is circulated, the pipes serving the double purpose of heating and stirring the sludge.

3. Drawing sludge from primary tank, heating it in a heat exchanger and passing it back into the digestion tank.

4. Heating alkaline supernatant water drawn off from the secondary tank and returning it to the primary tank. (This method has the advantage of helping to render the primary tank alkaline but adds to moisture content.)

5. Injecting steam into the bottom of the tank. (This method adds far too much to moisture content of sludge.)

Heaters for various of these methods are made by the Dorr-Oliver Co., Ltd., and Ames Crosta Mills & Co., Ltd.

The Dorr-Oliver "B.G.H." type of Primary Digester consists of a circular tank, having a conical bottom and a floating gasholder roof (see Figs. 111 to 114). The sludge is heated and circulated by means of the Dorr Type "B" Sludge Heating and Circulating Units located in chambers built on to the sides of the tank. Two sets of these chambers are required in the case of large digesters and one set on smaller units. In addition, a separate set of chambers is required for sludge draw-off and level control.

Sludge is pumped (usually daily) to the digester and passes into the tank through the recirculation return port at the bottom of the sludge heater chamber, the port being specially designed so that the direction and velocity at which the sludge enters the digester gives a rotary motion to the tank contents. The sludge feed displaces an equal volume of digested sludge, which passes through the sludge draw-off pipe from the central discharge point at the bottom of the cone to the sludge draw-off chamber. It then passes over an adjustable weir-penstock into the discharge chamber, before being delivered to the secondary digester or sludge drying beds. By adjustment of the weir-penstock, the sludge level in the tank can be controlled.

The gasholder as provided by Dorr-Oliver Co., Ltd., is spirally guided by roller carriages mounted directly on top of the tank walls, and is provided with a manhole, sight glasses, sampling tubes and an excess-gas-production relief assembly. It is also protected by a combined excess-pressure vacuum-relief device and alarm, which is connected to the gas line to the boiler-house or engines.

The heat for heating digestion tanks is usually obtained by heat exchange to circulating water from the exhausts of the sludge gas engine used for power production on the site. This may be supplemented by burning gas in a gas boiler.

In all cases where hot water is used for heating sludge, the circulated water should be softened to prevent deposition of boiler crust. It is not, however, possible to prevent boiler crust depositing on the outsides of heater coils as a result of the hardness of the sludge itself: where there are coils inside a tank this can produce a marked reduction of heat exchange after a period of years. The Dorr-Oliver "B" heater unit is arranged so that it can be lifted out for descaling. In installations designed by the author special cranes were provided for this purpose, acid baths for the removal of the crust and alkali baths to neutralise the acid.

The most frequent cause of trouble with heated digestion tanks is lack of temperature control. Digester temperatures should be taken and signalled to a control panel where they can be regularly observed and, preferably, recorded on charts.

Removal of supernatant water. The secondary digestion tanks serve the purposes of completing digestion to an economic limit, where necessary acting as storage between sludge production and disposal, and providing quiescent conditions to permit separation of supernatant water.

The aforementioned report of the Water Pollution Research Laboratory showed that the amount of dewatering achieved at various works was very varied. While at some more than 50 per cent of supernatant water was removed, at others dewatering was less than 10 per cent and at the majority less than 30 per cent. The differences were no doubt to some extent due to the nature of the sludge. But to obtain good dewatering it is necessary to give very close attention to the dewatering arrangements and at large works it is well worth having a man whose job it is to do nothing else than to check on the dewatering valves whenever they are in use. Poor dewatering may very well be the result of neglect in this respect.

The most common arrangement for the removal of supernatant water is the installation of several small-diameter valves connecting from various levels in the tank. The operation is to open a valve at a time and find where the supernatant water is, and then to draw it off until sludge appears. Ames Crosta Mills & Co., Ltd., manufacture a single rotating-arm decanting valve which does away with the necessity for several valves.

Supernatant water is returned to the flow of crude sewage for treatment.

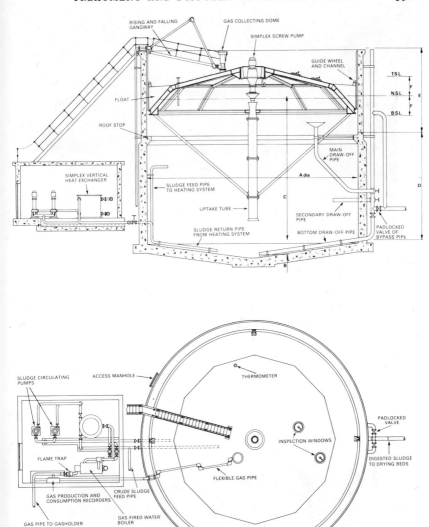

Fig. 115—Ames Crosta Mills "Simplex" digestion tank with floating roof and suspended circulating pump
By courtesy of Ames Crosta Mills & Co., Ltd.

Utilisation of sludge gas. Sludge gas consists of about 67 per cent methane and 33 per cent carbon dioxide and small quantities of other gases. As methane has a net calorific value of 895 B.Th.U. per cubic foot, the calorific value of sludge gas can be taken as being about 600 B.Th.U. per cubic foot. One British Thermal Unit (B.Th.U.) is the amount of heat required to raise 1 lb (or $\frac{1}{10}$ gallon) of water by 1 °F. (There are 3415 B.Th.U.

Fig. 116—Gasholder of the Dorr-Oliver "B.G.H." digester, Crossness sewage-treatment works

Fig. 117—Excess gas burner and primary sludge dewatering tanks, Beckton sewage-treatment works

in one Board of Trade Unit (B.T.U.), and one Board of Trade Unit is the same as one kilowatt hour (kWh).) Thus, allowing suitable efficiencies of prime mover, dynamo, etc., the calorific value of the sludge can be translated into electric power.

One pound weight of sludge gas at day temperature and atmospheric pressure occupies 15·4 cubic feet. Contrary to some opinion the weight of sludge gas produced is frequently about the same as the weight of solids destroyed by the digestion process. For example, in a series of large-scale experiments made at the Crossness sewage-treatment works the weighted average of the results for mesophilic digestion was 15·4 cubic feet of gas per lb of volatile matter digested.

Roughly, one-third of the heat available can be used in the production of horse-power, one-third economised by transfer of heat from cooling water and exhaust gases and one-third is irretrievably lost with exhaust gases, friction, etc. The heat economised from cooling water and exhaust gases is used (generally by hot-water circulation) for heating the sludge in the primary tanks, and is often sufficient for this purpose, although it is usually advisable to have stand-by gas boilers to make up any deficiency should there not be sufficient power plant running to produce all the heat required.

The main use of sludge gas, apart from heating the digestion tanks, is for power purposes, for the quantity produced is often sufficient to pump the whole of the sewage (where it is necessary for this to be done). At activated sludge works the power can be used for aeration purposes. Elsewhere it can be used for recirculation pumping in alternating double filtration schemes. Electricity is usually generated for use in these ways.

All engines should be isolated from gas mains by suitable flame traps.

Sludge gas has also been used for driving road vehicles. In this case the gas has to be purified, particularly with regard to removal of carbon dioxide,

TABLE 48
PERCENTAGE OF GAS PRODUCED

Days' detention	Percentage of gas produced in 30 days
1	16
2	30
4	50
6	65
8	76
10	83
12	88
14	92
16	94
18	96
20	97
25	99
30	100

after which it is compressed and stored in high-pressure cylinders. Compressed gas is drawn off from these cylinders and delivered by pipeline and flexible coupling to cylinders carried by the vehicles. This method has been used at Croydon and at Mogden.

Other uses for sludge gas. While sludge gas is used in this country for no other purposes than heating, power production and occasionally mixing with the local domestic gas supply,* it has been pointed out[157] that chemicals can be produced from methane and are produced from natural gas (which is principally methane) in America. Carbon black is probably the most simple product of methane, and it is produced from natural gas in America, and imported by Britain.

FIG. 118—Secondary digestion tanks, Crossness sewage-treatment works

Excess-gas burner. It is usual to dispose of any surplus gas, that cannot be utilised, by burning it in a suitable bunsen burner (although Dorr-Oliver Co., Ltd., have gas reliefs on their gasholders). This is quite a simple arrangement consisting of a vertical steel pipe with air inlet and gas jet at the bottom, similar to an ordinary laboratory bunsen burner on a larger scale. The orifice of the jet at the bottom is made large enough to pass, under the working gas pressure, the maximum quantity of gas likely to have to be discharged. The orifice is also arranged so that one or two smaller nozzles can be screwed on to reduce the orifice to about three-quarters of its maximum diameter. The vertical tube of the burner should have about three to four times the diameter of the largest orifice and the height should be suffi-

---

* It is dangerous to use sludge gas for domestic purposes unless mixed with coal-gas or otherwise given a distinctive odour. Also, used alone, it requires special burners.

cient to keep the flame well above ground-level. Such burners are lit by hauling lighted oily waste to the top with stainless-steel halyards and then turning the gas on from a safe distance; sparking-plugs just inside the top have also been used. The burner should be isolated from the pipework by a

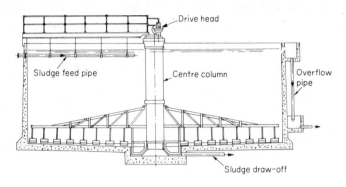

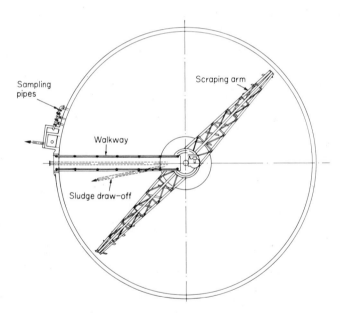

Fig. 119—Dorr secondary digester type "S"
By courtesy of Dorr-Oliver Co., Ltd.

suitable flame trap and located well away from trees and buildings. Table 49 gives the approximate discharges through the burner orifices. The burner should be able to discharge not less than 20 per cent of the maximum estimated gas production and not less than twice the maximum production of one digestion tank.

The above-described design of burner is probably the most reliable so far devised. Such burners do not blow out, do not deteriorate structurally to any great extent and give complete combustion of the gases with no production of carbon monoxide. Also, if tall enough, they carry the flame well away from ground-level. On the other hand, some burners not of the bunsen type have been known to blow out at frequent intervals and some of those involving structures in brick and firebrick have had structural failure. These other designs are liable to produce carbon monoxide and, unless provided with a tall and expensive flue, do not take the flame far enough away from ground-level for safety.

The burners as above described are also the least expensive provided they have the simple arrangements for control mentioned above. On the other hand, if they are given electric-pneumatic valves and push-button lighting from a special control board, they will then become forty times as expensive as they should be, to little or no advantage.

The burners should be well away from existing buildings or trees, 100 feet being the minimum distance: the writer found objectionable fumes in the air when on the top of some tanks down-wind from a burner about 500 feet away. The burner should be high enough to prevent the flame from blowing down: in another instance some scorching of grass was observed near a burner 40 feet high.

TABLE 49
DISCHARGE OF GAS THROUGH ORIFICES

Pressure of gas in inches of water	Cubic feet per minute per square inch of orifice
4	$42\frac{1}{2}$
5	$47\frac{1}{2}$
6	52
7	56
8	60
9	64

Gasholder. Whenever sludge gas is utilised it is necessary to have one or more gasholders, having in all a capacity of at least four, and preferably eight, hours' gas production. Separate gasholders can be used, but with some types of primary tank, such as the Dorr-Oliver "B.G.H." tank, the gasholder serves as gas collector and roof. In this case a special design of gasholder is desirable. First, the trusses should be so arranged that they do not dip below the surface of the sludge in the tank, for if they do, the rise and fall of the gasholder may be inhibited by the scum. The gasholder should be provided with windows fitted with wipers that can be worked from the outside so that the process and the operation of scum horns can be observed. The gasholders should be fitted with devices which automatically close the outlet pipes as they reach their lowest position and should also have anti-vacuum

TREATMENT AND DISPOSAL OF SEWAGE SLUDGE 357

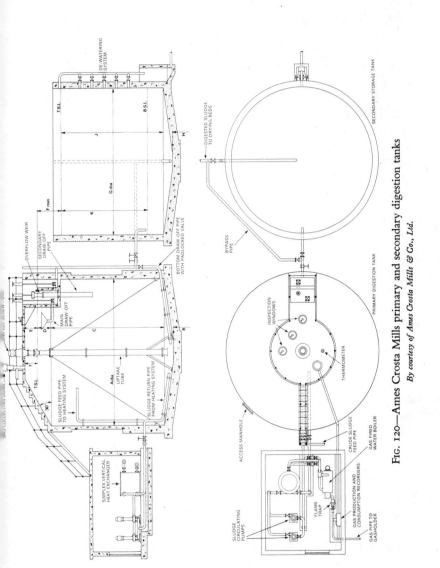

FIG. 120—Ames Crosta Mills primary and secondary digestion tanks
By courtesy of Ames Crosta Mills & Co., Ltd.

valves which must not have isolating valves. Spiral-guided gasholders are usual. Provision should be made for the removal of plates at various parts of the gasholder surface to facilitate opening of the tank for maintenance purposes.

Instrument panel. Sludge digestion is a process in which instrumentation is very desirable because, without proper control, the process may break down or there may be physical dangers. Control of temperature is very important and, particularly where there are many tanks, the temperatures of ingoing and outgoing sludge and sludge in the tank should be recorded for all tanks, also the rate of flow of sludge which should be integrated to give quantities. Where hot water is used for heating the sludge, quantities and ingoing and outgoing temperatures should be recorded so that the amount of heat used can be ascertained. The quantity of gas produced should be measured and recorded and either the amount discharged to waste or the amount utilised. The chemist should analyse the gas from time to time and there should be an automatic record of carbon-dioxide content. To avoid danger of negative pressure in the gas system, the levels of gasholder should be indicated and there should be visual and audible warnings of the approach of negative pressure. The quantity of supernatant water decanted should be recorded and the quantity of final sludge discharged to drying or other disposal.

Purging plant. So as to avoid the danger of explosive mixtures developing in tanks or pipework after they have been laid off and are to be returned to service they should be purged with deoxygenated air. For this purpose portable purging plant is available. This consists of a blower and oil-burning arrangement which converts oxygen to carbon dioxide. The same plant can be used for providing fresh air so that men can work in the tanks.

Methods of sedimentation involving digestion. In this country sludge digestion in separate digestion tanks is now usual, but methods involving digestion in sedimentation tanks or in chambers directly connected with sedimentation tanks have been used.

In the first place, there is the septic tank, a sedimentation tank sludged at very rare intervals on the over-optimistic assumption that the sludge lying at the bottom of the tank will digest. At one time such tanks were installed at municipal sewage works, but they were found to be unsatisfactory for other than the smallest works, for a number of reasons.

Septic tanks are still used for the treatment of sewage from isolated buildings in those circumstances where ordinary sedimentation tanks would not be practicable owing to lack of maintenance (see Chapter XXXIII).

Dr. Imhoff, of Essen, the inventor of the tank which bears his name, but which is also known as the "Emscher Tank," considered it highly undesirable for sewage to come in contact with decomposing sludge. He attached great importance to keeping the sewage fresh, and the sedimentation channels in his tank, which in some respects resembled the hydrolytic tank (see Chapter

XXXI), were accordingly made as small as possible consistent with the effective settlement of the suspended solids. An Imhoff tank has sedimentation channels, with a digesting chamber below, the former communicating with the latter through slots which run the whole length. The digesting chamber is carried up between the sedimentation channels, so that the gases may pass off freely.

Imhoff tanks have been installed in Great Britain, but they have never become generally popular. On the other hand, they have been widely used in Germany and in the United States, where they were once favourably regarded. In American installations much trouble has been experienced from foaming through the gas-vents.

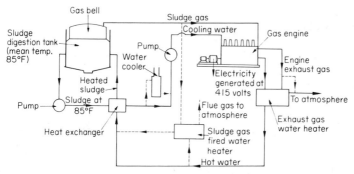

FIG. 121—Diagram showing utilisation of sludge gas and recovery of waste heat from engine exhaust and coolant

By courtesy of J. Campbell-Riddell Esq., City Engineers, Oxford, and The National Gas and Oil Engine Co., Ltd.

Utilisation of sludge as fertiliser. The utilisation of sewage is a subject on which lay writers of the would-be reformer type have sometimes been too enthusiastic and, without getting to know all the facts, have deplored the waste of "useful fertilisers" discharged to the sea or otherwise destroyed. In particular, the need for utilisation of sewage sludge has been stressed and the value of such natural manures, as against artificial manures, reiterated from time to time.

As those in the profession know, the real problems relating to sewage sludge and its utilisation as fertiliser are the best means of preparing sludge in order that it may be something more than a low-grade manure that can be injurious, the means of making it acceptable to farmers, and the precautions to be taken against the dangers to health that such utilisation of domestic sewage involves. The truth is that sludge is used as a fertiliser, but that ordinary sewage sludge (as against activated sludge and humus tank sludge) is not a good fertiliser, for, besides being of merely moderate manurial value, it is liable to contain mineral grease and chemicals which may injure the soil, together with weed seeds, also pathological bacteria which make it unsuitable as a manure for crops which are eaten uncooked. In this connection the following rules in force in California might well be kept in mind by British sewage-works operators:

(1) Raw sewage containing human excrement shall not be used for irrigating growing crops.

(2) Raw or undigested sludge shall not be used for fertilising growing vegetables, garden truck or low-growing fruit or berries unless the sludge has been kiln dried, bed dried or aged in storage for at least 30 days, well digested (i.e. odourless, readily drainable and containing not over 50 per cent of the total solid matter in volatile form), or conditioned to the satisfaction of the state board of health.

(3) Settled or undisinfected sewage effluents are banned for growing vegetables, garden truck, berries or low-growing fruit, or for watering vineyards or orchards where windfalls or fruit lie on the ground. Such liquid may be used for watering nursery stock, cotton, field crops such as hay, grain, rice, alfalfa, fodder corn, cow beets and fodder carrots, provided no milch cows are pastured on the land, when moist with sewage, or have access to ditches carrying sewage. Use is also permitted for growing vegetables used exclusively for seed purposes.

Certain kinds of sewage sludge, prepared in such a manner as to conserve manurial value and to inhibit the growth of harmful bacteria, are excellent manures and safe to use in most circumstances. But unfortunately, the best of these is practically unknown in this country, for the method of treatment is too expensive to adopt. English sewage-works managers and designers prefer such methods of sewage treatment and sludge treatment and disposal as are accomplished at the least cost to the ratepayer, particularly as regards capital outlay.

In this country the greater part of the information available on preparation of fertilisers from sludge is based on a few isolated examples where local authorities or private companies have made a special business of selling dried sludge composted with refuse, or otherwise treated as a fertiliser. Apart from these few special cases, nearly all the sludge used as fertiliser in Great Britain is either raw or digested sludge dried or pressed, and supplied to the farmer either as removed from beds or presses or broken and screened to make it easier to handle.

Some work has been done on composting sewage sludge with selected town refuse, but the method is falling into disuse for several reasons. It has been found that such compost, unless stored long enough for immobilisation of nitrogen, causes crop starvation. Such compost frequently contains broken glass: furthermore the product is of no better manurial value than the separate values of the sludge or the refuse and is definitely inferior to ordinary farmyard manure.

In America sludge utilisation has been given much more consideration, having been studied from the points of view of risk of infection and suitability of the various kinds of sludge for fertilising land. An important publication on this is Manual of Practice, No. 2, "Utilisation of Sewage Sludge as Fertiliser," prepared under the direction of the Committee on

Sewage Works Practice by the Sub-Committee on Sludge Utilisation for Fertiliser, Federation of Sewage Works Associations, U.S.A., 1946.[81]

The following is a summary of American opinion on the manurial values of different kinds of sludge:

Raw primary sludge. Raw primary sludge is not recommended as a fertiliser because it is odorous, also its grease content is higher than that of digested or activated sludge, and the grease may prove harmful in such quantity. Moreover, raw sludge in decomposing in the soil tends to bring about persistent acidity. Industrial waste, if present, may poison soil and crops. Raw sludge contains pathogenic bacteria to a greater extent than other sludges. Raw sludge is not humified for soil use, and may actually interfere with the development of a crop.

Liquid digested sludge. This is seldom used in the United States, and only occasionally in England.

Dewatered digested sludge. Digested sludge has little odour. Its appearance is good, and it is unobjectionable as a top dressing. Air-dried sludge may be stored without nuisance, although nitrogen is lost during storage. The storage, however, is an additional safeguard from the point of view of health. The sludge is humified, and the grease content is reduced.

Heat-dried digested sludge. The method is rare because the cost is higher than the quality of this variety of sludge justifies. It is, however, comparatively free from harmful bacteria, and can be mixed with chemical fertiliser as a conditioner.

Air-dried digested activated sludge. This is in most respects similar to dewatered (air-dried) digested sludge.

Heat-dried activated sludge. Heat-dried activated sludge is by far the best sewage sludge for use as a fertiliser, whether applied alone or mixed with other natural or artificial manures. It has the advantages of reduced bulk, easy application to the ground and little odour. It is unobjectionable as a top dressing for gardens. It has a good nitrogen content. It is believed to be free from harmful bacteria. The grease content is low. It is free from weeds. It must, however, be stored dry and under cover until wanted for use.

The general conclusions to be drawn are that sewage sludge is useful as a fertiliser if properly prepared and applied, but that its value varies with the type of sludge. Fresh sludge is not recommended. Digested sludge is considered to be of value as a manure and justifies transport up to a distance of perhaps twenty miles. But heat-dried activated sludge is recommended as a valuable source of nitrogen and safe to use with practically all crops. It would appear, however, that the cost of equipping and operating a plant for heat drying is very high, and therefore the method is not generally applicable.

The Porteous method of sludge dewatering. In the proprietary method of W. K. Porteous, Ltd., sludge is first passed through a disintegrator so as to break up any large solids, after which it is pumped through a heat exchanger in which the temperature is raised, and into a series of high-pressure vessels where heat is applied with the aid of a steam-jacket. When the sludge has been cooked for a sufficient period it is passed through the second jacket of the heat exchanger so as to give up its heat to the incoming raw sludge, thereby economising fuel to a very great extent. Next, the cooled sludge is settled in tanks from which approximately two-thirds of the volume is decanted as supernatant water. The thick sludge that is left after the removal of supernatant water is finally dewatered in filter-presses to a moisture content of about 40 per cent, the filter cake being further dried by storage in the air (see Fig. 122).

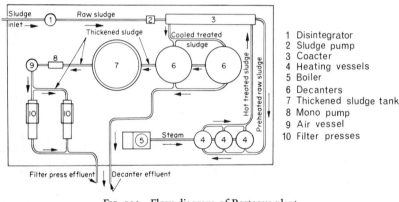

FIG. 122—Flow diagram of Porteous plant
By courtesy of W. K. Porteous, Ltd.

The dried filter cake from the Porteous process can be used as a fuel or as a fertiliser which, being sterilised by heat, is free from weed seeds and pathogenic bacteria.

Incineration of sludge. When there is no market for sludge as a fertiliser or for any other purpose, one of the alternatives to removing it from the site at the cost of cartage, or dumping it at sea, is to dry the sludge thoroughly and afterwards to burn it. Heat-drying of sludge and incineration have been used for some time in America, Europe and South Africa, but, so far, very little in Great Britain.

In the Nicholls Herreshoff "Multi-hearth" incinerator fairly-dry sludge cake is passed through a furnace consisting of a number of rotating hearths, one above another, and alternating with fixed hearths. The cake is raked so as to fall from the upper hearths to the lower, being first dried by the hot gases, then ignited and consumed. Gas or oil are used as fuel.

In the Raymond "Suspension-type" drier moist sludge is mixed with dried sludge, passed through a flash drier and then collected as dust in a cyclone dust collector. Part of the dried sludge is then returned for mixing with the wet sludge and part filled into sacks for use as fertiliser.

Dorr-Oliver Co., Ltd. have now installed a considerable number of sludge incinerators known as the "FS Disposal System." In this system there is a chamber known as the "reactor," at the bottom of which is a bed of silica sand kept in a state of turbulence by an upflow of air. Oil is injected and burnt in the sand until the whole becomes incandescent. The sludge to be burnt is cleansed of grit and dewatered by either centrifuge or vacuum filter —e.g. the "Coilfilter." The dewatered cake is then introduced directly into the bed of incandescent sand where rapid combustion takes place, the ash being blown out of the sand by the upflowing gases. These gases are cooled and cleansed by "scrubbing" in sewage-works final effluent.

Zimmerman process. The Zimmerman process is a method of oxidation of organic matter dissolved or suspended in water. The combustion produces heat in the same manner as if the dry matter were burnt in free air, and this heat, which is also necessary for the process, can be utilised for the production of the power required to drive the mechanism. To obtain the necessary temperature, the liquid has to be under considerable pressure and consequently the plant involves pumps and heat exchangers for economising the heat.

The sludge is heated in tanks to a temperature of 180 °F and pumped, first by a rotary pump and then by a high-pressure pump. Next, air is delivered into it at a pressure of 1200 lb per square inch and the sludge and air passed to a preliminary heat exchanger in which steam is used to raise the temperature during the start-up, then to a second heat exchanger which raises the temperature to 380 °F, and finally to a heat exchanger where the temperature is raised to 460 °F.

The hot sludge passes from the third heat exchanger into a pressure vessel. It enters this at the bottom and is oxidised as it passes upwards. Then it passes out at the top and down through an outer shell. Next the liquid, burnt products, gases, etc., at a temperature of 515 °F, pass through the second and third heat exchangers, dropping in temperature to 500 °F and 440 °F respectively, afterwards to be cooled by a water-cooled heat exchanger to 150 °F. The gases are vented to atmosphere and the settled ash pumped to lagoons.

It is understood that the process should be thermally self-sustaining if operated with sludge containing upwards of 3 per cent solids: with a solids content of 6 per cent and working under a pressure of 1200 lb per square inch it should be capable of doing without any added heat.

Sludge freezing. Laboratory experiments have been carried out on the addition of coagulates followed by very slow freezing, so as to produce water crystals which separate the solids. This makes the sludge easy to dewater. The method, however, has not been put into practice except for dealing with radio-active sludges and otherwise does not appear to be economically justified.

Flow of sludge in pipes. Researches in several countries have shown that the critical velocity at which laminar flow changes to turbulent flow is very high for sludge and very varied according to the nature of the sludge. It has frequently been stated that there is a lower critical velocity below which flow is laminar and sluggish, an upper critical velocity above which the flow is fully turbulent and in accordance with flow formulae for water, while between the two is a region in which flow may be either laminar or turbulent or a combination of both conditions. All experiments with which the author has been concerned have shown one critical velocity only for one diameter of pipe at which there is an abrupt and complete change between laminar and turbulent flow.

Table 50 is based on the average of values for the upper critical velocity for digested or semi-digested sludge as found at five overseas sewage works. This average is about 30 per cent higher than the average for the lower

TABLE 50

APPROXIMATE MINIMUM HYDRAULIC GRADIENTS FOR SLUDGE MAINS

Diameter (actual millimetres)	Critical velocity (metres per minute)	Gradient, 1 in:	Discharge (cubic metres per minute)
100	111·0	18	0·871
125	102·0	28	1·25
150	94·7	41	1·67
175	89·2	56	2·15
200	85·5	72	2·69
225	82·3	90	3·27
250	81·4	105	4·00
300	77·6	145	5·49
350	75·7	185	7·28
375	75·1	205	8·29
400	74·6	225	9·38
450	73·3*	270	11·66

critical velocity for the same works and about 15 per cent higher than the critical velocity found in the experiments with which the author was concerned. The table should therefore be reliable for use for most sludge-flow calculations. Where, however, very extensive sludge pipe-lines may have to be constructed it would be advisable for adequate experiments to be made with the sludge concerned.

Practice of some engineers is to design sludge mains so that flow is never less than the critical velocity: this not only simplifies calculations but ensures a self-cleansing condition.

---

* Minimum velocity for all larger diameters, about 70 metres per minute.

CHAPTER XXIX

PERCOLATING-FILTER TREATMENT

In the treatment of sewage by oxidation, the organic content is decomposed by serving as the food of aerobic micro-organisms. The process takes place in stages, different organisms taking part until finally stable inorganic salts are produced. The final effluent may then be considered unlikely to become offensive or a danger to fish life, provided that its biochemical oxygen demand is sufficiently low, taking into account the volume of pure river water into which the effluent is discharged.

Requirements of aeration. A normal sewage not containing sufficient inhibitory trade waste or other substance of a similar nature to interfere with the process of treatment will become oxidised rapidly in the presence of the necessary micro-organisms, and provided that the oxygen taken from solution in oxidising the organic content is replaced adequately. In favourable conditions a population of micro-organisms builds up in a new percolating filter or a newly filled aeration tank until a maximum condition is arrived at. If these organisms are kept well supplied with air, as in a well-constructed percolating filter, the required degree of oxidation can be effected in a fraction of an hour, or a few hours in an activated-sludge plant.

The process is essentially a biological one, not simple chemical reaction. This is a fact that has been overlooked by some chemists and engineers with the result that ridiculous suggestions and experiments have been made. The oxidation is effected by living vegetable matter and is in proportion to the quantity of healthy vegetable matter present. Just as a gang of workmen, who can do an amount of work when adequately fed and supplied with air, will not be able to work if starved of food or air, yet will not be able to do ten times as much work if given ten times the amount of food and air that they require, so it is with the micro-organisms. Broadly, the amount of aeration effected in either a percolating filter or an activated-sludge plant is in proportion to the weight of organisms and the time they have in which to perform their service.

In a well constructed and maintained percolating filter there is usually an adequate supply of air, but if the load of organic pollution is more than the organisms in the filter can oxidise the quality of the effluent must reduce. There is a limit to the weight of organisms that can develop in a percolating filter, this depending on the surface area of the particles of filter medium to which the organisms can cling.

In an activated-sludge plant it is possible to reduce the supply of air to less than that required by the organisms present in the tank. This reduces the health and activity of the organisms with the result that they become replaced by other organisms not suitable for sewage treatment and again the quality of the effluent suffers. On the other hand the provision of a greatly excessive quantity of air does not improve the effluent beyond the maximum aeration that the organisms can effect: it is merely a waste of expensive power.

In an activated-sludge plant it is possible to give partial treatment by greatly overloading the works. But intermediate between full treatment and partial treatment there is what is known as the "bulking zone" in which satisfactory treatment cannot be effected because the conditions are such that undesirable organisms prevail and there is bulking of the sludge.

It is, however, possible to vary deliberately the quantity of activated sludge in an activated-sludge works. If the sludge is healthy the density of the sludge liquor can be increased by crowding the organisms in the tank without detriment. Thus the capacity of an aeration tank is not critical and, as will be shown, can be varied inversely as the power required for aeration.

Sedimentation reduces not only inorganic suspended solids but also, by removing organic suspended solids, it reduces the B.O.D. of the sewage. The main purpose therefore of aeration is to oxidise the dissolved organic solids. In doing this the organisms build up their own physical structures converting dissolved solids to suspended solids which are removed in the humus tanks of an activated-sludge plant. It is for this reason that the amount of sludge produced at a sewage-treatment plant is greater than the quantity of suspended solids that arrive at the works. Unfortunately this is a matter that has not been given sufficient attention by research workers (see "Calculating quantities of primary sludge" in Chapter XXVIII).

The oxygen taken up by the organisms must be replaced and this occurs by solution from the air at the surface of the water. In a percolating filter there is a vast surface of the water that is trickling at very shallow depth over the particles of medium, and the rate of solution is more than adequate. In a body of water in a tank the surface exposed to the air is far from adequate unless something is done to increase it. There is much difference of opinion as to how and when the greater part of the aeration takes place. Some say that it is mostly at the moment of bursting of bubbles of air that have been injected into the tank or produced by splashing: others do not agree. The fact remains that oxygen can be replenished either by blowing air into the tank or by causing a disturbance at the surface as will be described in Chapter XXX. Both methods require power, the efficiency depending on the details of the method selected.

From the foregoing it will be seen that the requirements of sewage aeration are:

    1. The presence of a sufficient population of suitable oxidising organisms.
    2. A sufficient surface area of liquid exposed to the air.
    3. Sufficient turbulence to produce the degree of aeration required with the surface area available.
    4. A sufficient detention period for the organisms to effect oxidation.

If the surface area is reduced, the turbulence must be increased; if the aeration produced by surface area and turbulence is reduced, the detention period must be increased. For any particular method of treatment there would appear to be an optimum relation between these factors.

The percolating filter. In a percolating filter, settled sewage is trickled over the surface of particles of medium which soon become coated with a film of bacteria and other oxidising organisms, about 2–3 mm thick. It is the outer surface only of this film that is aerobic; underneath, and closer to the medium, anaerobic digestion is taking place. The majority of the organisms stay in the bed with the liquid flowing over them. The surface area is virtually the maximum possible, for the water trickles in a thin film; for this reason turbulence is not necessary. It follows that a very short detention period should suffice, and observations show that this is the case.

In Great Britain percolating filters are the most popular method of oxidation of sewage. Many years ago bacteria beds were found to be more economical than poor-quality or expensive land for the treatment of sewage, less liable to cause nuisance and occupying less space. The earlier form of bacteria bed, the contact bed, was superseded by the percolating filter because the latter was found to be doubly efficient for the same cubic capacity and capital cost, and also to require less attention.

A percolating filter (which is not a mechanical filter as its name might suggest) consists of a bed of medium, such as broken stone or clinker, which should not be "graded" but be of even particle size so as to contain a large proportion of voids. In addition to the oxidising bacteria, there are numbers of other organisms. Growths of algae and other forms of vegetable life occur on the surface where there is sunlight, and these tend to increase, particularly in rapid flows of sewage, and would, in a short time, choke the beds were they not controlled by a variety of animal organisms, particularly worms and larvae, which feed on the surface growth, breaking it up and transforming it, so that it is discharged with the effluent in the form of small particles.

In the investigations of Dr. Reynoldson[203] the surface growth was the blue-green alga, *Phormidium*, and the chief organisms responsible for removing surface growth were enchytraeid worms. By careful experiments Dr. Reynoldson demonstrated that the spring off-loading—that is, the breaking up and discharge of vegetable matter as humus, which occurs in the spring—was due to the activity of these worms, which increased suddenly with the warm weather, the worms coming to the surface and attacking the growth. Dr. Reynoldson proved also that, as the broken particles of vegetable matter passed through the beds, the worms followed them down, feeding upon them.

Required properties of medium. A section of research which has received some, but not sufficient, attention is the investigation of the comparative qualities of different filter media. It is generally agreed that the desirable qualities are:

> High surface area in proportion to volume.
> Rough texture to assist adhesion of bacteria.
> High proportion of voids to volume, to allow free passage of air and effluent.
> The particles should be cubical, not flaky.
> Durability in all weather conditions and in the presence of corrosive effluents.
> Low cost.

Different materials have these qualities in varying proportion. There are some materials which, physically, are very good but which are expensive; there are others which are less satisfactory but which are of lower cost. The designer of a percolating filter scheme has to decide on the most suitable material available in the district in which the works are to be constructed. He should select a material which will be durable under working conditions, but he should avoid the temptation to buy the most expensive material offered, for it often proves that a filter more efficient in terms of result in proportion to cost can be secured by using a large volume of cheap material rather than a small volume of expensive material.

The difficulty when choosing filter medium is to assess the difference of value of two materials in the absence of accurate information as to their performances. The designer who wishes to use what he may consider to be a material of low performance but low cost is advised to increase the capacity of the filter to a volume greater than that required for an average material, but not to increase the capacity to such an extent that the cost of the filter is greater than it would be with an expensive material.

A British Standard Specification for filter medium was issued in 1948. This includes tests for size, shape and durability of media.

It is unwise to require that percolating-filter media shall pass the sodium sulphate (durability) test of British Standard 1438. This test is far too severe and is known to rule out some very satisfactory percolating-filter media which, while they fail the test badly, will stand up to normal weathering indefinitely. It should be mentioned that the American specification on which the sodium sulphate test was based has now been relaxed.

Another fault of the British Standard specification is that it does not include the larger grades of media which are useful particularly in high-rate filters.

The grading of media is important and a satisfactory compromise has to be arrived at between a small-grade medium which will give very good efficiency for a short time but later become choked with ponding at the surface and a coarse medium which is free from ponding but of which a greater quantity is required to give the same degree of treatment.

The treatment is effected by the bacteria on the surfaces of the particles of medium and as there is a greater surface area in a cubic yard of 1-inch medium than, say, in a cubic yard of 3-inch medium the 1-inch medium is, in suitable circumstances, capable of treating that much more sewage. On the other hand, the spaces between the particles of 1-inch medium are smaller, permitting less ventilation and augmenting the risk of chokage and ponding.

For most purposes media of $1\frac{1}{2}$-inch or 2-inch grade, according to British Standard Specification 1437, should be satisfactory. Finer grades are generally not to be recommended and $2\frac{1}{2}$-inch to $3\frac{1}{2}$-inch grade could be suitable for very strong sewage, for treatment in high-rate filters or for the top layer of any filter. A layer of very large material of 4-inch or 6-inch diameter is usual at the bottom of the filter to assist ventilation and drainage.

Apart from the differences of grade at the surface and at the bottom the material of the filter should be of constant size throughout not, as has often been provided in error, increasingly fine towards the surface. The effect of this grading, as in a waterworks filter, is to filter out solids and cause chokage. It is also of importance that the sizes of the particles in any grade of medium should not be too varied, for if there are any particles much smaller than the largest particle they fill in the spaces interfering with ventilation and drainage. Generally the smallest particles should not have a diameter of less than two-thirds of the diameter of the largest particles in the same part of the bed.

Capacity of percolating filters. Percolating filters are still designed more or less in accordance with the findings of the Royal Commission, because the Royal Commission investigations into percolating filters were extensive, and formed a reasonable basis for design. Nevertheless, it must be complained that design has tended to degenerate to the application of a rule of thumb, for many engineers when designing treatment works for newly sewered areas commence with two assumptions:

1. that the sewage will be strong;
2. that the rate of flow of sewage per head of population will be high.

In the majority of instances, if one of these assumptions is right, the other must be wrong, for, as far as domestic sewages are concerned, a high rate of flow means dilution and, consequently, a weak sewage.

The Royal Commission classed filter media as coarse, medium size, fine and very fine; medium size being 1-inch to $\frac{1}{2}$-inch diameter, fine material $\frac{1}{4}$-inch diameter and very fine material $\frac{1}{8}$-inch diameter and smaller. The Commission's findings relating to fine media do not need to be discussed, for nearly all present-day percolating filters are constructed of material varying in size from about 1-inch diameter to 3-inches diameter or even larger.

The Final Report of the Royal Commission, which was a summary of their recommendations, does not give any recommendations as to capacity of percolating filters. But in the Fifth Report the findings of the Commission are summarised, the relevant information being as follows:

> ... percolating filters of coarse material, 6 feet deep, can treat a strong septic tank liquor (strength 73 to 85) at a rate of 100 to 125 gallons per cube yard per day, or a liquor of half this strength at the rate of 200 to 250 gallons...
>
> ... 80 gallons per cube yard of a strong liquor of strength 113 or 150 gallons of a weak liquor of a strength 58.
>
> ... As regards the question whether it is easier to treat upon percolating filters of coarse material one volume of strong septic liquor or two volumes of weak, the organic impurity being the same in both cases, there is little difference either in the actual work done by the filters or in the quality of the effluents, at rates of working up to 200 and 400 gallons per cube yard per day.

The above figures give the rates at which sewage can be treated per cube yard of material and a satisfactory "Royal Commission effluent" be obtained, provided the rate of flow remains constant from day to day, varying only throughout the day. It was, however, the practice at the time that the Royal Commission was sitting, and it still remains the practice today, to install sewage-treatment works of sufficient capacity to treat a rate of flow of three times the average dry-weather flow. The Royal Commission found that:

> The rates of filtration which we give could generally be doubled in wet weather. Thus, assuming it was desired to provide enough filtering material to filter in storm times up to three times the dry-weather flow of sewage, the amount of material required would be 50 per cent more than the amount which would be required for the dry-weather flow, calculated on our figures.

From the above quotations it will be seen that the findings of the Royal Commission were that the amount of water in a sewage does not affect the required capacity of percolating filter, but that the latter depends on the amount of organic matter to be oxidised. This is in agreement with some American practice, for American designers have used the principle of allowing so many cubic feet of material per pound weight of biochemical oxygen demand per day, regardless of the quantity of water in which the organic matter is transported.

From the Royal Commission's findings it is possible to calculate the average of the opinion stated and to set out the quantities of media required for the treatment of different strengths of sewage. There are three ways of approaching this section of the design of percolating filters and of making use of the information available. These are as follows:

1. When the works are for a newly sewered area and the actual strength of sewage is not known, but the sewage will be mainly domestic, the capacity of percolating filters can be based on the population, regardless of the amount of water in the sewage. In this case about 0·533 cubic yard (0·408 cubic metre) of medium are required per head of population.

2. When, as in the case of an existing works, the strength (McGowan)

is known, the capacity of filter, in terms of gallons per day per cubic yard of medium, may be determined in accordance with Table 51.

3. While in the past sewage-treatment works have been designed according to the strength of the sewage as estimated by McGowan's formula, this practice has been confined to Great Britain. Elsewhere—e.g America—works are designed on the B.O.D. value of the crude sewage, and the tendency now is for English engineers and chemists to follow overseas practice. When the five days B.O.D. (parts per million) is known, the capacity of the percolating filters can be calculated by allowing 120 cubic feet per lb weight ($7\frac{1}{2}$ cubic metres per kilogram) of B.O.D. per day.

Lb B.O.D. per day is:

$$\frac{\text{Gallons per day} \times \text{B.O.D. (parts per million)}}{100{,}000}$$

In the case of mainly domestic sewages, the writer recommends an allowance of 0·12 lb B.O.D. per head per day. Tests made by the Water Pollution Research Laboratory at Stevenage New Town on flow from a purely residential area of which full particulars of population were known showed that the load was 0·099 lb B.O.D. per head per day. For a large industrial town sewered on the combined system, and including its associated residential areas, a figure of about 0·17 lb B.O.D. per head per day is about normal in England. These figures might be compared with the American figures given by Metcalf and Eddy,[177] which are as follows:

Separate sewers without industrial wastes, 0·115 lb B.O.D. per head per day.

Combined sewers without industrial wastes, 0·139 lb B.O.D. per head per day.

Other figures understood to be in common use in America are:

For cities with separate sewers, 0·14 lb B.O.D. per head per day.
For cities with combined sewers, 0·16 lb B.O.D. per head per day.

Presumably these include some proportion of trade waste.

Different capacities are necessary when the preliminary treatment is other than continuous-flow sedimentation or septic-tank settlement, or when the percolating filters are operated according to the principles of alternating double filtration or recirculation, or when the filters are enclosed with a roof and given artificial ventilation. If, in place of continuous-flow sedimentation, the sewage is chemically precipitated in continuous-flow tanks, the quantity that can be treated per cubic yard of medium can be increased by about $33\frac{1}{3}$ per cent at least.

The foregoing capacities of percolating filters were arrived at before synthetic detergents introduced a new problem. There is, however, no evidence that the recommendations of the Royal Commission need revision, for percolating filters appear to be able to treat sewages containing synthetic detergents without the difficulty that the activated sludge systems sometimes have.

Alternating double filtration, recirculation and enclosed filtration. For long it had been known that the rate of dosing percolating filters in terms of gallons per day per unit of superficial area of bed had a marked effect on performance. An early assumption that very regular fine-spray distribution should be the most efficient was disproved by experiment, for it was

TABLE 51

LOADING OF PERCOLATING FILTERS AFTER CONTINUOUS FLOW SEDIMENTATION
(WORKS DESIGNED TO TREAT UP TO THREE TIMES DRY-WEATHER FLOW)

Crude sewage strength (McGowan)	Gallons per day dry-weather flow per cubic yard of filter medium*
30	222
35	190
40	167
45	148
50	133
55	121
60	111
65	102
70	95
75	89
80	83
85	78
90	74
95	70
100	67
110	60
120	56
130	51
140	48
150	44
160	42
170	39
180	37
190	35
200	33

found that better results were obtained with short-period heavy discharges at infrequent intervals and that, consequently, rectangular beds with distributors that gave infrequent heavy discharges were more efficient than circular beds with rotating distributors running continuously.

Further investigations into the special methods of alternating double filtration, recirculation and enclosed filtration have shown that it is possible to greatly improve on the performance of an ordinary percolating filter about 6 feet deep and given fairly regular distribution on its surface.

* Divide by 168·2 to give cubic metre of sewage per day per cubic metre of medium.

These new methods were, at first mysteries, undoubtedly successful but no one knew why. The author is, however, convinced that the secret of their success lies in the fact that each and all of them has, by heavy surface loading, the effect of bringing the whole of the medium into use leaving few or no dry patches or surfaces free from organisms. This opinion was strengthened by the facts that, where there was already a high rate of flow per superficial yard of bed, the new method produced a moderate degree of improvement only, but where the initial rate of flow was low, the new method effected a great improvement; also that tests made by the Water Pollution Research Laboratory showed that the maximum wetting of material in a bed occurred when the flow was 900 gallons per day per superficial yard of bed at about which rate the maximum B.O.D. removal in pounds per day per cubic yard of filter medium reached its maximum (see Table 52).

TABLE 52
GALLONS RETAINED IN A CUBIC YARD OF MEDIUM

Gals per day per yd sup.	Gals retained in a cu yd of medium, indicating degree of wetness
200	7·47
400	8·75
600	13·12
900	**18·4**
1,200	14·8
1,600	13·05
2,000	12·82

To return to the above-mentioned methods, the findings of the research workers were:

E. V. Mills, in "The Treatment of Settled Sewage in Percolating Filters in Series with Periodic Change in Order of the Filters: Results of Operation of the Experimental Plant at Minworth, Birmingham, 1940 to 1944,"[179] states:

> From results of the continuous operation of the large-scale experimental plant at Minworth for four consecutive years it has been shown that settled sewage can be treated by the process of alternating double filtration at four times the rate of treatment at which similar sewage can be treated by single filtration.

This opinion was based on full-scale experiment, but as practical experience at a number of works has yet to be gained, most authorities who are changing their works over to alternating double filtration have in mind a rate of treatment per cubic yard per day of about twice that applicable to ordinary percolating filter treatment.

The same writer, in "The Treatment of Settled Sewage by Continuous Filtration with Recirculation of Part of the Effluent,"[180] states:

> It is concluded from these experiments that the recirculation process is comparable in efficiency with alternating double filtration, if effective action can be taken

to counteract the effects of obstruction of the subsurface layers of the filtering medium which occurred in the recirculation process in the winter.

The experiments to which E. V. Mills' paper relates were on recirculation of effluent in equal quantity to the rate of flow of sewage. It is not easy to make direct comparisons between normal percolating filters and the various recirculation methods used in America, because the various companies manufacturing types of filters have been interested in giving publicity to the results achieved by their own products. However, examination of results at military installations* suggests that it would be safe to apply twice the rate of flow permissible with normal deep filters by using deep filters on recirculation at a rate of equal parts of effluent to sewage.

Details of these methods are described later under the head "High rate filtration."

Alexander Hunter and T. Cockburn, in "Operation of an Enclosed Aeration Filter at Dalmarnock Sewage Works,"[116] state:

> On an average over four years 1940-43 inclusive the Enclosed Filter when compared with our Open Filters has treated with an equal degree of purification twice the volume of tank liquor per cubic yard.

Depth of filter. As the loading rate of superficial area of bed is important it is clear that unless there are arrangements for recirculation the depth of filter is important. The Royal Commission considered the normal depth of a percolating filter to be 6 feet although increased depth could improve performance. It is now known that this depth is uneconomical unless there are arrangements for recirculation. It is also known that, with most methods of distribution, the flow does not spread over the whole area of the bed until it has penetrated to a depth of 1 or 2 feet, with the result that one can consider the first 6 or 12 inches of depth of percolating filter as being wasted as far as treatment capacity is concerned.

Depth is not infrequently restricted by the amount of head available but in no case should filters be made less than 4 feet 6 inches deep and, broadly, as depth is reduced, so should capacity be increased to make up for loss of efficiency.

Shapes of percolating filters. Percolating filters are usually either circular or rectangular, the type adopted depending on the site. Octagonal filters have been constructed. Circular filters are most commonly employed because they permit the use of the simplest distributing mechanism, the revolving-sparge distributor. Rectangular filters are preferred for two extremes—the very small filter, used for isolated buildings, which employs a tipping-tray mechanism for distribution of tank effluent, and for comparatively large works on congested sites. The main advantage of the large rectangular filter is that it economises land, for circular filters involve some wasted space between the beds.

* "Sewage Treatment at Military Installations—Summary and Conclusions." By the N.R.C. Subcommittee on Sewage Treatment. *Sewage Works Journal*, January 1948.

TABLE 53
EFFECT OF SIZE OF PERCOLATING-FILTER MEDIUM
FILTRATION WITHOUT POST SETTLEMENT

Rate, gals per cu yd per day*	B.O.D. of tank effluent, pts per 100,000	Loading lb per cu yd per day	Effluents, lb B.O.D. removed per cu yd per day					
			Granite	Raschig rings				
			$1-3''$	$\frac{3}{4}''$	$1''$	$1\frac{1}{2}''$	$2\frac{1}{4}''$
172	18·6	0·340	0·224	0·260	0·241	0·217	0·215
343	18·4	0·631	0·186	0·407	0·345	0·266	0·204
686	18·9	1·296	0·608	0·615	0·573	0·651	0·615
1362	22·6	3·078	1·352	1·587	1·538	1·476	1·488

Dosing. Rotating distributors for circular beds are actuated by the reaction of the water as it leaves the holes in the sparge pipes. At times of minimum flow the reaction may not be sufficient to rotate the distributors, which, accordingly, stand still and dribble. To overcome this trouble it is usual to install some kind of mechanism which produces a periodic flush equal to the maximum rate of flow. This usually takes the form of a dosing tank with automatic siphon. If a dosing tank is installed, the flow during slack periods becomes a series of short bursts of sufficient intensity to set the arms in motion.

In a large plant dosing tanks may be omitted if it is required to economise head as far as possible. In such works, at times of small flow, the whole of the flow may be passed to one or two filters, thereby making certain that no dribbling occurs. But when the head is available, that little extra required for dosing should be utilised and tanks constructed.

The minimum capacity of a dosing tank is generally taken as $\frac{1}{2}$ gallon per yard super of filter-bed served, but for very small filters this capacity should be increased to as much as 2 gallons. The amount of head lost depends on the size and type of siphon, being generally more for larger flows. Low-draught siphons for small flows are manufactured to draw from a tank with a working depth as little as 9 inches, or in some cases 6 inches. But for general purposes deeper tanks are preferable, in that they have a smaller surface area and smaller area of flat floor to be periodically cleansed. Where particularly shallow tanks are used it is permissible, and, in fact, advantageous, to include the capacity of the effluent channels of the sedimentation tanks in the dosing capacity. Then the channels fill up to a level sufficiently below the weirs for the sewage to give a clear fall and, on the siphon discharging, the channels empty rapidly, thereby cleansing themselves. Finally, the last of the effluent is drawn by the siphon from the sump in the dosing tank, and the cycle repeated.

* Divide by 168·2 to give cubic metres of sewage per day per cubic metre of medium.

The best arrangement is to have a separate dosing tank for each filter. If, however, for the sake of economy, a dosing tank is to supply, say, four filters, the siphon must be capable of discharging the total flow at the maximum peak rate of flow that will be fed to those four filters; but as there is the possibility that one out of the four beds will have, at some time, to be rested, the filter arms of the remaining three beds must be capable of discharging the total maximum rate of flow under the head available when the dosing tank is full. This gives the size of siphon and the discharge rate of arms for the particular condition under discussion. In order that no dribbling shall occur,

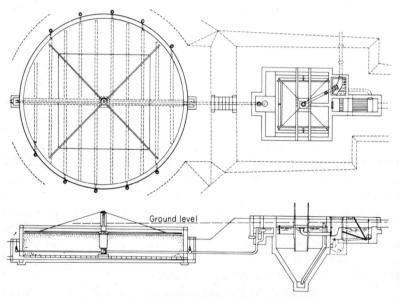

FIG. 123—Sedimentation tank and percolating filter
By courtesy of Adams-Hydraulics, Ltd.

the discharge from the four filters should not be less than dry-weather flow when the dosing tank is on the point of emptying. Thus it is obvious that the design of dosing tanks and the arrangement of siphons are not things to be applied by rule of thumb.

Where several dosing tanks all receive their flow from one supply, there should be a means by which the flow can be weir-controlled to each tank, as otherwise the siphon action of one will affect the rest, so that they do not receive their full share of the flow. The simplest arrangement is to have one dosing tank for each individual battery of settling tanks, as weir-feeds mean further loss of head.

Some small distributors have dosing tanks housed in a central column. When these are used, there must be no other dosing tank and each distributor must be separately weir-fed, or distribution will be very unequal.

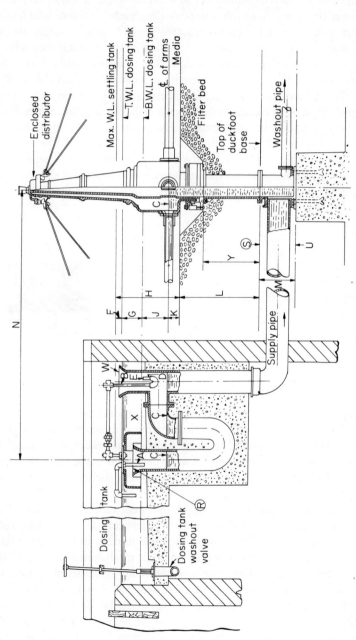

FIG. 124—Section of dosing tank and percolating filter
By courtesy of Adams-Hydraulics, Ltd.

In Fig. 125 are illustrated the component losses of head between settling-tank weir level and top medium level of a circular filter. First, there is the free fall over the weirs, A, to top water-level in channel, a figure which can be cut fine in some cases, but which generally might be taken as being about 3 inches or more. Then there is the friction loss in the channels themselves when running at the maximum rate of flow, B. Third, and usually the greatest of the component losses of head, there is the working depth of the dosing tank, C. The maximum head of the filter arms will be C, D and E, and this should give the maximum rate of flow to the beds, usually three times dry-weather flow plus allowance for resting. The minimum rate of flow occurs when the head is only D plus E, and at this head the flow should be 1 to $1\frac{1}{2}$ times dry-weather flow. D is the allowance for friction in feed-pipes and through valves and bends at the maximum rate of flow, and E the residual head required for discharge through the holes in the arms and for producing

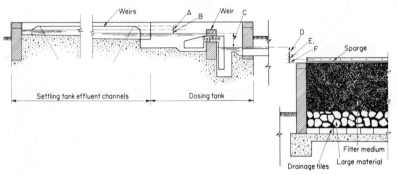

FIG. 125—Illustrating loss of head from sedimentation tank to percolating filter

the motion of the arms themselves. This last can, in some cases, be as little as 3 inches. F is the distance between the holes in the arms and top surface-level of the bed. This can be reduced to about 3 inches if necessary in the smaller beds, but in large beds the diameter of the arms may be sufficiently great to require a larger figure. Where head does not matter 6 inches might be taken as a general rule.

While small percolating filters may be operated with an overall loss of head from sedimentation-tank weirs to material level of 2 feet or under, even as little as 15 inches when there is no dosing, the distributors for rectangular beds require heads up to 3 feet 6 inches when the mechanisms are driven by the flow, and even as large as 2 feet or more when there is electrical operation.

Excessive fall from settling tanks to filters does not matter, provided that the rate of flow is not influenced by the fall, as would be the case if the feed-pipe to the filters had a sucking effect on the dosing siphon. Where, however, the siphon discharges to an open chamber from which the feed-pipe connects, and the siphon is properly sized, there can be any amount of fall from this point to the filters.

Distributors for circular beds. There is considerable difference in the design and quality of the various makes of percolating-filter distributors although the general appearance is similar. The vast majority consist of a cast-iron central column with ball bearings and watertight seal and a number of radiating sparge arms, usually two or four in number, suspended by guy wires. Leakage between the incoming feed and central column is prevented by a mercury seal, glands, multiple water-seal or, in one instance, a device incorporating the venturi throat principle. In some designs there is a weir in the central column so that two arms are fed continuously and the other two come into operation on increase of flow.

It has been usual to make the sparge pipes of galvanised iron and support them with galvanised wires and strainers. But protective or non-corrosive materials have come into use.

The distribution holes are, in the better-quality machines, fitted with bronze bushes and distributors and are spaced, being closer together nearer the delivery than at the centre, so as to give even distribution over the whole of the bed during normal running. The holes in the different arms are staggered so that no two holes are at the same distance from the centre of the bed. So as to cause rotation of the arms the holes discharge horizontally opposite to the direction of rotation, but in circumstances where there is no recirculation or other means of adjusting the rate of flow per unit area of bed the holes in one arm out of four can be made to discharge in the opposite direction so as to reduce speed of rotation.

At the end of each sparge pipe there is a removable cap for cleaning purposes.

Because of the very great difference of quality in the various makes of distributor it is not good policy to obtain distributors by competitive tender. The engineer should use the kind of distributor which he has found most satisfactory and recommend to his client that this and no other should be used.

Feed distribution. The feed-pipe system to a series of percolating filters should, to a certain extent, follow the bifurcation method, as described for the feed-pipes and channels of settling tanks (see Chapter XXVII), but in this case it is not so important as in the case of tanks. Although it is advantageous if the flow is more or less equally distributed to the filters when all valves are open, it is not necessary that the distribution should be exact, for a number of percolating filters, although constructed alike in every detail, will not necessarily be found of equal efficiency. While one may be capable of treating satisfactorily large quantities of effluent, another may, for no apparent reason, give poor nitrification to the same flow. It therefore becomes necessary for the flow to be regulated in accordance with the capabilities of the filters, and this being the case, exactly equal distribution of flow can be an unnecessary refinement.

To ensure self-cleansing conditions, the velocity of flow in the feed-pipes containing settled sewage should be not less than 2 feet per second at the time of peak rate of flow during the day, although theoretically these mains should not contain any heavy sludge-producing solids. The figures may have to be modified according to circumstances. For example, sometimes it may be necessary to reduce velocity so as to economise head. In calculating head losses it should be borne in mind that the various junctions, bends and valves in percolating-filter feed systems involve considerable velocity-head losses.

The effluent pipes and channels leading from percolating filters to humus tanks should be designed to give velocities of not less than $2\frac{1}{2}$ feet per second at the peak daily rate of flow, for although humus is light material easily swept along by the flow, the filters, particularly when new, give off quantities of dust or detritus washed out of the medium. The flow should be sufficient to carry this to the humus tanks.

Once the final effluent has been settled in the humus tanks, it should contain very little solid material and practically none likely to settle in the channel. Velocity then becomes of small importance and, if there is need to economise head, channels may be made large. Nevertheless, the velocity should be more than two-thirds of a foot a second at some time during the day, otherwise some sedimentation is to be expected. Where falls are adequate, velocity should be kept as high as practicable for the sake of economy in channel construction.

Construction of beds. Circular filters are constructed with floors usually of 6 inches of concrete, which generally need not be reinforced, and external walls to retain the medium. At one time it was common to construct external walls of open brickwork to assist ventilation, but this was found not to have any real advantage, as it assisted aeration at the edges of the filter only, the main flow of air being inwards through the under-drains and upwards through the medium, and also they were found to have various disadvantages, increasing insect nuisance and being more liable to failure than solid brickwork.

Walls consisting of large or boulder clinker, or of dry rubble walling, were also at one time popular, but are not now common, because boulder clinker is not so easy to obtain, and the skilled labour for dry walling is becoming rare. A badly built dry wall of any material is liable to fail, even though the pressure due to such material as clinker, which is light in weight and has an angle of repose of about 35° to the horizontal, is not high.

Where filters are nested together they should be divided one from another by a dwarf wall.

The falls of the floors are usually not very steep, about 6 inches being sufficient for a bed 100 feet wide. The fall may be towards the centre radially, towards the outside radially or straight across from one side to another, the choice of method being dependent on local conditions.

With the first arrangement a central manhole is constructed, to which the under-drains gravitate. The use of a central manhole in this manner is the most satisfactory solution of the collecting together of effluent under-drains when the beds are dug deep below ground-level and an external channel would be unduly expensive. This method is to a certain extent popular for general use, probably because of the slight advantage, at least on paper, of having an access manhole at the centre column of the distributor. It has, however, disadvantages for general purposes. In a large filter, under-drains, if purely radial, become too widely spaced at the far ends and too closely spaced at the centre. In fact, if they are to be at right spacing at the far ends there is no room at the centre for them to be placed side by side, and some form of "herring-boning" becomes necessary. Second, as a result of this arrangement of the under-drains, the aeration of the beds is irregular, being unsatisfactory at the outside, unless special ventilating-pipes are provided. The construction of a central manhole is comparatively costly, wasteful of head and requires effluent drains laid below floor-level.

The method of laying under-drains radiating from the centre to an external peripheral channel is probably best for the purposes both of drainage and aeration. It affords access to the ends of all pipes, which can therefore be rodded if necessary, and the bed is aerated equally from all sides. This method, however, is slightly more costly than some, owing to the length of external channel required, but it does not necessitate excessive length of under-drains. It is the best design for filters constructed wholly above ground-level on a level site.

The most economical construction for general purposes is that involving cross-flow of under-drains, in which the drains are laid parallel and discharge to a channel which is equal in length to one-quarter of the circumference of the bed. The fall of the floor should, of course, be in the same direction as the fall of the land. This design requires the minimum length of drains for the degree of drainage and aeration obtained, the minimum excavation on a sloping site and the minimum cost of channel construction. It is particularly applicable to the sloping site.

Rectangular filters. The construction of rectangular percolating filters is generally similar to that for circular beds, with the exception of the differences of feed and effluent channels necessitated by the difference of form and the type of distributor. The supply channels run longitudinally down the sides or centre of each bed and are constructed of concrete, brickwork or steel troughs. These channels support the rails on which travel the distributors. The distributors draw their supplies from the channels with the aid of siphons.

The treated effluent is discharged by under-drains or drainage tiles into culverts, which are usually constructed underneath the feed channels, and which are large enough both to give a good air-flow for aeration of the beds and to give access for men in order that cleansing may be performed without difficulty.

Rectangular distributing mechanisms vary considerably in type, some being self-driven by the rotating of water-wheels carried on the sparges, some are driven by electric motors, and some are cable-driven, either by electric power or water-power. The earlier types of mechanisms gave some trouble, but most modern devices are satisfactory in all weather conditions.

An important point to be borne in mind when designing rectangular beds is that the distributors for such beds usually move slowly, discharging at heavy momentary rates compared with rotating distributors for circular beds. Within limits this adds to the efficiency of the bed, but if momentary discharge is excessive there is a danger of discharging tank effluent on to the filter at such a rate that it passes right through the bed in a matter of seconds,

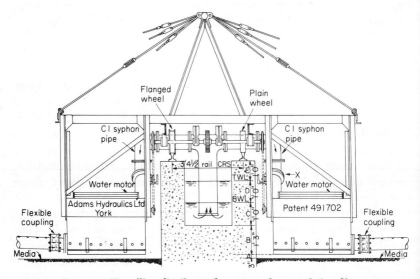

FIG. 126—Travelling distributor for rectangular percolating filter
By courtesy of Adams-Hydraulics, Ltd.

receiving insufficient treatment. Furthermore, seeing that rectangular distributors are expensive, the designer may be tempted to make his beds unduly long, thereby reducing the number of distributors and increasing the rate of discharge per distributor. For satisfactory working, rectangular beds should be comparatively short—say not more than 250 feet, unless the beds are divided into bays, each of which is served by an individual distributor.

Humus tanks. Humus tanks are generally similar to primary sedimentation tanks, but differ in detail mainly because they have to collect humus, which, although easily settled, is liable to gas and rise to the surface if not expeditiously removed. Pyramidal-bottomed tanks are largely used, although they have the disadvantage of involving deep excavation, being usually low in level relative to the general level of the works. At larger works mechanically swept circular tanks are not uncommon.

While with sedimentation tanks one has to consider carefully rates of upward flow, sludge capacities and positions of inlet, this is not so with humus tanks. With the pyramidal humus tank it is necessary only to design to be within the limit of a maximum rate of $7\frac{1}{2}$–9 feet per hour upward flow. A capacity of four hours dry-weather flow is most commonly adopted, which is more than the Royal Commission recommendation. Length of weir is not important, in that the rate of flow is large compared with the size of tank, and so tanks may be made over quite a large range of sizes, with efficient working.

The channels of the humus tanks should be at least as large as those of the sedimentation tanks: too often designers make the channels large or small in accordance with the size of tank, regardless of flow. Humus tanks should be fitted, with adequately deep scum-boards well away from the weirs, for there is always a distinct possibility of gassing and floating humus, even when the tanks are given regular attention.

Pyramidal tanks are sludged under hydrostatic head, in every way similar to sedimentation tanks, but they should be given even more regular attention. The humus may be discharged to separate drying beds, where it may be preserved because of its high manurial value, or if there is no demand for humus, it may be passed to digestion tanks or returned to the sedimentation tanks and resettled with crude sludge.

The quantity of humus for the year must be the quantity of sludge that the primary sedimentation tanks have failed to settle less the suspended solids escaping with the final effluent plus the suspended solids produced from dissolved solids in the percolating filters. On the average the humus amounts to about one-third of the sludge produced at the works during the year. But the quantity is very variable and allowance should be made for high rates of flow during the spring off-loading. The moisture content of humus is very high, averaging in the region of 99·23 per cent (see beginning of Chapter XXVIII).

High-rate filtration. When percolating filters are dosed at high rates, the amount of work done in B.O.D. reduction is increased, but nitrification may be reduced. Too great an overload of a normal filter will cause ponding, and eventually a reduction of performance. But by special methods of operation, percolating filters can be made to do more work and remain efficient. On the one hand, a filter can be made to partially treat a large quantity of effluent; on the other, full treatment can be given on a reduced quantity of medium. Moreover, if the rate of flow is greatly increased it has a washing effect which prevents ponding.

In America the term "high-rate filtration" applies particularly to heavy loading of percolating filters to produce partially-treated effluents such as may be permissible where discharge is into high dilution. In Great Britain there is little opportunity for the discharge of partially treated effluent, and the only methods involving heavy loading of filters that have been used to any extent are those which achieve Royal Commission effluents—i.e. recirculation or the Dorr "Biofiltration" process; alternating double filtration; and, to a much lesser extent, enclosed filtration.

The Dorr Biofiltration process. The following description of the Biofiltration or recirculation process is included by courtesy of the Dorr-Oliver Company, Ltd.

Years of experimental work have been devoted to the search for a simple method of increasing the dosing rate on normal percolating filters, with a view to curtailment of initial cost and reduction of the large area required. Subsidiary objects of research have been the reduction of nuisance from adult psychoda flies and reduction of odours.

From original investigations by Jenks in the United States in the 1930s on the effect of recirculation of filter effluent through a preceding clarifier, coupled with the use of very shallow filters, the Biofiltration Process has been developed to the point of having become a standard method of treatment in America, South Africa and elsewhere, utilising dosing rates up to and in excess of 1000 galls/cu yd/day on a raw sewage D.W.F. basis.

The essential feature of the Process, which is patented, is the recirculation of a relatively large volume of filter effluent, and the contact of this recycled effluent with the incoming flow in a clarifier large enough to give effective contact time. The filter itself does not effect complete purification but seems to act as a rapid oxidising agent, whose effect is developed in the clarifier owing to the interaction of the recycled flow with the raw flow, coupled with dilution. This explains the success of shallow filters, since oxidation rather than nitrification is the object; furthermore, the shallow filter is much easier to ventilate and owing to the smaller head on the pumps, the power load for recirculation is much reduced below that which would be feasible otherwise.

The Process can be single stage or two stage, according to the degree of purification desired, and the recirculation flow quantity is closely related to the amount of 5 day B.O.D. desired to be removed. Furthermore, the clarifier contact time can, as a variation, be given as successfully in a clarifier which follows as in one which precedes the filter.

It is important, however, that the settling tanks employed for the detention zone should be of the mechanically cleaned type—inadequate cleaning of settling tanks can defeat the whole design.

In practice, Biofilter plants are flexible, shock-proof in their capacity to accept sudden overloads, odourless and substantially free from fly nuisance. In this latter connection they obtain the usual benefit from the presence of the psychoda fly larvae, but the majority of the adult flies, in the winged stage, are drowned owing to the very high rate of flow through the filter.

The power load for purification by the Biofiltration process naturally varies according to the degree of treatment attained and site conditions. For complete treatment, however, it could be estimated as 10–15 H.P. per million gallons treated, as against 25–30 H.P. per million gallons by the activated sludge method. The following flow sheets are generally in use (see Fig. 127):

(a) Single-stage partial treatment

This flow sheet, with the units adequately sized, will give some 50 to 60 per cent removal of 5-day B.O.D. Its chief application lies in the field of reconstruction of existing overloaded conventional type sewage plants, whose filters or aeration tanks can be made to accept a greatly increased flow, provided that the pollution load is partially removed by Biofiltration. In reconstruction schemes, it is frequently impossible to provide additional head for expansion of process between existing sedimentation tanks and subsequent treatment. A feature of this flow sheet, since the flow is pumped to the Biofilter and runs back to sedimentation at

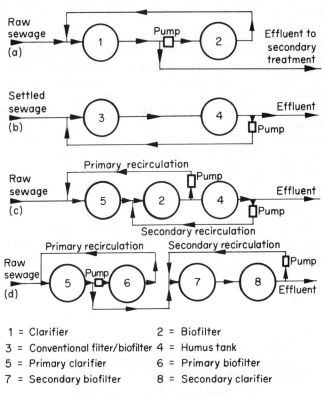

1 = Clarifier   2 = Biofilter
3 = Conventional filter/biofilter   4 = Humus tank
5 = Primary clarifier   6 = Primary biofilter
7 = Secondary biofilter   8 = Secondary clarifier

FIG. 127—Flow diagrams of Dorr "Biofiltration" process
By courtesy of Dorr-Oliver Co., Ltd.

the original level, is that no additional head is required, and the original hydraulic scheme of existing plants can be maintained. Recirculation quantity is usually two or three times the dry-weather flow.

(b) Single stage complete treatment—on settled sewage

While it is difficult to predict actual 5-day B.O.D. removal, the capacity of existing conventional-type filters can be at least doubled by recirculation of final humus tank effluent ahead of the filters. Humus tanks must be made adequate for flow plus recirculation. This flow sheet is also useful for extension of existing overloaded plants, particularly when their primary sedimentation tanks are too modern to replace, or of a type unsuitable for incorporation in a scheme embodying flow sheet 'a.' Recirculation quantity need not exceed dry-weather flow.

(c) *Single-stage complete treatment—on raw sewage*

This flow sheet employs primary and secondary sedimentation with single-stage filtration and double recirculation. Overall 5-day B.O.D. reduction of 75 to 85 per cent is obtained on normal sewage, and though an effluent of Royal Commission Standard is not to be expected, the degree of treatment is adequate for discharge into large bodies of water or estuaries where plain sedimentation is in itself insufficient. Total recirculation quantity is usually about three times dry-weather flow, divided into primary and secondary stages.

(d) *Two-stage complete treatment—on raw sewage*

This flow sheet is really a combination of 'a' and 'b,' and will give an effluent of Royal Commission Standard very economically in comparison with Aeration Plants or conventional filter plants. By adoption of a moderate dosing rate in the secondary Biofilter and extension of this unit to conventional depth, an advanced degree of nitrification can be obtained. This may not always be necessary or even desirable, and if consistently satisfactory removal of B.O.D. will suffice, then full Biofilter dosing rates and shallow filters may be employed for the secondary stage as well as in the primary stage. Recirculation quantity is usually about three times dry-weather flow, as in flow sheet 'c.'

The equipment for the Biofiltration Process is of conventional type, but the following points are stressed:

Clarifier should be of centre feed radial flow design, and must be equipped with mechanism for continuous sludge removal. Scum skimming is required on primary units.

Filters must be adequately designed from the point of view of drainage capacity (in view of the high flow rate which may be from 15–100 times normal flows) and ventilation.

Distributors require to be specially designed for the high dosing rates. Dosing tanks are not required since flow is continuous.

Recirculation pumps operate against low head, say 10′ 0″—11′ 0″ manometric. They are usually of large capacity and in general pumps of axial-flow type are the most suitable, in view of their high efficiency.

In conclusion, the value of the process can be summed up as follows:

Very large saving in area and installation cost owing to the high permissible dosing rates and small size of filter employed.

Small offset in cost for recirculation pump power load.

Flexibility in operation, ready absorption of shock loads, skilled supervision limited to mechanical maintenance.

Suitability for existing plant reconstruction as well as new plant designs.

Alternating double filtration. The process of alternating double filtration was developed in the first place for treating waste waters from dairies and cheese factories; it is true that it had been used rather tentatively before that for treating waste waters from beet-sugar manufacture, but it was not really developed in its present form until the work on milk was done. Milk wastes, particularly those containing whey, give rise to an unusually thick growth of biological film, containing a high proportion of fungal hyphae, on the surface of a percolating filter, and the process was tried in the first place in an endeavour to reduce the growth of this film, and so to enable larger quantities of waste water to be treated in plant of a given size. This effect in keeping down the growth of biological film is no doubt an important feature in the use of the process for the treatment of sewage.

In this respect the process of alternating double filtration is similar in effect to recirculation. In this process the quantity of effluent to be pumped equals, of course, the quantity of sewage delivered to the works, and is not varied to two or three times, as it can be the case in recirculation.

The process involves two batteries of percolating filters arranged in series, but normally constructed at the same level. The effluent from the sedimentation tanks passes through the first battery of filters, and thence to humus tanks, after which it is pumped to the second battery of filters and gravitates to a second battery of humus tanks.

After the sewage has flowed through the works in accordance with the above circuit, the operation is altered, the first battery of percolating filters and their humus tanks taking the place of the second battery, which in turn become primary filters and tanks. This involves an arrangement of pipes and valves which permit the easy alternation of function of filters. The period of alternation that is most convenient and, at the same time, most satisfactory, can be found by experiment.

Alternating double filtration schemes should generally be so designed that they can be run as recirculation schemes whenever desired. This makes them more flexible, in particular in that it permits adjustment of flow per unit of surface area to the optimum.

Economics of recirculation and alternating double filtration. Taking into account capital and running costs, these methods are probably the most economical to use in many circumstances.

Enclosed percolating filters. Enclosed filters are deep percolating filters of coarse material which are totally enclosed, having a roof to protect them from the weather. Ventilation is ensured by a fan, which may be placed in the apex of the roof. The effect of enclosing the medium and excluding the light as well as the cold encourages psychodae and similar organisms to flourish throughout the year, and inhibits the usual growths that are the cause of ponding. Other growths of an unusual kind, however, have been known to develop, but the enclosed filters are capable of passing greater flows than normal, and are more efficient than the ordinary percolating filter. On the other hand, the roofs are expensive structures which balance-out much of the saving.

Operation and maintenance of percolating filters. Percolating filters require little attention, and that is one of the reasons for their popularity. The most regular work to be done is the clearing of the sparge pipes. This is usually a daily labour. The mechanical arrangements need little attention;

the rotating mechanism at the centre, the water-seal, etc., may require attention according to the type installed, but with a good device this should seldom be necessary.

A wash-out should connect from the centre column, by which each individual distributing pipe may be separately emptied without interfering with the flow to the pipes feeding the other beds. This should not discharge to the effluent, but flow should be returned for treatment.

Galvanised sparge pipes have advantages, in that they require less frequent renewal than other types. The support wires are the components which probably need renewing most often, and it is worth-while having spare sets of these in stock, ready made up to length.

The former practice of resting beds periodically is no longer favoured because it causes dry-out and loss of film. Another possibility which can, in some cases, be considered is a mean by which beds may be flooded for the purpose of reducing the activity of psychoda flies. Objections have been raised to this method of dealing with psychodae, but it might be pointed out that it is simply turning the beds over to the condition of contact beds for a short period of time. Where flooding is not possible psychoda control may be accomplished with the aid of insecticide chemicals.

Effluent polishing. In cases where, for special reasons, it is necessary that an effluent of better than Royal Commission standard is essential, the most effective method is not to increase the sizes of aeration and settlement units but to ensure by other means that the suspended-solids figure of the final effluent is kept low, for generally a low suspended-solids content tends to ensure a low B.O.D. value also. This is known as effluent polishing. The methods that have been applied are micro-straining, sand filtration, land filtration and irrigation through grass plots.

Micro-straining, for which the machinery is made by Glenfield & Kennedy, Ltd., is straining through fine stainless-steel gauze or plastic fabric which is stretched over rotating drums. The filter fabric is kept clean by back-washing with clean water or treated effluent. It has also, of course, to be cleaned from time to time to prevent organic growth. Micro-strainers on the average reduce the solids content of an effluent to one-third. Thus, allowing a factor of safety, they could be relied on to transform a humus tank effluent containing 30 parts per million to a final effluent having 15 parts per million.

Rapid sand filters, generally similar to those used for water treatment (see "Rapid filters" in Vol. I, Chapter VII), have been used to treat 140 gallons of final effluent per square foot per hour under a maximum head of 9 feet. The filters had to be back-washed every 8 or 10 hours when the loss of head had reached $6\frac{1}{2}$ feet, the amount of back-wash water being 9·5 per cent of the flow. This achieved a suspended solids content of 4·6 parts per million.

Where the subsoil is suitable, land filtration may be very effective (see Chapter XXII). Slow sand filters can also be used. It is not usual to back-wash these and, when the operating head exceeds about 2 feet, they are cleansed by being drained, the sludge scraped off and fresh sand added as necessary. They are particularly suitable for small works.

The above-described effluent-polishing processes should always be preceded by normal humus-tank settlement, otherwise they would become uneconomical or call for undue attention. They are not methods to use at sewage works where an effluent of Royal Commission standard will suffice.

In addition to the foregoing methods, land filtration followed by short-period settlement can be very effective where suitable land is available, and calls for less maintenance. It is a method that can be, and is, used after other methods of aeration at works designed to produce Royal Commission effluents.

At the works of the Birmingham Tame and Rea District Drainage Board, humus-tank effluent was treated on grass plots, an area of 1 acre per 4000 persons served being allowed. To allow for maintenance and reconditioning, the area was divided into not fewer than three plots with separate supply systems fed by concrete channels with side weirs. The length of each plot was about 300 feet and the fall of the land 1 in 60. The maintenance was little: each plot was dried out and mown twice in the growing season to prevent the development of weeds. The result of this process was, at the Langley Mill Works in 1956, to reduce a humus-tank effluent of 16·2 parts per million B.O.D. and 17·0 parts per million suspended solids to a final effluent of 7·6 parts per million B.O.D. and 8·1 parts per million suspended solids.

The author has used areas of grassland to remove solids from effluents that had not been settled in humus tanks, and, thereby, obtained superb effluents but with larger areas of land than recommended by the Ministry. The Ministry recommendation is a loading of 150,000 gallons per acre per day.

The dissolved solids in sewage effluents can be reduced by passing the treated effluent to an algae pond where algae are grown for the purpose. It is necessary to encourage a continued and rapid increase in algal growth which can be aided by the addition of carbon dioxide to the effluent, and to harvest the live algae so as to remove the solids before decay sets in. This is done by draining the ponds in turn and scraping the algae from the sides and bottom. The method also reduces B.O.D. by virtue of the oxygen produced by the algae and effects some reduction in bacteria.

Where methods of effluent polishing are used careful consideration should be given to any arrangements for dealing with storm water, for it would be unreasonable to spend money on plant to produce a superb effluent and at the same time discharge quantities of much more polluting storm water from the storm tanks at the sewage works or overflows elsewhere on the sewerage system.

CHAPTER XXX

ACTIVATED-SLUDGE SYSTEMS

In percolating-filter treatment and in land treatment the oxidising organisms are attached to particles of medium or of earth respectively, and the sewage to be treated flows over them in a thin stream. In all the activated-sludge processes the oxidising organisms are gathered together in freely floating flocculi in the "mixed liquor" of sewage and recirculated activated sludge, and are not permitted to come to rest or to become attached to any part of the tank-work, as any deposition tends to cause septicity, which can upset the process.

Just as a percolating filter has to build up its population of organisms, the body of activated sludge has to build up over a period of many days and the concentration must be preserved. The activated-sludge flocculi, which develop by growth while feeding on the organic content of the sewage, are separated from the mixed liquor in the final sedimentation tanks. The treated effluent is passed on to the outfall: the activated sludge is returned in measured quantity to the aeration tanks, where it continues to do its work of oxidising the sewage. An excess of activated sludge, usually referred to as "surplus activated sludge," is built up, separated, discharged to the flow of crude sewage and settled out in the primary sedimentation tanks with the raw sludge or may be separately thickened and passed to sludge disposal.

Activated-sludge systems. Activated-sludge systems are classed under two heads: the diffused-air system, in which aeration and mixing of the sludge and sewage are effected by injecting air into the bottom of the tank, and the mechanical-agitation or surface-aeration system, in which aeration of the sewage is effected by splashing at the surface and disturbance of the sludge is produced by some form of stirring.

The diffused-air system has been marketed for many years by Activated Sludge, Ltd., who have always used diffuser plates or domes of a porous type for diffusing the air into the mixed liquor. Wm. E. Farrer, Ltd. have a similar process in which jet aerators are used. Under the head of "surface aeration" are included the "Simplex" system, which is the proprietary design of Ames Crosta Mills & Co., Ltd. and the "Kessener" system developed by Dr. Kessener of Holland.

Requirements of the process. Effective aeration of sewage by an activated-sludge process requires a sufficient population of aerobic organisms, efficient aeration of the mixed liquor so as to provide these organisms with oxygen and sufficient detention period for the oxidation to take place. It is also necessary for there to be a degree of turbulence to prevent local settlement

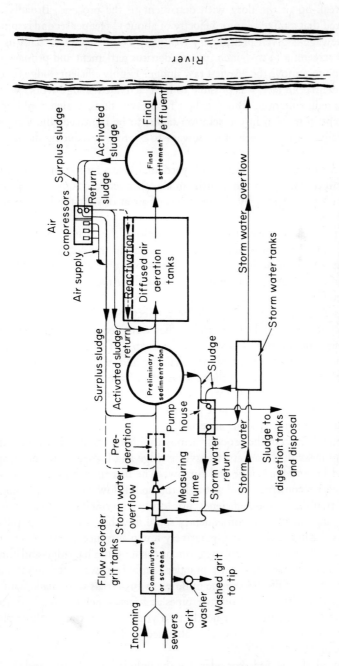

Fig. 128—Flow diagram of typical diffused-air plant
By courtesy of Activated Sludge, Ltd.

of sludge leading to septicity: on the other hand, excessive turbulence can cause break-up of the flow with detriment to the process. Broadly, it is considered that a mixed-liquor velocity of about $1\frac{1}{2}$ feet per second is suitable.

The sewage is given preliminary treatment before aeration. This usually includes screening ($\frac{1}{2}$ to $\frac{3}{4}$-inch spaces), detritus settlement and primary sedimentation to remove the heaviest raw sludge. Highly efficient sludge removal is not essential, in fact, it is possible to have activated-sludge works without primary sedimentation. But it is best for heavy solids, which could not be easily kept in motion in the aeration tanks, to be removed, and sludge removal reduces the strength of the sewage to be treated, reducing the required size of aeration plant.

The diffused-air system. The particular characteristic of the diffused-air system that makes it different from all other activated-sludge processes is that aeration is effected by the injection of air at or near the bottom of the aeration tank.

To a practical engineer, the observed results of the various means of treatment are of more importance than theory, and in this connection it can be said that the diffused-air method is at least one of the leading methods of activated-sludge treatment on both sides of the Atlantic, being particularly preferred for large works.

Diffused-air aeration has usually been effected in long tanks or channels. Formerly it was thought that cross-baffles were necessary to prevent short-circuiting; but it is now known that the degree of turbulence produced by aeration is ample to prevent short-circuiting and, moreover, experience has shown that cross-baffles had the reverse effect from that which was desired.

Research and experimental work on the diffused-air method of aeration in the activated-sludge process have proved that the efficiency of aeration can be increased with reduced aeration periods.

In the early days of the process various arrangements of diffuser were tried, including individual pockets, transverse ridge and furrow, saw-tooth, spiral flow and longitudinal ridge and furrow. The first three arrangements were expensive to construct, and though the spiral-flow tank was cheaper in construction the aeration efficiency was lower with spiral flow than with any other arrangements, and consequently spiral flow has never been popular in the United Kingdom and comparatively few plants incorporate this system.

During the period 1930 to 1957, use of the longitudinal ridge-and-furrow type of tank became standard practice, with furrows 5 feet wide and a continuous line of diffusers in the middle. In 1958 trials were made with flat-floored tanks with lines of diffuser pipes at 2 feet 6 inch centres, and domes spaced 12 inches apart. After nearly two years of operation of three plants, proof of increased efficiency was obtained and flat-floored tanks became standard practice.

Activated Sludge, Ltd., consider that the bubbles must be fine and that this necessitates a low air volume of 2 to 3 cubic feet per minute per square foot of diffusing area. For this they claim increased efficiency of aeration. The diffusers should be separated to reduce coalescence of bubbles and the dispersion of bubbles should be spread over the area of the tank as much as is practicable.

Present-day flat-bottomed aeration tanks can be any width to suit the site and need not be multiples of a fixed figure such as 5 feet. The depth of tank can be anything between 6 and 10 feet to suit local conditions. With shallower tanks the volume of air is increased for the same horse-power.

FIG. 129—Dome diffusers mounted on air main
By courtesy of Activated Sludge, Ltd.

The pressure through the diffusers should be kept at a minimum by keeping the diffusers clean. For this reason air is filtered prior to being passed through the diffusers, the best results being obtained with electrostatic dust removers. With satisfactory air filtration there should be no appreciable pressure rise over a period of nine or more years.

Aeration-tank capacity. In Great Britain it has generally been considered that, for a sewage of average strength giving a primary sedimentation-tank effluent with a B.O.D. of 200 parts per million, a detention period of seven to nine hours is satisfactory. (A crude sewage having a strength (McGowan) of 100 or a five-day B.O.D. value of 350 parts per million should, after primary settlement, give a tank effluent in the region of 200 parts per million.) These capacities include any aeration-tank capacity intended to be used for reaeration of activated sludge, for which about 20 per cent of the total aeration capacity is reserved in some British plants. This capacity is so

FIG. 130—Flat-bottomed aeration tank
By courtesy of Activated Sludge, Ltd.

arranged that it can be used for either sewage aeration or reaeration of activated sludge. This feature is unusual in America.

It has frequently been commented that American aeration periods are shorter than those used in Great Britain and it has been assumed that this is because American sewages are generally weaker and final effluents of lower standard. What can be overlooked are that the inputs of power for aeration purposes are, on the average, greater than those used in Britain by the same

FIG. 131—Arrangement of dome diffusers in flat-bottomed aeration tank
By courtesy of Activated Sludge, Ltd.

amount as the aeration, and final sedimentation-tank capacities are lower. Thus, if the American figures are compared with the British with the aid of the formula given at the end of this chapter, it will be found that the same results are to be expected.

Broadly, from the operator's point of view, a long detention period is preferable, for it gives a margin of safety. It is usually easy to increase horse-power when loads are high or sewage proves difficult to treat, and to economise in horse-power at other times; whereas if a plant is designed to a short detention period it may be difficult or impossible to deal with peak loads. On the other hand, the shorter detention periods are probably more strictly economical, particularly at the present time, when high rates of interest have to be paid on money borrowed for construction of works.

Proportions of channels. There should be not fewer than four and preferably six or more separate aeration channels in any diffused-air plant. The depth should not exceed 12 feet and, according to current opinion, be between 8 and 10 feet. The width and length of the tanks depend on the size of the plant and are not otherwise considered to matter where tanks are of the flat-bottomed type, not the spiral-flow type. But, if the channels are wide, the inlet and outlet arrangements must be carefully designed to ensure even flow throughout.

The spiral-flow principle is now virtually obsolete as regards aeration tanks in Great Britain. But it is used for supply and sludge-return channels which are aerated so as to permit low velocities of flow and minimum head loss.

Methods of diffusion. Air is introduced by two kinds of diffusers— porous diffusers and jet aerators. In the United States various kinds of jet aerator are used, the most effective probably being that in which a downward jet of water impinges on the surface of a metal bowl in which air is released. The air is broken into small bubbles which fly off from the rim of the bowl. In another type, recirculated effluent is used as a vehicle for injecting air into the channels in the same manner as water is injected by steam into a boiler. These methods require water (effluent) to be pumped as well as air, considerably adding to horse-power demand.

Diffusers can be flat plates, tubes or domes formed of porous material. Those generally preferred are tiles made up of particles having sharp edges because such sharp edges permit small bubbles to be released, whereas bubbles grow to considerable size on a smooth surface before they break away.

Flat-plate diffusers, consisting of tiles fixed by screws in cast-iron boxes and surrounded by rubber gaskets, were at one time used for all small plants in Great Britain, but are not installed in new works and are now supplied only for replacement. The plates were fed individually or in groups of two or four by screwed-barrel pipes brought down from air mains carried along the tops of the walls of the channels. These screwed-barrel pipes, when of steel, involved considerable maintenance costs, and it has been found economical to replace existing steel pipes with polythene pipes.

Plate diffusers when new will pass about 2 cubic feet of air per square foot per minute with a loss of head of about 3 inches water gauge. In course of time they may become clogged and should be removed for cleaning. This can be done by burning the tiles which, in many instances, is all that is necessary. If they have become choked by iron oxide from the pipe-lines this can be removed with 30 per cent hydrochloric acid. Organic solids can be removed by 50 per cent nitric acid or by caustic soda.

Diffusers in the form of long tubes are used in America, but not in Great Britain. The favoured British diffuser is now the "dome" diffuser of Activated Sludge, Ltd., which is a mushroom-shaped tile pressed down on to a plate by a single gunmetal screw and sealed by a rubber ring. This makes the diffusers much easier to change than the plate diffusers, each of which is fixed by four screws. A further advantage of the dome diffuser is that the complete diffuser units of plate, dome, etc., are secured to P.V.C. mains that run along the bottoms of the aeration channels. Thus, the dome diffusers considerably reduce the amount of maintenance costs and make some reduction in the initial cost of air mains.

Dome diffusers are in two sizes, 4- and 7-inch diameter. The 4-inch diffusers, which are used for aerated detritus tanks only, pass from 0·3 to 0·5 cubic foot of air per minute: the 7-inch diffusers, generally used for aeration tanks, etc., pass 0·5–1·0 cubic foot of air per minute, 0·64 being the approximate working figure and 0·8 the preferred maximum. Removable diffusers are available for tanks or channels which cannot be emptied.

In practice, the amount of air used varies from $1\frac{1}{2}$ to $2\frac{1}{2}$ cubic feet of air per gallon of sewage treated. For design purposes air blowers or compressors, pipework and diffusers should allow for a maximum air flow of $2\frac{1}{2}$ cubic feet per gallon of sewage, where the sewage is mainly domestic, and more where difficult trade wastes have to be dealt with.

Another rule for the volume of free air required is to allow 700 to 800 cubic feet per lb of B.O.D. removed in tanks 10 feet deep as an average figure. But again, allowances must be made for maximum conditions.

In the spiral-flow system the tank or channel is flat-bottomed but should have chamfered or rounded junctions between the walls and the floor. The diffusers are arranged close to one side of the tank so as to cause a spiral rotation as the mixed liquor of sewage and activated sludge passes along the channels. This naturally means a closer concentration of the diffusers on the air pipes.

Air supply. Air must be delivered from the compressors at a pressure sufficient to overcome frictional resistance in the pipes and diffuser tiles, and the pressure of the sewage on the diffusers. The pressure of the sewage depends on the submersion of the diffusers being 0·43 lb per foot of depth. The resistance of the diffusers is 9 inches head of water at 0·64 cubic foot per minute of air when the diffusers are new. Loss of air pressure in mains can be calculated by Tables 32 and 33.

Air should be compressed by rotary blowers only, not by reciprocating compressors that require internal lubrication. The efficiency of the machines varies according to the size (see Table 55) and naturally the most efficient

TABLE 54
RECOMMENDED VELOCITIES IN AIR MAINS

Internal diameter in inches	Free air velocity in feet per second	Free air discharge in cubic feet per second
4	25	2·2
6	33	6·5
8	40	14·0
9	43	19·0
12	52	41·0
15	60	73·5
18	68	120·0
21	76	183·0
24	82	257·0
27	89	354·0
30	96	472·0
33	102	606·0
36	108	765·0
39	112	930·0
42	120	1155·0

TABLE 55
APPROXIMATE EFFICIENCIES OF LOW-PRESSURE AIR-BLOWERS

Capacity of compressor in cu ft of free air per minute	Overall efficiency per cent	
	Roots blowers	Turbo blowers
1000	53	
2000	58	
3000	62	
4000	64	
5000	66	
6000	67	71
7000		72
8000		73
10,000		74

will be purchased if the size is practicable. Some blowers are noisy, but this can be overcome by careful design. In any case silencers should be considered necessary.

As previously mentioned air filtration is essential, otherwise dust will eventually clog the diffusers which will then have to be taken out and reburnt or cleaned with chemicals. The electrostatic filter is now generally used for very large plants. Alternatives for smaller plants are filters containing mats of metal, glass wool, organic wool or hair covered with viscous oil. Air mains should be provided with blow-out pipes at the end of each section for removal of condensation water and oil or removal of large quantities of water should the mains be accidentally flooded.

Sludge recirculation. In the early days of activated-sludge treatment, activated sludge was recirculated in modest quantities. But the tendency over the years has been to increase rates of recirculation. Latest practice for British diffused-air plants is to allow for a maximum recirculation equal to 100 per cent of the dry-weather flow, and normal rates of recirculation of two-thirds of the maximum are recommended. This is in line with tendencies in the United States, where rates of recirculation of activated sludge vary, according to Babitt, from 20 to 100 per cent of the dry-weather flow. In design the maximum figure should always be allowed.

Incremental feeding and tapered aeration. The modification of the activated-sludge process known as "incremental feeding," "incremental loading" or, as it is termed in America, "step aeration" is the feeding of the settled sewage into the aeration tanks, not all at the inlet end (where all the recirculated sludge is fed in), but at various points along the aeration channels even to as much as two-thirds of the distance towards the outlet end. Step aeration is usually allowed for in design, but it need not be used and can be applied when considered desirable by the operator.

"Tapered aeration" is the arrangement of diffusers so that more air is injected near the inlet end of the channel than near the outlet end. Practice has been to apply about 45 per cent of the total air to the first third of the detention period, up to 30 per cent to the second third and the remainder to the last third. This practice is based on the tapering rate of oxygen demand due to the fact that more work is done in reducing B.O.D. in the first stages of treatment than in the following stages. British operators are not, however, all convinced that tapered aeration is desirable and as, in the diffused-air system, it is not practicable to change the number of diffusers in operation, there is much to be said for arranging diffusers in equal numbers throughout the length of the aeration channels and leaving it to the operator to effect any desired tapering of aeration (or the reverse) by valve-control of air-flow.

Apart from the dissolved air in the mixed liquor, the oxidising organisms appear to be able to store oxygen and use it later. An example of this is in the contact bed which is emptied completely to aerate the organisms, then filled with settled sewage and allowed to stand for some time while the organisms, with the aid of the oxygen they have acquired, aerate the sewage.

The author's study of tests made on existing surface-aeration activated-sludge works shows that the activated sludge continues to aerate sewage in the final sedimentation tanks at the same rate as in the aeration tanks although there is no aeration of the sewage in the former.

Fig. 132—Aerial view of Maple Lodge sewage-treatment works: a diffused-air plant

By courtesy of M. A. Kershaw, Esq., General Manager, Colne Valley Sewerage Board; and Activated Sludge, Ltd

Operation. Activated-sludge plants may be operated in more than one manner. Within limits, capacity of aeration tanks, rate of injection of air and the proportion of activated sludge may be varied. For example, an inadequate tank capacity may be remedied by an increased rate of air injection or by the combined effect of an increased rate of air injection and the use of a higher proportion of sludge. Moreover, methods of operation vary according to the kind of effluent required. Formerly plants were operated so as to obtain stable, well-clarified effluents containing comparatively little oxidised nitrogen compounds (nitrite and nitrate). Now, the tendency is for the purification to be carried further, and high-quality effluents containing 10–20 parts of nitric nitrogen per million parts of effluent are obtained. On the other hand, high-rate partial treatment has been effected by giving very short-period aeration to mixed liquor and very long reaeration to activated sludge.

Generally, the higher the proportion of sludge, the more rapid the purification of the sewage; but the advantages to be obtained by the use of high proportions of sludge cannot be expected unless the rate of injection of air to the mixed liquor is sufficiently high to meet the oxygen requirements of both the sludge and the sewage.

Materially lower rates of air injection normally slow down the rate of purification of the sewage, render the process unduly sensitive to adverse conditions—e.g. overloads—and tend to encourage the development of filamentous growths in the sludge, which may lead ultimately to rapid increases in the bulk of sludge (bulking) and poor settlement in final sedimentation tanks.

Sludge-volume index. As a means of ascertaining the condition of the activated sludge during treatment so as to be able to decide on any changes that should be made in the proportion of sludge to sewage, etc., in the mixed liquor, regular tests of sludge-volume index have to be made. The sludge-volume index is defined as the volume in millilitres of sludge which, after thirty minutes settling, contains 1 gramme weight of dry solids. From 1920 onwards various methods of determining sludge index have been proposed, including those of Theriault, Donaldson and Haseltine. Today the Mohlman sludge-volume index is the most extensively used for determining the condition of the activated-sludge process at any time. The procedure is as follows:

A sample of mixed liquor is taken from the aeration tank, and from this a 1-litre graduated cylinder is filled and allowed to settle for 30 minutes and the percentage volume occupied by the settled sludge is recorded. Next, the suspended-solids content of the original sample is determined. Then:

$$\text{Sludge-volume index} = \frac{\text{Per cent settling by volume in 30 min}}{\text{Per cent suspended solids}}$$

A good settling sludge should have an index below 100; one with poor settling characteristics may have a value in the region of 200. A rising sludge index is an indication of trouble developing in the process of treatment. The sludge-index test is not too reliable, for the rate of settlement is easily influenced by the way in which the sample is handled.

Bulking. The term "bulking" is applied to a condition (now becoming rare) in which the sludge increases in volume very rapidly and to a great extent. For example, in one day the quantity of sludge may have increased by as much as 100 per cent or more. The effect of this is to render final sedimentation difficult, with the production of a bad effluent.

Microscopic examination of the sludge when bulking occurs has revealed the presence of extensive filamentous growths—e.g. *Cladothrix*, and of types of protozoa not normally present in the sludge. There is evidence to show that trade wastes containing carbohydrates—e.g. brewery waste, milk wastes and waste containing starch—when present in sewage in undue proportions, favour the development of such growths.

If bulking occurs the first thing to suspect is operation, for bulking seldom occurs when works of adequate capacity and oxygen input are producing a Royal Commission effluent and also when works are being used, as sometimes in America, to give partial treatment in a condition which in England would be considered overload. It is an intermediate condition known as the "zone of bulking" when bulking is likely to occur, when the sludge-volume index will rise, the activated sludge cease to settle satisfactorily and a bad effluent will result.

To avoid bulking in works designed to give full "Royal Commission" treatment, the load should not exceed 0·3 lb B.O.D. per day per lb of volatile solids in the aeration tanks.

Final sedimentation tanks. The final sedimentation tanks for an activated-sludge plant need to be highly efficient, because they have to settle out very large quantities of light flocculent sludge and produce an effluent which satisfies the Royal Commission standard. Fortunately, activated sludge settles comparatively easily, and can be removed efficiently in upward-flow or semi-upward-flow central-inlet tanks.

These tanks should be designed to treat a rate of flow equal to three times dry-weather flow plus maximum rate of flow of returned sludge, and, at this rate of flow, the upward flow should not exceed 7 feet per hour. The capacities of the tanks should generally be not more than six or less than four and a half hours' dry-weather flow.

The sludge must be removed continuously, otherwise the level of sludge will rise in the tanks until sludge passes over the weirs. If hopper-bottomed tanks are used, the sides should be sloped at 60° to the horizontal, or steeper.

A much-favoured type of mechanically raked final sedimentation tank is one with a conical bottom sloping at 30° to the horizontal and having simple bars or trailing chains to disturb the sludge sufficiently to make it gravitate to the central outlet. Where the nature of the site makes these tanks undesirable, flat-bottomed tanks may be used with echelon or curved blades.

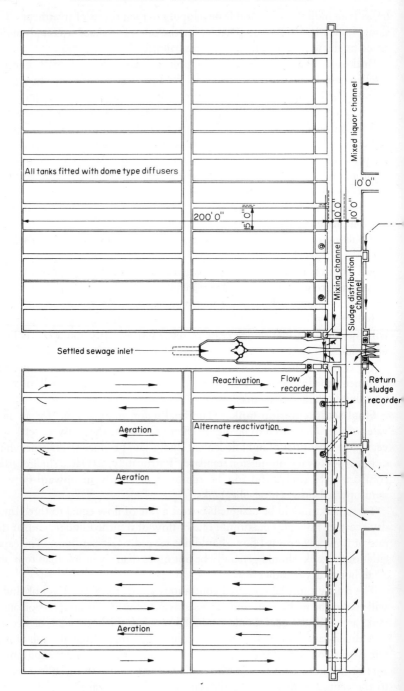

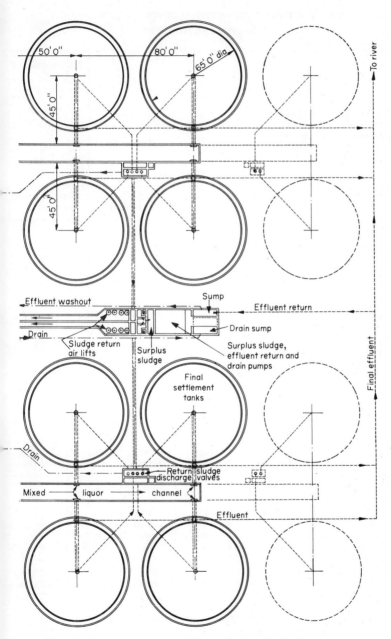

FIG. 133—Lay-out of activated-sludge plant, diffused-air system
By courtesy of Activated Sludge, Ltd.

TABLE 56

MOGDEN SEWAGE-WORKS—A TYPICAL DIFFUSED-AIR SCHEME

SUMMARY OF OPERATING DATA, 1936–46

	Million gallons per day
Sewage flows:	
Dry-weather flow, 1936	40·0
Dry-weather flow, 1946	58·0
Maximum rate of flow received	455·0
Average flow received	62·4
Average flow treated by activated sludge	60·59
Average flow given sedimentation only (storm water)	1·89

	Number	Hours' detention period
Activated-sludge plant data:		
Aeration units in use for mixed liquor	9·99	6·60
Re-aeration units in use for sludge	1·36	0·90
Total	11·35	7·50

Air supply, cubic feet per minute	48,950·0
Air per square foot of aeration-tank surface, cubic foot per minute	0·165
Air per gallon, cubic feet	1·16
Air per lb B.O.D. removed: cubic feet	711·0
Horse-power per million gallons of daily flow	34·0
Horse-power per lb B.O.D. removed	0·47
Volume of return sludge, per cent of sewage	33·0
Suspended solids in mixed liquor, parts per 100,000	236·0
Surplus sludge, m.g.d.	0·92
Surplus sludge, per cent solids	0·81

	Raw sewage	Settled sewage	Activated-sludge effluent	Total percentage removed
Analytical results, parts per 100,000:				
Oxygen absorption from acid KMnO$_4$, 4 hours at 26·7 °C	8·01	4·65	1·13	—
Ammoniacal nitrogen	3·80	3·96	1·53	—
Albuminoid nitrogen	0·81	0·46	0·085	—
Nitrous nitrogen	—	—	0·29	—
Nitric nitrogen	—	—	1·24	—
Chloride (as chlorine)	10·3	10·3	9·9	—
B.O.D., 5 days at 18·3 °C	30·7	17·2	0·83	97·3
Suspended solids	23·8	6·9	0·7	97·1

Number of daily composite samples found putrefactive after incubation, 6 days at 26 °C	None
Number of samples incubated	3497

Reproduced from "West Middlesex Main Drainage: Ten Years' Operation." By courtesy of C. B. Townend, Esq., C.B.E., M.I.C.E.

Ames Crosta Mills & Co., Ltd., supply a mechanism by which the sludge is drawn off by vertical pipes and delivered to a central box connecting to the sludge outlet.

As activated sludge must be drawn off continuously, it is usually withdrawn under hydrostatic head over telescopic weirs, the levels of which can be adjusted to control the flow.

Activated sludge is very voluminous, often having a moisture content in excess of 99 per cent, and although it is inoffensive when fresh, it rapidly deteriorates and soon becomes septic. Therefore it should be returned without delay to sewage re-aeration channels or aeration tanks where it will be aerated.

Surplus sludge. The quantity of surplus sludge can be approximately estimated by the formula:

$$D = \left[\frac{A(100 - E)}{100} + B - C\right]\frac{100}{E}$$

where: A = dry solids in crude sewage in lb per day
B = dry solids produced by activated-sludge process in lb per day (about 0·2 lb per lb B.O.D. of crude sewage)
C = dry solids lost with final effluent in lb per day
D = dry solids in surplus sludge in lb per day
E = percentage efficiency of primary sedimentation.

This calculation, which is explained by the diagram illustrated in Fig. 134, is not strictly true, for it depends on the assumption that all types of solids are settled with equal efficiency in the primary tank.

As a check on the above calculation, it can be said that the average quantity of dry solids in the surplus sludge of a diffused-air plant is about 0·4 lb per lb B.O.D. of crude sewage.

It has been suggested that the quantity of surplus sludge solids is much greater for a surface aeration plant than for a diffused-air plant. But the writer thinks that this view may be due to the publication of a few erroneous figures for large works. An appreciable part of the solids content of surplus sludge is sewage solids which the primary sedimentation tanks failed to settle: the quantity of these cannot exceed 100 per cent.

The quantity of sludge depends very much on the moisture content, which averages in the region of 99·33 per cent but can vary widely. In practice, it is desirable to allow pipework, pumps, etc., of sufficient size to pump quantities of surplus activated sludge equal to 25 per cent of the dry-weather flow of sewage. During normal running the quantity of surplus sludge approximates to 1 gallon per day per head of population, unless the sewage is strong by virtue of trade wastes. Surplus sludge is returned to the flow of crude sewage for settlement with the primary sludge in the primary sedimentation tank or can be separately dewatered (see Chapter XXVIII).

INKA system. The INKA process is a method of diffused-air activated-sludge treatment which was developed in Sweden and embodied in a number of European sewage-treatment works. In Great Britain it has been taken up by Dorr-Oliver Co., Ltd.

The process differs from the diffused-air process of Activated Sludge, Ltd. in that the aerators consist of stainless steel or plastic grids having 2·5 mm

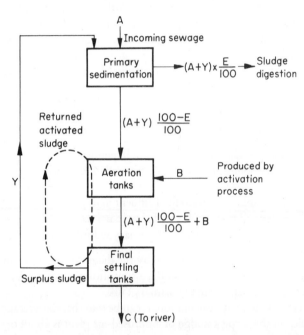

FIG. 134—Diagram illustrating approximate calculation of quantity of surplus activated sludge

Y = Dry solids in surplus sludge (equals D in above formula)
A = Dry solids in crude sewage
B = Dry solids produced in aeration
C = Dry solids lost to river
E = Percentage efficiency of primary sedimentation

diameter holes on the underside and which are submerged at a depth of 2 feet 6 inches. The aeration grids are arranged to one side of a spiral-flow tank, extending from the wall to a vertical baffle of corrugated fibreglass, the purpose of which is to assist the spiral flow. Tank depths vary from 10 to 15 feet, and widths from 10 to 30 feet. The width of the aeration grid is 6 feet 8 inches for tanks up to 20 feet wide and 7 feet 4 inches for wider tanks. The depth of fibreglass baffle varies from 4 feet 6 inches to 9 feet 6 inches for tanks from 10 feet to 15 feet deep respectively.

Air is blown in at a working pressure from 35 to 40 inches water gauge at $5\frac{1}{4}$ cubic feet of air per gallon of sewage treated.

Mechanical agitation or surface aeration. In the surface-aeration methods the procedure is to disturb the contents of the aeration tanks at the surface by skimming off mixed liquor and throwing it through the air to fall again within the confines of the tank. Taking off the liquor at one part of the surface causes upward flow, which must be maintained at a velocity sufficient to prevent any settlement. The drops of mixed liquor are to some extent aerated as they pass through the air, but in falling on the surface they cause violent surface aeration and bubbles of air are forced downwards.

At one time it was believed that, because of the difference in method of aeration, the surface-aeration processes required longer detention periods than did the diffused-air process. It now appears that the differences between the methods are only in matters of detail and that the detention periods for surface aeration can be similar to or even less than those required for a standard diffused-air plant, provided that sufficient disturbance of the mixed liquor is caused.

To cause disturbance requires power, which costs money, and therefore the tendency of investigation has been aimed at finding how the most disturbance can be produced with the least expenditure of power.

Simplex system. Ames Crosta Mills & Co.'s Simplex system is a surface-aeration method, which has always had considerable status in Great Britain, but which, by virtue of detail modifications greatly improving efficiency of power utilisation, has recently challenged most other methods of activated-sludge treatment.

In this method the mixed liquor is drawn from the bottom of the aeration tank through an uptake tube and sprayed over the surface by an aerating "cone" which is a specially designed centrifugal impeller of considerable efficiency for lifting large quantities against low heads. In very small works for villages or institutions there may be one tank only, square on plan and hopper-bottomed with a single uptake tube, and final sedimentation may take place in pockets arranged in the corners of this tank (see Fig. 155). In larger works the hopper-bottomed pockets, with their uptake tubes and aerating cones are arranged in lines or long channels and the aerating cones are driven by line-shafts in numbers of up to six per electric motor, the motors being arranged just outside the area occupied by the tanks.

The outlet weir of the Simplex aeration tank is a double weir in the form of a trough of variable level and with telescopic outlet pipes. This makes it possible for the water-level of the tank to be adjustable so as to vary the submergence of the cones and the power input. For this reason the installed horse-power of the motors needs to be about 25 or 30 per cent higher than the calculated normal running horse-power.

Recirculated activated sludge is introduced at the inlet ends of the channels: sewage is introduced in stages down the channels by the incremental-feeding

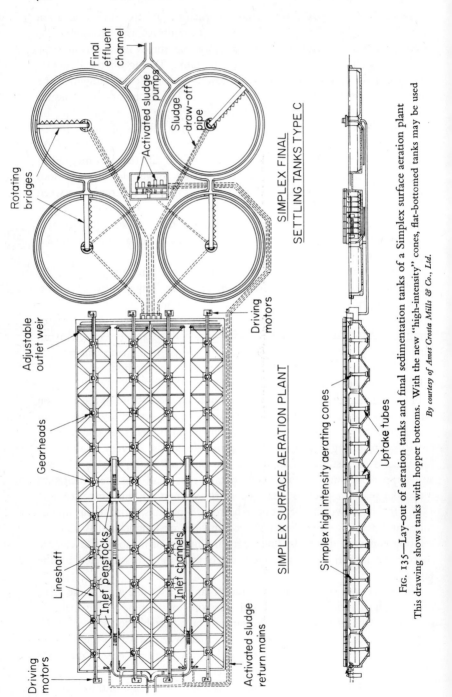

Fig. 135.—Lay-out of aeration tanks and final sedimentation tanks of a Simplex surface aeration plant. This drawing shows tanks with hopper bottoms. With the new "high-intensity" cones, flat-bottomed tanks may be used

By courtesy of Ames Crosta Mills & Co., Ltd.

method (see Fig. 135). By the provision of cross baffles at the inlet end, one or more hopper-bottomed pockets of each line can serve for activated-sludge re-aeration when required.

In all the earlier plants the uptake tubes were arranged in pockets having hopper-bottoms sloped at angles generally between 30° and 45° to the horizontal, as will be seen by reference to Table 57.

TABLE 57

DIMENSIONS OF TYPICAL SIMPLEX AERATION TANKS

Capacity imp galls	A	B	C	D	E
25,000	20' 0"	6' 0"	5' 3"	7' 7"	12' 10"
30,000	22' 0"	6' 0"	6' 0"	7' 3"	13' 3"
35,000	22' 0"	6' 0"	6' 0"	8' 11"	14' 11"
40,000	24' 0"	8' 0"	6' 0"	8' 4"	14' 4"
45,000	24' 0"	8' 0"	6' 0"	9' 8"	15' 8"
50,000	26' 0"	8' 0"	6' 9"	8' 9"	15' 6"
55,000	26' 0"	8' 0"	6' 9"	9' 10"	16' 7"
60,000	28' 0"	8' 0"	7' 6"	8' 10"	16' 4"
65,000	28' 0"	8' 0"	7' 6"	9' 10"	17' 4"
70,000	30' 0"	8' 0"	8' 3"	8' 10"	17' 1"
75,000	30' 0"	8' 0"	8' 3"	9' 8"	17' 11"
80,000	32' 0"	8' 0"	9' 0"	8' 7"	17' 7"

A = Length and width of pocket
B = Length and width of flat bottom of hopper
C = Depth of hopper
D = Depth of vertical sides above hopper
E = Overall water depth, or $C + D$

It was the introduction of the modern high-intensity cone which made the great improvement in the Simplex system, for it rendered possible the reduction of aeration capacity and the use of tanks with flat floors.

On the basis of results with high-intensity cones at various sites, Ames Crosta Mills & Co., Ltd., published a design curve which is reasonably well expressed by the formula:

$$D = 0.16 \, \text{B.O.D.}_1^{\frac{2}{3}}$$

Fig. 136—Primary sedimentation tanks, Crossness, feeding to aeration tanks (Fig. 137)

where: D = aeration tank detention period in hours' dry-weather flow

$B.O.D._1$ = biochemical oxygen demand of settled sewage in parts per million.

While this curve may give reasonable detention periods, the author does not consider that it is strictly true, as will be shown in the following pages. Experiments on Simplex plant at Manchester showed that by increasing the rate of rotation of the cones or otherwise increasing the horse-power applied, further reductions of detention period could be made and that performance of the plant depended on horse-power and the capacity of the aeration tanks *plus that of the final sedimentation tanks.*

Activated sludge is recirculated in modern Simplex plants at similar rates to those now employed in diffused-air plants—i.e. at two-thirds to three-quarters of the dry-weather flow. It is usual to allow for a maximum rate of recirculation equal to the dry-weather flow in the design of a plant. In the experiments at Manchester this rate was generally applied.

Other methods. In addition to the foregoing, a number of other modifications of the activated sludge process have been proposed, the essential feature of all of them being the agitation of the sewage with a greater or less proportion of activated sludge.

In the Brush Aeration system, introduced by Dr. Kessener of Holland, the mixture of sludge and sewage is agitated and aerated by means of a partly submerged cylindrical brush revolving at a high rate. The brush is fixed longitudinally on one side of the aeration channel, and in rotating causes the mixture of sludge and sewage to be flung outwards across the channel in the form of spray. This brings about both aeration of the mixture and circulation

Fig. 137—"Simplex" aeration tanks, Crossness, fed from primary sedimentation tanks (Fig. 136)

of the channel contents. In order to facilitate circulation the bottom of the channel is rounded, the side of the channel opposite to that on which the brush is mounted is inclined inwards towards the base, and near to the sloping wall is fixed a board, submerged $1\frac{1}{2}$ to 2 inches below the liquid surface, which serves to deflect the flow downwards (see Fig. 139).

Fig. 138—Simplex high-intensity aerating cone
By courtesy of Ames Crosta Mills & Co., Ltd.

In a plant at Hamoir, Belgium, designed to treat 37,000 gallons of dairy waste water per day, brushes 26 inches diameter and 52 feet in total length were installed. These, rotating at 65 to 70 revolutions per minute, dipped about 0·6 inch into the liquid. In some of the more up-to-date plants the brushes consist of stainless-steel combs and rotate at 140 revolutions per minute. This system is used chiefly in Holland, for the treatment of industrial wastes and to a limited extent for the treatment of municipal sewage.

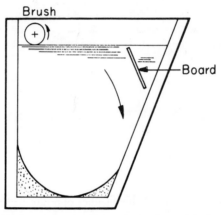

Fig. 139—Cross-section of tank illustrating the principle of the Kessener process
By courtesy of the late Mr. W. T. Lockett

Comparison of the activated-sludge methods. Side-by-side experiments involving the diffused-air method, the Simplex method and varieties of the bio-aeration methods or Sheffield system have been made at Crossness, Croydon and Manchester and, on these, the engineers concerned have based their conclusions as to the type of plant to be used for future extensions. What is particularly noticeable is the difference of opinion that resulted from these tests. After the Crossness experiments a modified Sheffield plant was installed to treat part of London's flow at the Beckton Works: nevertheless the major construction at a later date at Crossness was a full-scale Simplex plant and the works now being designed to take the whole of the flow at Beckton are of the Activated Sludge, Ltd. diffused-air type. It is understood that the Simplex system was favoured at Manchester but the large extension at Croydon is again a diffused-air plant.

The difficulty of comparing the system side by side is mainly one of fair operation of each of the methods under test. The Simplex system cannot be upset very easily, for the horse-power cannot be varied without difficulty and therefore tends to remain constant and there is little to deteriorate as a result of neglect. Broadly, the plant can be upset only by bleeding-off too much surplus sludge so as to have a mixed liquor of the wrong density, or by reducing the recirculation of sludge in the extreme. The diffused-air plant

ACTIVATED-SLUDGE SYSTEMS 413

can also suffer from these causes but it will also appear at a disadvantage if horse-power loads are made excessive either by blowing more air than is necessary for the process or by permitting the diffusers to become clogged by neglect. Both of these factors have influenced figures that have been studied from the point of view of what plant to use and, therefore, all results should be very carefully examined and questions asked by anyone who is wishing to make a decision on the type of plant to install.

It should also be borne in mind that figures of performance do not give indications of comparative costs of construction and of running costs, which include not only power demands but also labour and difficulties of operation.

The writer has attempted to arrive at some means of comparing the performance of different plants at various sites *in terms of tank capacity and power input* and, on the examination of data, has arrived at what he considers is a representative empiric formula. The following was the method of approach.

When, for any series of experiments, B.O.D. of settled sewage was plotted to exponential ordinates and abscissae against detention period in aeration tanks required to produce a standard final effluent, the relation was found to be:

$$\text{Detention period} = \frac{\text{B.O.D.}}{X}.$$

Similarly, when B.O.D. was plotted against horse-power per million gallons per day, the relation was:

$$\text{Horse-power per m.g.d.} = \frac{\text{B.O.D.}}{Y}.$$

From this one may derive the formula:

$$\text{m.g.d.} \times \text{B.O.D.} = K \times \sqrt{C} \times \sqrt{\text{H.P.}}$$

where: m.g.d. = million gallons per day
B.O.D. = biochemical oxygen demand of primary tank effluent in parts per million
K = a constant
C = tank capacity in million gallons
H.P. = total aeration horse-power.

When, however, the results of all the series of experiments were plotted to the above formulae, C being taken as the capacity of aeration tank only, it was found that the value of K varied for each series, and on inspection it was clear that the variation was related to the ratio of final sedimentation tank capacity to aeration tank capacity.

Accordingly, B.O.D. load was plotted against the product of total horse-power and *total capacity of aeration plus final sedimentation tanks*, again to exponential ordinates and abscissae. This gave a remarkably straight line for the averages of the results for various conditions, and all individual data points conformed reasonably well and with such regularity that agreement could not be accidental.

Thus, in the foregoing formula, C became capacity, in million gallons, of aeration tanks plus final sedimentation tanks.

The findings of this investigation are not altogether unexpected, for reduction of B.O.D. (partly at expense of dissolved-oxygen content) must continue while activated sludge is in contact with sewage in the final tanks and also, at a reduced rate, afterwards. Furthermore, increased final sedimentation-tank capacity must improve sedimentation which, again, reduces B.O.D. It would therefore appear that final sedimentation-tank capacity cannot be neglected as has often been done in aeration calculations.

In the design of activated-sludge works it is usual to allow for a margin of safety and aim at an average final B.O.D. below Royal Commission standard. Sometimes a standard of 10 parts per million is specified and sometimes 12 or 15. Also when comparing works it is not always possible to find comparable final effluent values. To make adjustment for the B.O.D. value of the final effluent, the following empiric formula was suggested. This appears to satisfy conditions for all the data that have been examined so far.

$$K = \frac{M.G.D. \times \sqrt{B.O.D._1} \times \log_{10}\left(\frac{B.O.D._1}{B.O.D._2}\right)}{\sqrt{C} \times \sqrt{H.P.}}$$

where: $M.G.D.$ = million gallons per day
$B.O.D._1$ = biochemical oxygen demand of primary sedimentation tank effluent in parts per million
$B.O.D._2$ = biochemical oxygen demand of final treated effluent (the formula is applicable to values between the limits of 8 and 30 only)
K = "performance index" dependent on efficiency of power utilisation, difficulty of treatment of sewage, etc. (the value varies between 3·3 and 4·3 for most good works)
C = capacity of aeration *plus final sedimentation* tanks in million gallons
$H.P.$ = aeration electric horse-power (excluding sludge recirculation horse-power, etc.).

This formula is applicable only where nitrification is negligible and B.O.D. not greatly reduced below Royal Commission standard, but, with those limitations, when applied to long-term average results of carefully collected data, appears to give good indication of the comparative performance of treatment plants treating the same sewage at the same works. The higher the value of K the better the performance.

It must be understood that this formula is purely empiric. It cannot be extrapolated unreasonably and it cannot be explained. It suggests that an increase of horse-power can make up for a deficiency of capacity and vice versa, which the tests certainly suggest. But it remains a fact that there is a limit to the reduction of capacity necessary to house the organisms that effect the oxidation and the excessive increase of horse-power input beyond that necessary to supply the organisms with adequate oxygen is simply waste of power.

The treatment is, of course, not effected by the concrete tanks or the air blown into the tanks but by the organisms living in those tanks and consuming the oxygen. The quantity of organisms depends not only on the tank capacity but the density of the sludge, which itself depends on the air input and general health of the organisms. It is possible to have a large tank capacity and/or a dense sludge with a long "sludge age" and produce a good effluent. It is also possible to effect partial treatment with an excessive load. But intermediately is the condition in which bulking occurs with a change in the character of the sludge and the quality of the final effluent will fall off seriously. For general English practice an adequate volume of healthy sludge is required.

P. S. S. Danson and S. S. Jenkins (*Sewage Works Journal*, Vol. XXI, No. 4, July 1949) found that oxygen uptake (in laboratory experiments) was directly related to the weight of the activated-sludge solids and to the time during which they worked. The figures were in accordance with the formula:

$$x = 22{,}500 \, yt$$

where: x = oxygen uptake in micrograms
y = activated sludge solids in millilitres
t = detention period in hours.

About 0·525 gramme of oxygen is consumed for every gramme B.O.D.$_5$ removed.

Advantages and drawbacks of the process. The activated-sludge process is not without its drawbacks. In first cost it compares favourably with a normal percolating filter installation, but its working expenses are heavier. It requires skilled management. For these reasons its adoption on small works has not been extensive: it is more readily accepted for the treatment of the sewage of large communities.

The increasing use of synthetic detergents, which have a harmful effect on all sewage-treatment processes, is particularly objectionable at activated-sludge works, where depths of foam 4 or 6 feet deep have occurred over the aeration tanks. The foam is liable to be picked up by the wind and become a nuisance in the neighbourhood and also to involve a danger to health owing to its bacteria content. Where the foam settles and dries out it leaves a black, greasy deposit. Foam also forms in the outlet channels of final sedimentation tanks and floats in masses down the watercourse into which the effluent is discharged.

Foaming can be reduced or eliminated by the use of small quantities of suitable chemicals applied at the point where it is tending to occur, or may be suppressed by spraying the surface with treated effluent.

Probably the most effective method is the application of $1\frac{1}{2}$ to $3\frac{1}{2}$ parts per million of oils such as the Vacuum Oil Co.'s "Foamrex" or of Shell Chemical Co., Ltd., "Sewage Anti-foam." The chemicals should be applied only when foaming becomes excessive. The equipment for application consists of remote-controlled mechanical lubricators delivering to about eight points along each aeration channel through $\frac{1}{8}$- to $\frac{1}{4}$-inch diameter copper, P.V.C. or nylon tubing, there being a separate tube for each point of delivery. If of plastics, these tubes are best in single lengths without joints.

For economy the chemicals should be purchased in bulk, stored in large tanks and gravity-fed to ball-valve controlled cisterns locally supplying each mechanical lubricator.

In recent years a further deterrent to the adoption of the activated-sludge process for medium-size works has been the discovery of the alternating double-filtration and recirculation methods of operating percolating filters. These new processes can reduce the capital cost of percolating filter schemes by a large percentage, and while they add to the running costs by the amount required to cover pumping power, this, in the case of alternating double filtration or recirculation of effluent in equal proportion to the rate of flow of sewage, is considerably less than the cost of power required for aeration etc., at an activated-sludge works.

On the other hand, during the same period of development of processes for the anaerobic fermentation of sludge at temperatures above that of the atmosphere and the advances made towards collection of the gas produced and utilisation of the gas as fuel in gas engines for power production, by providing suitable means for the disposal of the surplus sludge, and by providing a cheap source of power, have done much to enhance the value of the several systems involving the use of activated sludge.

In addition to a relatively low first cost, the activated-sludge process possesses the following special advantages:

1. Small area of land for plant installation.
2. Freedom from odour nuisance.
3. Entire freedom from nuisance from filter flies.

CHAPTER XXXI

METHODS OF SEWAGE TREATMENT OF HISTORICAL INTEREST

BEFORE the methods of sewage treatment now in use had become firmly established, other methods of various types were tried, recommended and, in some cases, largely used until their defects led to their abandonment. Although most of these methods are of no apparent value, knowledge of them is desirable for all interested in sewage treatment. There are also some methods of comparatively recent introduction which are rarely used. The methods include processes for the removal of suspended solids, aeration of clarified tank effluent and other procedures which reduced the strength of the final effluent.

Septic tanks. Septic-tank treatment came to the fore as a result of Cameron's experiments at Exeter. The theory behind the septic tank was that, if a sedimentation tank were made sufficiently large for the sludge to be stored for a very long period, practically the whole of the organic matter in the sludge would be digested, producing gas and soluble material, so that virtually none would have to be removed as sludge. It was also thought that all pathogenic bacteria would be destroyed and that anaerobic digestion would make the settled effluent easier to treat on land, in contact beds or on percolating filters.

As a result of this belief, very many septic-tanks were constructed at municipal sewage works, and some of these were of considerable size. For example, the septic tanks at Manchester measured 300 feet by 100 feet by 6 feet deep. But it was found that the theory that complete digestion would take place was unsound: about 25 per cent digestion was more usual. Moreover, the flat-bottomed tanks were not easy to sludge, and the sludge, left in position for a long time, became very dense and difficult to manipulate. For these reasons septic tanks became virtually obsolete for municipal purposes, and many tanks were being demolished or converted to other purposes shortly after the First World War.

Septic tanks are, however, still used very largely for the settlement of sludge at small works serving isolated buildings, where frequent sludging of continuous-flow sedimentation tanks is impracticable. Any digestion of solids that may occur is no longer taken into account by designers with much experience of this class of work. (For details see Chapter XXXIII.)

Imhoff tank. The Imhoff tank, in which sedimentation and sludge digestion are effected in upper and lower compartments of the same tank respectively, while almost obsolete in Great Britain, is still used overseas, being popular in Europe and the United States. For this reason it is described towards the end of Chapter XXVIII.

Travis hydrolytic tank. The hydrolytic tank was designed by W. Owen Travis, who held the view that the deposition of colloids to form sludge was dependent on the exposure of the suspending liquid to surfaces such as the sides of the tank or grids of laths fixed in the tank, and was a purely physical phenomenon. The hydrolytic tank was designed to make use of this principle and achieve as complete settlement as possible, particularly of colloids. The tank was circular with a central manhole containing penstocks for drawing off sludge and three annular rings, the outer ring being divided into four portions. In these rings were fixed the "colloiders," that is, wooden frames having 1-inch by 1½-inch vertical slats arranged at 6- to 9-inch centres. By a complex arrangement of weirs and openings, the sewage was caused to flow from chamber to chamber, passing through the colloider and depositing its sludge.

Although causing sewage to flow through a grid of vertical laths or rods is similar to the process of flocculation by the movement of a "picket-fence" through stationary liquid, a method which was used some decades ago for sludge thickening but found to be unreliable, the Travis tank did not achieve much popularity, and the design went out of use.

Fieldhouse tank. This was a design of sedimentation tank in which the incoming sewage was made to mingle with the contents of the tank by outward diffusion as it travelled towards the outlet weir. Like the Travis tank, it was circular in form, consisting of a central conical-bottomed tank into which the sewage was fed and an outer ring with several hopper-bottoms and a common peripheral weir. The sewage passed from the inner tank to the outer ring through holes in the vertical upper portion of the circular wall between them.

Dibdin slate beds. These were tanks filled to a depth of 3 feet with horizontal layers of slate, the layers being kept 2 inches apart by slate distance-pieces. They were intended to be filled with crude sewage once a day and then thoroughly drained in the manner of contact beds.

Contact beds. Contact-bed treatment was one of the methods considered by the Royal Commission on Sewage Disposal. But the Royal Commission's investigations showed that percolating filters of the same capacity and cost as contact beds would treat twice the quantity of sewage and with less attention. As a consequence, contact beds went out of general use.

Contact beds were tanks filled with medium of the same type as used in percolating filters. But instead of settled sewage being distributed over the surface and allowed to percolate through well-aerated medium, the bed was filled with sewage, which was retained for a time, then drawn off by gravity to a second bed at a lower level for similar treatment, while the first was left to take up oxygen from the air that had moved in on the outflow of the liquid. In some instances there were three stages of treatment. It was necessary to have not fewer than two beds in each stage, so that one could be filled while the others were emptying or standing. The control of the beds was by manual operation of penstocks, but operation by automatic siphonic gear was also used.

Electrical treatment. In 1888-9 Webster invented an electrolytic process whereby water, and sodium, magnesium and other chlorides were decomposed by an electric current into their constituent elements. The chlorine and oxygen, which were liberated in a nascent state, quickly combined with the organic matter, with the formation of innocuous compounds. The earlier experiments were made with platinum electrodes, but these were too expensive for practical use. After many months of exhaustive experiments, carried out on a large scale (half a million gallons per day) at Crossness, it was thought that plates of cast-iron were satisfactory.

A gallon of ordinary sewage required on an average one ampere of current for ten minutes, the organic matter in solution being reduced by 61 per cent. At this rate the power required to purify a million gallons of sewage per day was estimated at 26 horse-power. Nineteen horse-power effected a purification of 50 per cent. The iron electrodes were gradually eaten away, the average loss being 2 grains per gallon of sewage treated. (At the present rate of flow at Crossness, this would amount to about 4500 tons of iron per annum!) In the course of some similar experiments carried out in Paris the number of organisms was reduced from 5,000,000 per cubic centimetre to 600.

This process was favourably reported on by Sir Henry Roscoe, and high expectations were formed of it. It was tried on a large scale at Salford, and later at Bradford, Yorkshire, but it never came into general use.

It received, however, a good deal of attention from the Niersverband, one of the youngest of the German river authorities. Experiments showed that a purification equal to that effected by a complete mechanical sedimentation plant, and two-thirds of that from an activated sludge plant, could be obtained by the Webster process in from ten to thirty minutes.

A few years after Webster carried out his experiments, a somewhat similar process was introduced by Hermite, at Lorient, France. Hermite, however, instead of electrolysing the sewage itself, passed a current of electricity through sea-water, liberating chlorine. The electrolysed sea-water was then added to the sewage. The Hermite process was tried at Worthing, but made little or no headway at the time. Ten years later, however, a company was formed in this country under the name of Oxychlorides, Ltd. to develop a process of decomposing sea-water (or a solution of salt) in a specially designed electrolyser and utilising the resultant liquid for the treatment of sewage. The process was tested on a working scale at the Guildford sewage-works.

Electrolytic methods have also been tried in America. A process very similar to that of Hermite was invented by Albert E. Woolf, of New York, and was adopted by the Health Department of New York City in the spring of 1893 to treat the sewage from some thirty houses at Brewster, N.Y., before discharge into the Croton River, one of the sources of New York's water supply. The plant was burnt down in 1911.

In the Landreth "Direct-oxidation" process electricity and lime were employed. The electric current liberated at the electrodes oxygen and hydrogen, which, it was claimed, brought about the destruction of pathogenic bacteria and the reduction of the nitrogenous organic matter to albuminoids, peptones and amino-compounds, which were subsequently oxidised to nitrites, nitrates and carbon dioxide. The lime furnished an alkaline medium, which lowered the electrical resistance of the sewage and rendered the electrodes passive, thus greatly reducing the quantity of iron dissolved from them. The electrodes consisted of mild-steel plates, spaced $\frac{3}{8}$-inch apart, with paddles revolving slowly between them, with the twofold object of keeping the plates clean and bringing the sewage into intimate contact with the gases produced by electrolysis while these were still in a nascent state.

A plant, serving a population of 7000, and dealing with 1,000,000 U.S. gallons per day, was laid down in 1918 at Easton, Pa., near the centre of the city, and was favourably reported on by a Committee of the Franklin Institute and by the Engineering Division of the Pennsylvania State Department of Health.

Lanphear, supervising chemist at the Worcester, Mass., sewage works, compared the results from the "direct-oxidation" process with those obtained in everyday working by lime precipitation at Worcester. He arrived at the conclusion that the results obtained by the Landreth process at Easton were probably due to the excessive quantity of lime employed (3720 lb per million U.S. gallons) rather than to the electrical treatment of the sewage. Fuller, of New York, formed the same opinion, but the City Engineer of Allentown, Pa., where a part of the sewage had been treated for a year by a direct-oxidation plant, controverted Lanphear's statement and spoke strongly in favour of the process. After a seven-day test of the plant, made a few years later, the consulting engineers reported that the process was not suitable for dealing with the sewage of the whole city, the purification effected being inadequate, the operating charges high and the quantity of sludge produced excessive.

Other electrical methods of purification have been tried in America. Between 1913 and 1915 seven electrolytic sewage disposal plants were laid down in the State of Oklahoma. Of these one was never used; five were abandoned after being in operation from eighteen months to six years; and one-third part of the seventh was in use for eight hours per day. Among the causes assigned for the abandonment of the process were the difficulty in disposing of the solids and in keeping the plates clean and broken parts replaced, the high cost of repairs, the lack of skilled attendance and the failure to effect any appreciable purification.

Of recent years there has been some revival of treatment by electrolysis including a method developed by Dr. Ernst Foyn which was exploited by Elektrokemisk A/S, Oslo. This method was developed to deal with pollution in Oslofjord and in 1958 the Oslo authorities built what was described as a semi-commercial plant. Tests showed that the very high power input of one kWh per 220 gallons treated was required, and there is little question that this would make the method uneconomic for use in Great Britain.

Miles' acid process. This process, which was suggested about 1900 by Mr. George W. Miles, a chemist, of Boston, Mass., aimed at the decomposition of the soluble soaps and the liberation of the fatty acids. Sulphuric acid might be used for the purpose, but Miles preferred sulphur dioxide gas. In experiments carried on with sewage from two sewers at New Haven, Conn., 840 lb and 1356 lb of sulphur dioxide were used per million imperial gallons. After treatment with this gas, the sewage was allowed to settle for four hours. The treatment removed from 61 to 66 per cent of the total suspended solids, 90 per cent of the settleable solids and more than 99 per cent of the bacteria. As the result of the experiments, the process was recommended for adoption by the City of New Haven.

It was originally estimated that the receipts from the sale of the grease and tankage recovered would repay from 40 to 50 per cent of the cost of treatment, but it was found that the low value of these products rendered the cost of the process almost prohibitive. It is said that the method gives a better effluent than that from an Imhoff or a plain sedimentation tank.

Treatment with zeolite. This process, which was developed at the Guggenheim Laboratory in New York, consisted essentially of three stages:

1. the removal of the suspended solids by precipitation with iron compounds and lime;
2. the removal of the basic nitrogen compounds, together with attendant groups containing carbon, hydrogen, sulphur, etc., by an exchange reaction using a preferred type of zeolite;
3. the regeneration of the zeolite, and consequent concentration of the basic nitrogen compounds in the salt solution, and the subsequent recovery of ammonia from this solution.

In this final stage of the purification the basic nitrogen, which was mostly in the form of ammonia, was exchanged for the sodium in the zeolite, resulting in a high concentration of ammonia in a very small volume of zeolite. The zeolite was then regenerated by a back-wash of water containing 20 per cent of salt, the ammonia being transferred to this solution. Sixteen thousand pounds of zeolite were required per million gallons of sewage treated. It was said that the suspended solids were reduced to practically nil, the biochemical oxygen demand to 5 parts per million and the bacteria by 99·1 per cent. The cost of the necessary plant was estimated at about two-thirds that of an activated sludge plant to do the same work.

The sludge from the purification tank was pumped to a vacuum filter, in which the moisture was reduced by about 80 per cent. The organic matter in the filter cake was then destroyed in a rotary kiln incinerator, consisting of two sections—a drying section and a burning section. The resulting ash was treated with sulphuric acid to regenerate the ferric sulphate, which was again used for coagulation.

The process was tried on a small scale in New York City, and on a larger scale at the North Side Sewage Works at Chicago, where the results were said to compare favourably with those obtained by the activated sludge process.

Chemical-mechanical treatment (Laughlin process). In this process, used at Dearborn, Michigan, the sewage was first screened and mixed with lime and paper pulp, prepared by grinding in pebble mills. On leaving the mixing-chamber it received a dose of ferric chloride or ferric sulphate. It then passed through two circular "clarifiers," or precipitation tanks, each 60 feet in diameter and holding about 200,000 U.S. gallons, in which it remained just over an hour. It was admitted at the centre of the bottom of the tanks, through which it flowed outward and upward. At the top of each tank there was a filter, consisting of a 3-inch layer of crushed magnetite sand, resting on a perforated bronze screen, which covered about 50 per cent of the surface of the tank. The sewage passed upward through these filters at the rate of about 3 U.S. gallons per square foot per minute, and then overflowed by a weir which surrounded the tank. Means were provided for chlorinating the effluent when necessary.

The magnetite was cleaned by a mechanism which consisted essentially of a solenoid which travelled over the filter at the rate of 4 feet per minute. A timed relay switch alternately closed and opened the electric circuit eighteen times per minute. When the relay closed the solenoid picked up a section of the sand clear of the screen, and when the circuit was opened the sand dropped back into place. This operation scrubbed the sand, the dirt from which was washed by a current of settled sewage from under the filter and returned to the centre of the tank.

The amount of purification effected by this process was said to be far greater than that brought about by simple sedimentation, and to approach the purification obtained by an activated sludge or percolating filter plant. The cost of the Dearborn installation was said to be very little more than that of a plain sedimentation plant of the same capacity, and the working expenses lay between those of simple sedimentation and activated sludge treatment.

The Sheffield system, or bio-aeration. One of the first to realise that the activated-sludge process may be carried on without compressed air was the late John Haworth, Chemist to the Corporation of Sheffield. As the result of his observations on the self-purification which went on in polluted rivers, Haworth came to the conclusion that the necessary oxygen might be picked up by the sewage from the atmosphere, provided that the surfaces exposed to the air were constantly changed. He accordingly constructed an artificial river by dividing an existing tank into narrow channels, 4 feet wide and 4 feet deep, connected at alternate ends, so as to form one continuous channel nearly three-quarters of a mile in length. In the centre of each length he placed a paddle-wheel, driven by an electric motor. These paddle-wheels propelled the sewage along the channels at a velocity of about $1\frac{3}{4}$ feet per second, and at the same time broke it up into a succession of waves.

Further disturbance of the surface took place when the direction of the flow was reversed at the ends of the channels. The first and last channels were connected together, so that the whole system formed one continuous circuit, round which the sewage travelled again and again. Fresh sewage was continuously admitted at one point, an equal quantity of purified effluent being drawn off at another, remote from the first. The effluent so drawn off was passed through settling tanks. The greater part of the sludge which was deposited in these tanks was pumped back to the aeration tanks, the surplus being removed.

The first large-scale plant on these lines was laid down in 1920 at Sheffield to deal with the sewage from the Tinsley district, about a quarter of a million gallons per day, one-third of which consisted of the trade wastes from large steel works. After three months' working very strong sewage from the city was added, the volume treated being gradually increased to half a million gallons per day. The sewage then took about seventeen hours to go through the aeration tank, during which it passed along the whole length of the channels some thirty times.

Although the Sheffield system was one of the activated-sludge processes, Haworth used the term "bio-aeration" to apply to this particular method. This term should not be used for all activated-sludge methods or be confused with bio-filtration (recirculation in percolating filters), or with bio-flocculation or bio-precipitation. (The last two terms have been given such widely different meanings by authorities on both sides of the Atlantic that they are best avoided altogether.)

In the Sheffield system the capacity of the aeration channels was usually in the region of sixteen to twenty-four hours' dry-weather flow. The results of work carried out by J. H. Edmonson and S. R. Goodrich, in which paddles of an improved type were used, indicated that it is possible to introduce more air into the mixed liquor stream, by impact and rotation of the paddles (that is to say, by using the paddle-wheels as "aerators" in addition to their original purpose of providing velocity), without any appreciable increase in power requirements, and that a substantial reduction in the required capacity of aeration channels could thereby be effected.

However, the Sheffield system does not appear to be capable of such modification as would be necessary to achieve such high efficiency of power utilisation as would enable it to compete with other methods. While some plants still exist, it is probable that this system will fall into disuse.

CHAPTER XXXII

TREATMENT OF TRADE WASTES

THE treatment of trade wastes has been the subject of textbooks[65,226] and can only be described in outline in the space of a chapter.

Where trade wastes have to be dealt with, the fullest possible information should be obtained on both their volume and their composition. Such information is sometimes difficult to obtain, for trade wastes are discharged at many points, and their volume and composition vary greatly from hour to hour, day to day and seasonally.

Trade wastes in large proportion to domestic sewage often modify its character to such an extent as to necessitate special modes of treatment. While their nature may influence the choice of the preliminary treatment, a given volume of trade waste will not, as a rule, call for any more tank capacity than an equal volume of domestic sewage of the same strength. Some organic trade wastes can be treated alone in the same manner as sewage, the aeration units requiring to be no larger than would be needed for the treatment of a domestic sewage of the same strength. On the other hand, some trade wastes are germicidal in nature, others encourage vegetable growth; some, such as citrus fruit wastes, cannot be treated by bacterial aid unless a nutrient for the organisms is added, and others, while they can be treated when diluted with domestic sewage, render treatment difficult.

Where different trades are carried on in a town the effluent from one will often neutralise that from another, in which case the mixed liquids will be easier to treat than either of them separately. In some cases, too, the work of the sewerage authority may be lightened, with profit to the manufacturer, by the recovery at his works of valuable substances which, through inadvertance, had been allowed to escape with the waste. Gas companies, for instance, were formerly in the habit of letting their coal-tar run to waste, but when they were compelled to treat it they found it a source of valuable by-products.

The Royal Commission on Sewage Disposal devoted a section of their Fifth Report to a consideration of the effect of trade effluents on sewage purification. All the trade effluents on which they received evidence interfered with or retarded the purification to some extent, but they were not aware of any case in which the admixture of trade refuse made it impossible to purify the sewage. They cited the following cases in which some modification in the mode of treatment had to be adopted.

The sewage of Burton upon Trent consisted mainly of brewery refuse, the domestic sewage constituting only about one-fifth of the whole. A large quantity of lime (40 or 50 grains per gallon) was added to the sewage, which was then treated on land.

The sewage of Bradford contained a large proportion of wool-scouring liquors, and was treated by "cracking" with sulphuric acid as a preliminary to filtration.

At Rochdale the sewage contained wool-scouring liquor and the waste water from fellmongers and rubber-recovery. After precipitation with sulphuric acid and Aluminoferric it was dealt with either in filters or upon land. This sewage was successfully treated, without any neutralisation, by septic tanks followed by percolating filters.

The sewage of Wolverhampton was largely composed of the waste water from galvanising works, which contained a great deal of ferrous iron in acid solution. It was treated with lime and irrigation.

Part II of Volume III of the Commission's Seventh Report contains the evidence received with regard to the treatment of trade wastes generally, and Part I of the Ninth Report deals with the disposal of liquid wastes from manufacturing processes. These wastes are classified as follows:

Liquid wastes which are polluting by reason of the suspended matter which they carry with them—
    Coal-washing waste.
    Waste waters from tin mines.
    Waste waters from lead and zinc mines.
    Waste water from china-clay pits.
    Stone-quarrying waste.
    Stone-polishing waste.

Liquid wastes which are polluting mainly because of the suspended matter which they carry with them; but also to a considerable extent because of the dissolved impurities which they contain—
    The cotton industry.
    The woollen industry.
    The manufacture of paper and cardboard.

Liquid wastes which are polluting mainly because of the dissolved impurities which they contain—
    Waste liquor from wool scouring.
    Brewery waste.
    Steep water from maltings.
    Waste liquor from the manufacture of sulphite.
    Cellulose.
    The metal industries.
    Fellmonger's waste.
    Tannery waste.
    Dairy waste.
    Margarine waste.
    Waste liquor from shale-oil distillation.
    Spent gas liquor.

Since 1915, when the Commissioners made their Final Report, many changes have taken place, but their observations as to the methods suitable for the treatment of the various classes of wastes mentioned still hold good. There are, however, other wastes the treatment of which is now receiving a good deal of consideration.[226]

Gas liquor. Gas liquor is produced in the distillation of the ammoniacal liquor formed in the manufacture of coal-gas. Its composition varies according to the type of retort used, but it always contains a great deal of ammonia, sulpho-cyanides and phenols, and absorbs a large amount of oxygen. It is among the most intractable of all the waste liquids which find their way into the sewers. As little as 0·5 per cent of it is said to have caused difficulty in the treatment of sewage, and as much as 8 per cent is sometimes present. So high a proportion will be met with only where the gas-works from which it proceeds supply an area much larger than that from which the sewage is derived, but proportions of 2 or 3 per cent are not uncommon. In these concentrations gas liquor can usually be dealt with in an ordinarily efficient sewage works, and does not call for any modification in the mode of treatment.

Milk wastes. The waste waters from dairies and creameries are rich in organic matter and very liable to undergo fermentation, with the formation of lactic, butyric and other organic acids which give it a particularly objectionable odour. When milk wastes are treated alone or are in large proportion to other wastes or domestic sewage, the works need to be of acid-resisting construction, because lactic acid produced by the souring of milk destroys cement and concrete.

Beet-sugar wastes. The beet-sugar industry has been carried on for many years not only on the Continent and in America, but also in this country. The factories are usually situated, not in towns, but in rural areas, often on the banks of some hitherto unpolluted stream. The quantity of water used in the process is very large—from 3 to 4 million gallons per thousand tons of beet—and the waste is highly polluting.

The Water Pollution Research Board has for many years carried on an intensive investigation into the purification of these wastes. By far the greater part of the water is employed for washing the beets and cooling the condensers, and may be reused again and again, leaving about one-eighth of the total volume to be purified. Sedimentation tanks, followed by percolating filters, have given promising results. The Board has now reached the conclusion that this difficult problem has at last been satisfactorily solved.

New problems. The problem of trade wastes is continually changing with the changes in industry. The new development of light industry, for example, has involved an increase in the number of chromium and other plating plants which have necessitated the installation of neutralising tanks and treatment to precipitate the chromium compounds in insoluble form.*

---

* J. H. Spencer. "The Treatment of Chromium Plating Wastes." *Proceedings Institute of Sewage Purification*, 1939, Part 1.

Storage and/or disposal in sealed capsules in deep sea. These are methods applicable to radioactive wastes. Radioactive isotopes deteriorate with time and those which have a short "half-life" (the time in which they lose one half of their activity) may be rendered harmless if stored for a sufficient time. The method of disposal in sealed containers in deep sea is applicable to those which cannot be rendered harmless.

Radioactive liquid wastes can be rendered less harmful by removal of radioactive sludge or by absorbing the dangerous substances with activated charcoal.

Control of trade wastes at points of origin. Under the *Public Health (Drainage of Trade Premises) Act*, 1937, local authorities can control discharge of trade effluents and set limits on the periods during the day when any effluent may be discharged, the maximum quantity that may be discharged per day and the maximum rate of flow at any time. Moreover, they may insist on the control of temperature of effluent and require it to be discharged in a neutral condition, and they may insist on the exclusion of any condensing or cooling water. Constituents liable to injure or obstruct sewers or to render treatment difficult or unduly expensive must be eliminated if the local

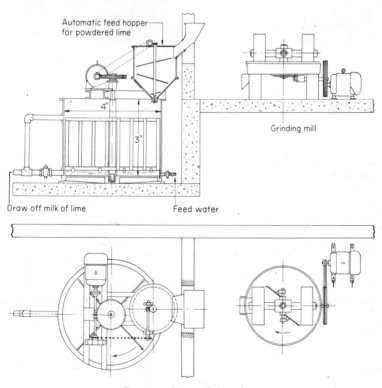

FIG. 140—Lime-mixing plant
By courtesy of Hartleys (Stoke-on-Trent), Ltd.

Fig. 141—Labyrinth for chemical treatment of sewage
By courtesy of J. H. Garner, Esq., Chief Inspector, The West Riding of Yorkshire Rivers Board

authority so requires. In the case of sewers discharging to harbours, etc., without treatment, objection can be raised to trade wastes liable to injure or obstruct navigation.

The types of treatment plant most commonly required at point of discharge to sewers, etc., are settlement tanks for the removal of heavy sediments which occur in the wastes, or for precipitation of insoluble compounds produced by chemical treatment; grease- or oil-traps for the removal of grease or oil lighter than water, and catchpits to settle out either heavy solids or liquids; screens for the interception of large floating particles; tanks for holding up the flow over prohibited periods, or for balancing purposes, or for cooling purposes; and tanks or labyrinths for chemical treatment.

Some indication of the types of wastes liable to be frequently encountered and the methods of treatment applicable are given in Table 58, which is prepared from the Appendix to M. A. Kershaw's paper "The Inspection and Control of Trade Waste Effluent Discharges."[145]

Screens and settlement tanks for treating wastes are similar to those described for the treatment of sewage at small works and grease-traps are generally similar to those required for the interception of grease from kitchens (see Vol. I, Chapter XX). Grease and oil removal can be described in Chapter XXIV.

The most frequent form of neutralisation is that of acid with the aid of lime and/or caustic soda. Lime is applied in most cases in the form of milk of lime, which is mechanically fed into the flow of waste and thoroughly mixed therewith in a labyrinth. In some cases acid is neutralised by passing it through a contact bed of limestone chippings (which become coated and ineffective). Labyrinths are tortuous channels designed to produce turbulent flow and complete mixing in a small space (see Figs. 140 and 141).

Charges for treatment of trade wastes. Local authorities are empowered to charge for the reception of trade wastes into the sewers. Such charges are usually made in accordance with the cost of sewage treatment in the district concerned and the estimated fair proportion of this cost chargeable for the trade waste in accordance with its quantity and strength in terms of biochemical oxygen demand or strength (McGowan), suspended-solids content and difficulty of treatment. Opinion can vary considerably on what is a fair charge in any particular circumstances. It is understood that the Middlesex County Council has charged according to what is known as the Mogden formula:

$$\text{Pence per 1000 gallons} = 1 + \frac{S}{60} + \frac{M}{75}$$

where: S = suspended solids in parts per *hundred thousand*
M = strength (McGowan).

This formula has been widely adopted elsewhere, although the values should be adjusted according to local costs. In 1957, owing to the decreased purchasing power of the £, cost as estimated by the above formula had to be increased by the factor 1·25. (See Vol. I, Table 3.)

Chemical-resisting construction. When acids are present in trade wastes and are not very dilute, precautions have to be taken so as to prevent the destruction of clay-pipes, cement joints, brickwork and concrete. This involves the use of acid-resisting materials, most of which are expensive.

For the construction of drains and sewers special salt-glazed ware pipes and fittings with chemically resistant properties are available (B.S. 1143). These are manufactured from a quality of clay, such as Dorset clay, which renders them resistant to acid, and are tested by hydraulic pressure to 40 lb per square inch. They should not be confused with chemical stoneware, a very expensive material used for the conveyance of chemicals in factories, and jointed by means of a variety of special (e.g. bolted, flanged) joints.

Chemically resistant stoneware pipes are jointed in special jointing materials selected according to the nature of the waste. One of the peculiarities of acid-resisting cements, etc., is that while they may be resistant to an acid in high concentration, they may be easily broken down by an alkali or injured by heat, and therefore not only the nature of the waste but also its temperature and the periodic variation of its reaction have to be considered.

TABLE 58

SELECTION OF PRODUCTS MANUFACTURED ON A TRADING ESTATE, EFFLUENTS PRODUCED AND TREATMENT METHODS APPLIED BEFORE DISCHARGE TO SEWER

Type of product	Product	Type of trade waste effluent	Treatment of units
Inorganic, non-metallic	Pre-cast concrete slabs and kerbstones	Cement and grit, particles in suspension, traces of oil	Settlement
Metallic	Precious metal refining	Dissolved acid fumes and other chemical waste	Neutralisation with lime and caustic soda
	Electro-medical equipment	Metal cleaning, quartz cleaning and hydrofluoric acid	Neutralisation and settlement
	Lead carbonate, paints and lead nitrate. Metal hardening	Lead carbonate in suspension, cyanide solution and acid wastes	Neutralisation and settlement
	Alloy metal work, castings, forgings and engineering	Chromate, silica and anodising waste, oil and grease	Dilution, grease traps, and neutralisation
	Motor-car assembly	Metal cleaning and repairs	Oil traps and catchpit
Foodstuffs	Jams and honey. Confectionery	Sugar solution and fruit pulp	Grease traps, screens and catchpits
	Margarine	Machinery and floor washings	Grease traps
	Canned food products	Washing and cleaning vegetables	Screens and catchpits
Organic compounds other than foods	Cosmetics	Vat washings	Grease traps
	Soaps, sulphonated. Alcohols	Soap lyes	Neutralisation and settlement
	Bitumen, tar and pitch solution and emulsions	Mixing tank and floor washings	Catchpits
	Paints, varnishes and synthetic resins	Resin scum, drum cleaning and solvent recovery	Catchpits, neutralisation, filtration and precipitation
	Insulating materials	Caustic waste	Catchpits and neutralisation
	Wallpaper and colour printing	Colour waste with clay in suspension	Catchpits
	Rubber and latex	Organic chemical waste and latex solution	Cooling and catchpits, and acid coagulation
	Capsulated medicines	Gelatin and glycerine solution	Cooling and catchpits
	Glues and starch products	Vat and barrel washings	Catchpits
	Knitting, felting and dyeing	Dye liquor, soapy water with fibrous suspended matter	Screening, cooling and neutralisation

Tanks for neutralising or otherwise treating chemical wastes may be constructed of special brickwork, or of concrete lined with chemical-resisting asphalt, or with thin, flexible protective coatings. Bricks for this purpose need to be impervious to water and to have a smooth surface. Good-quality Staffordshire Blue Bricks will serve, but for most high-class work best pressed Accrington "Nori" bricks are often preferred because of their very smooth surface and high density. These, carefully jointed in chemically-resisting material, form an inner skin or facing to the tank-work, but may be backed by ordinary brickwork, providing that leakage is entirely prevented. To make sure that there is no leakage, in the best-quality work the facing of chemically-resisting brickwork is backed by a chemical-resisting asphalt about $\frac{3}{4}$ inch in thickness or a sheet of plastic.

A variety of materials manufactured by Prodorite Ltd. is useful either for jointing chemical-resisting pipes or for constructing chemical-resisting tank-work. The following are the approximate qualities of some of the materials:

Semtex S.X. 141. This is a mixture of rubber latex and high-alumina cement which, when used as a cement mortar for jointing pipes, will withstand dilute acids. It is, however, liable to go soft and spongy if the acid strength is at all high.

Semtex F.P.C. This is a coating for application to the interior surfaces of tanks. It is slightly flexible.

Cement Prodor HFR. This is a super potassium silicate cement which will resist all acids except hydrofluoric. It has low porosity and great adhesion.

Cement Prodor SWD. This cement will tolerate a 30 per cent solution of sulphuric acid, a 100 per cent concentration of nitric acid, but will not tolerate any alkali. The maximum temperature it will resist is about 1000 °C.

Cement Prodor SWK. This material is similar in most respects to Cement Prodor SWD, but it will withstand sulphuric acid of any concentration.

Prodorphalte. This mastic material will stand a concentration of 30 per cent sulphuric acid, concentrated hydrochloric acid or 5 per cent caustic soda. It will not tolerate a temperature higher than 35 °C if exposed, but it will withstand a temperature of 65 °C when used as a backing to acid-resisting brickwork.

Prodorkitt. This again is a mastic material with properties similar to Prodorphalte as regards heat resistance. It will, however, withstand a 60 per cent concentration of sulphuric acid, concentrated hydrochloric acid or a 20 per cent concentration of caustic soda.

All the above materials will tolerate higher percentages of acid or alkali applied for short periods only.

Sulphur compound. Sulphur compound will tolerate up to 80 per cent concentration of sulphuric acid, 40 per cent concentration of nitric acid or 1 per cent concentration of alkali. It will withstand a temperature of about 90 °C.

Asplit. Asplit is a corrosion-resistant cement of the synthetic resin type resistant to acids (with the exception of those of higher strength), alkalis and solvents. This cement may be used at temperatures up to 180 °C. There are three grades, Asplit A, Asplit C.N. and Asplit O. This cement can be used in contact with foodstuffs.

Prodor-glas. This is a phenolic resin-based coating bonded to metal by a stoving procedure.

Prodorfilm. This is a resin-based coating, polymerised by chemical reaction, which can be applied on site. Products which can safely be brought into contact with either of the two foregoing include alcohol, edible oils, fruit juices and molasses.

Prodorcote. This is an epoxy resin.

Polyvinyl chloride (P.V.C.). This is available in sheets or tubes. The hard variety is preferred for chemical-resisting purposes. Softness is induced by the addition of a plasticiser which may come out with ageing. P.V.C., if not plasticised, is resistant to oils, some acids and alcohol absolute except at high temperature.

Polyethylene (Polythene). This is available in flakes for pressing into special shapes, thin or thick sheets, tubes and valves. It tends to oxidise in sunlight and if exposed to light should have an opaque filler. It is resistant to several acids but not to oils or strong alcohol.

Nylon. This is available in fibres, rope, sheets and tubes. It will withstand animal, vegetable and mineral oils but not acids. It has great strength and durability and can be made into gearwheels, etc.

Polytetrafluoroethylene (P.T.F.E.). This has great resistance to most chemical substances. It has an exceptionally low coefficient of friction against itself, making it useful for unlubricated bearings. Rods and tubes are available or it can be applied as a coating.

Of metals, chemical lead, which is very pure lead, is sufficiently resistant for use with mild alkalis and the most dilute acids, but not nitric or hydrochloric acid. Joints should be by lead burning and not by soldering. Stainless steel, in suitable grade, will resist acids and alkalis, but not hydrochloric acid or strong sulphuric acid. It is attacked by various chlorine compounds, perhaps electrolytically. Cast-iron containing silicon has chemical-resisting qualities.

Rubber-lined or plastics-lined pumps are available for pumping chemicals.

CHAPTER XXXIII
SEWAGE DISPOSAL FOR ISOLATED BUILDINGS

THE disposal of sewage for isolated buildings is work that is outside the experience of many public health engineers practised in designing sewage-treatment works for towns. The small sewage works involved may often be included in architects' contracts, yet architects, more often than not, have very little knowledge of sewage treatment of any kind. As a result the design and construction of such works tends to be left to contractors who generally may be assumed to have no interest in the production of inoffensive effluents.

It has thus come about that by far the majority of small sewage-treatment works are designed not to treat sewage effectively but to give the appearance of treating sewage. Furthermore, following upon this established practice, some committees or other producers of advisory publications have described how to design and construct works which look like sewage-treatment plants but which never could produce a Royal Commission effluent.

The writer, in investigating conditions at a number of large country houses, converted or to be converted into schools or other institutions, found that twelve of these premises had sewage-treatment works incorporating percolating filters of which none was producing a satisfactory effluent, and sixteen had cesspools none of which was sanitary or complying with the by-laws operative at the time.

It must be understood that this was not a particularly bad sample but quite representative of conditions throughout the country. Moreover, several of these works were very bad indeed. In one instance the septic tank was derelict, the percolating filter distributors derelict, and the filter overgrown to such an extent and for so long that two trees of 9-inch girth were growing through it. The sewage works of a school consisted of nothing but a so-called "septic tank," the effluent from which flowed into a near-by field where it was drunk by cows. As milk from this herd was used in the school it was not surprising that there was an epidemic of diarrhoea at about the time of the investigation. In three instances there were quite ineffective treatment plants and the drains from the premises crossed public highways and land under different ownership to discharge into streams, in one case over a mile away. The effluent from one quite ineffective sewage-treatment plant was discharged into public bathing pools in London and analysis of the water in them showed varying degrees of contamination. Where cesspools overflowed (nine cases), this was indicated by either sewage fungus or faeces and toilet paper in the watercourse. There were seven cases of leaking or deliberate discharge to the subsoil. In one instance soakage from sub-irrigation killed a hedge, in another the sewage could be seen running along a ditch beside the highway.

In all these instances new works had to be constructed to secure sanitary conditions. In the case of premises housing several head of population and where no part of the works was capable of repair it was decided to install the type of treatment plant illustrated in Figs. 145 and 150. Thereafter nuisance was generally avoided and in some instances superlative final effluents were obtained. Contrary to some opinion it *is* possible to give satisfactory treatment to sewage at very small treatment works and without involving exorbitant costs.

When compared with the sewerage of towns and sewage treatment at municipal disposal works, the treatment of sewage from isolated buildings is always problematical. The chief reason for this is that, whereas at the large sewage works the manipulation of the plant is the responsibility of an experienced manager and staff, the maintenance of sanitary conditions at the works serving an isolated house or group of houses involves an unpleasant duty which is very liable to be neglected. Thus, methods of sewage disposal for isolated buildings are selected in accordance with the difficulties of securing satisfactory maintenance, and details of design are based on the assumption that the small treatment works are more liable to be neglected than otherwise.

Methods of disposal. The methods of disposal of sewage from those buildings or small communities which cannot be served by main drainage include both conservancy methods and the water-borne system. Under the former heading come earth-closets and pail- or chemical-closets. The water-borne system of drainage may be used in conjunction with a cesspool in which the sewage is stored (which is a form of conservancy), afterwards to be removed by tank vehicle at the hands of the local authority or a contractor; the disposal of sewage without treatment by an outfall into a tidal water; and finally, some form of sewage treatment.

Conservancy methods. Conservancy sanitation has been defined as sanitation by keeping refuse matter in privies, pails, earth-closets and cesspools for its periodic removal.

No conservancy method at present in use can compare with main drainage either as a sanitary method or from the point of view of amenity. Nevertheless, there are occasions when main drainage is impracticable, and then a suitable conservancy method has to be considered. In these circumstances great care should be taken that the method selected is, in fact, in accordance with the dictates of public health and legal requirements and not some expedient which has become more or less accepted in localities where standards of sanitary education or technical qualifications are low.

The types of conservancy sanitation which are legally permitted are those involving the use of earth-closets and cesspool drainage, but they are legal only if they conform to statutes and regulations.

The use of earth-closets and cesspools respectively are limited by the following sections of the *Public Health Act*, 1936:

43. (1) Where plans of a building or of an extension of a building are, in accordance with building byelaws, deposited with a local authority, the authority shall reject the plans unless either the plans show that sufficient and satisfactory closet accommodation consisting of one or more waterclosets or earthclosets, as the authority may approve, will be provided, or the authority are satisfied that in the case of the particular building or extension they may properly dispense with the provision of closet accommodation:

Provided that—

(i) unless a sufficient water supply and sewer are available, the authority shall not reject the plans on the ground that the proposed accommodation consists of or includes an earthcloset or earthclosets; and

(ii) if the plans show that the proposed building or, as the case may be, extension is likely to be used as a factory, workshop or workplace in which persons of both sexes will be employed, or will be in attendance, the authority shall reject the plans, unless either the plans show that sufficient and satisfactory separate closet accommodation for persons of each sex will be provided, or the authority are satisfied that in the circumstances of the particular case they may properly dispense with the provision of such separate accommodation.

37. . . .

(3) A proposed drain shall not be deemed to be a satisfactory drain for the purposes of this section unless it is proposed to be made, as the local authority, or on appeal a court of summary jurisdiction, may require, either to connect with a sewer, or to discharge into a cesspool or into some other place:

Provided that, subject to the provisions of the next succeeding subsection, a drain shall not be required to be made to connect with a sewer unless—

(a) that sewer is within one hundred feet of the site of the building or, in the case of an extension, the site* either of the extension or of the original building, and is at a level which makes it reasonably practicable to construct a drain to communicate therewith, and, if it is not a public sewer, is a sewer which the person constructing the drain is entitled to use; and

(b) the intervening land is land through which that person is entitled to construct a drain.

In the above the circumstances in which a local authority may insist on connection with their sewers and the provision of water-closets are clearly set out. There remain circumstances where cesspools are permitted, and other circumstances in which the provision of earth-closets or chemical closets is acceptable. But to be considered proper sanitary provision, an earth-closet or chemical closet must fall under the definition in the *Public Health Act* and comply with *Building Regulations*.

* The site of a building is the land covered by the building, and does not include the curtilage (author).

Earth-closets and chemical closets. An earth-closet is defined in the *Public Health Act*, 1936, Section 90, as:

> ... a closet having a *moveable** receptacle for the reception of faecal matter and its deodorisation by the use of earth, ashes, or chemicals, or by other methods;

Thus, a chemical closet of a suitable design can be, in law, an earth-closet.

The *Building Regulations*, 1965, require that any earth-closet which is not a chemical closet shall be so constructed that it can be entered from the external air only or from a room or space which itself can be entered directly from the external air only. No earth-closet, or chemical closet shall open directly into a habitable room, kitchen or scullery or room in which any person is habitually employed in manufacture or business. An earth-closet which can be entered directly from the external air shall have an adequate ventilation opening as near the ceiling as practicable and one which cannot be entered from the external air shall have a window or similar means of ventilation to the external air capable of opening to not less than one-twentieth of the floor area. Earth-closets shall be so situated that they shall not be liable to pollute any source of water used or likely to be used for domestic purposes.

The regulations also require that the floor of the closet shall be non-absorbent, and, if the closet is entered from the external air, shall be not less than 3 inches above the adjoining ground and fall towards the entrance at not less than 1 in 24. The receptacle shall be non-absorbent and so constructed and located that the contents cannot escape by leakage or otherwise or be exposed to rainfall or other drainage. The receptacle and fitting shall be so constructed and arranged that maintenance shall not constitute a nuisance or be a danger to health. *No part of the receptacle of an earth-closet or chemical closet shall discharge to a drain.*

Chemical closets. Chemical closets of the simple pail type, if of suitable materials and capacity, can be legally considered to be earth-closets provided they comply with *Public Health Act* definition. Warning should, however, be given that from time to time chemical closets have been marketed which do not conform to the definition of any type of closet in the *Public Health Act*, and which on many counts offend against *Building Regulations* and the interests of public health.

The ordinary pail chemical closet is essentially a vessel provided with a seat and lid and sometimes having other features more of sales interest than practical value. The material should be strong and non-corrosive, and the design such that cleansing is easy.

Many kinds of chemical can be used, but broadly there are three types in use:

1. Coal tar preparations with oil seal.
2. Coal tar preparations without oil seal.
3. Formaldehyde preparations.

* Author's italics

The chemicals for use with oil seal are intended to effect a degree of disinfectant action; the oil seal is to prevent evaporation from the surface with the possibility of objectionable odour. The preparations without oil seal rely on the disinfectant to effect deodorisation.

The chemicals should not only deodorise excreta immediately but remain effective as they become diluted for a period of at least a week. They should

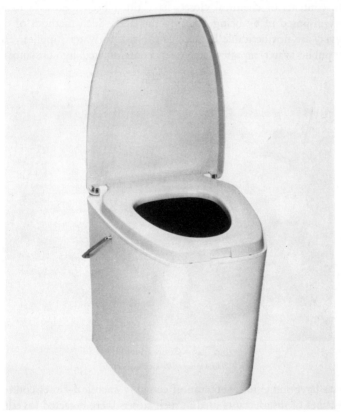

FIG. 142—"Elsan" chemical closet
By courtesy of Elsan, Ltd.

not permit the formation of obnoxious gases and should be odourless, and of a colour that will obscure the contents of the closet. In view of the possibility of splashing, the fluid should not be injurious to the skin.

The chemicals should be of a type not to cause deterioration of the materials of which the closet is made.

One of the difficulties in the use of chemical closets is that the contents tend to splash the user. So far, no devices have been invented that will prevent splashing other than in connection with some designs of closet which are ruled out as being contrary to the *Public Health Act,* 1936.

Where earth or chemical closets are common, it is highly desirable that there should be a regular service for emptying them. Suitable cesspool-emptying vehicles can be used for this purpose: for example, the Dennis cesspool emptier (see Fig. 143) can be provided with a hopper into which the contents of earth or chemical closets can be tipped and rapidly drawn into the tank of a vehicle by suction.

The contents of chemical closets have often been used as a manure or merely disposed of by being dug into the land. Such methods of disposal, however, are not desirable because of the risk to water supplies. Recently some public water supplies have been contaminated by substances which

FIG. 143—Cesspool emptier
By courtesy of Dennis Bros., Ltd.

analyses have suggested were almost certainly chemical-closet contents: the usual types of disinfectants used in such closets were detected together with sewage contamination (see also end of Vol. I, Chapter IV).

For estimating the provision of chemical closets or disposal of contents, the following information should prove useful. The average deposit of faecal matter amounts to 135 gm per day at 75 per cent moisture content. The quantity of urine has been stated as being, for the average man, 1500 gm of water per day containing 72 gm of solid matter. This amounts to 0·36 gallon per head per day or $2\frac{1}{4}$ gallons per person per week, to which should be added about 1 gallon per closet per week for chemicals and dilution water. Effective capacities of chemical closets available vary from 3 to $5\frac{1}{2}$ gallons, from which it will be seen that if closets are emptied once a week (a not uncommon period where the local authority undertakes the service) the largest size of closet is hardly adequate for a household of two persons and, generally, for domestic purposes two closets should be provided. It will also be observed that the recommendations of paragraph 5 of the schedule attached to "Statutory Instrument No. 966, Shops and Offices The Sanitary Conveniences Regulations, 1964" are inadequate: with the minimum provision recommended, the closets would have to be emptied every day!

Cesspool drainage. In common parlance, a cesspool is a tank the purpose of which is to store sewage between periods of disposal by tank vehicle or other means; but according to Section 90 of the *Public Health Act*, the term cesspool means also any tank for the reception or disposal of foul matter—i.e. a sedimentation tank or a septic tank.

The *Building Regulations*, 1965, require that any cesspool, including a settlement tank, septic tank or other tank for the reception of foul matter from a building shall be so constructed as to be impervious to water both from inside and outside. It shall be sited so as not to render any spring, stream or well used or likely to be used for domestic purposes liable to pollution. There shall be means of access for cleansing without carrying contents through any building used for residence or trade or to which the public has access. The cesspool shall not be so close to a building used for residential or business purposes or to which the public has access as to be liable to become a nuisance or danger to health.

The *Regulations* require that any cesspool not a settlement or septic tank (which presumably means a tank used as part of a treatment plant) shall be of a depth that permits complete emptying, covered to exclude surface water or rain-water, fitted with a suitable manhole cover, adequately ventilated and without any outlet for overflow or discharge other than that provided for emptying or cleansing. Such a tank must have a capacity below the level of the inlet pipe of not less than 4000 gallons (about 18·2 cubic metres). This last requirement is a step in the right direction for it will, in many instances, make cesspools more expensive to construct than proper sewage-treatment plants.

Any settlement tank or septic tank shall be of suitable depth and adequate size for the purpose and in no case of a capacity less than 600 gallons. It shall be covered and adequately ventilated with means of access for inspection and cleansing or, if not covered, it shall be fenced in.

While the *Public Health Act*, 1936, uses the term "cesspool" to mean not only a watertight tank without outlet but also the septic tank or other sedimentation tank of a small sewage-treatment plant, the two things are very different and need separate consideration. The true cesspool is a tank intended to store without leakage or overflow the whole of the sewage from a building until such time as it can be completely emptied, preferably with the aid of a tank-vehicle which can take the contents to some sewage-treatment works or other satisfactory place of disposal. Such tanks do exist and some are properly maintained as intended by the *Public Health Act*, but they are extremely rare.

Cesspools are, in fact, a survival from a former age when sanitation was in its infancy and the common use of baths, lavatory basins and water-closets was only coming in. The term cesspool in fact suggests a pond into which all filthy matter was discharged and allowed to accumulate—a condition of things which could not be tolerated at the present day. The intention of the definition in the *Public Health Act, 1936*, was that cesspools should be completely watertight vessels in which sewage could be stored until it could be removed in a sanitary manner, and from which no leakage or overflow liable to cause a nuisance or danger to health should be permitted. Such a watertight cesspool is, however, usually impracticable in present-day circumstances, for the amount of foul water discharged from an ordinary small dwelling-house amounts to about 700 gallons per week, or roughly the contents of an average-size cesspool-emptying vehicle. Thus, if cesspools were emptied by contract the householder would have to pay upward of £150 a year for this service. If, on the other hand, a local authority had undertaken cesspool-emptying and all the householders insisted on their rights, under the *Public Health Act*, to have their cesspools kept from overflowing, that local authority would have to build up an immense fleet of cesspool-emptying vehicles such as does not exist anywhere.

It will be seen from the foregoing that, as a general rule, the use of cesspools is quite impracticable. Nevertheless, cesspools do exist in large numbers throughout the country, and those who are not conversant with the facts may wonder how this can be. The reasons are these:

1. While local authority building inspectors may insist on cesspools being rendered watertight and being tested for watertightness, it is common practice for builders secretly to knock holes in the bottoms of cesspools after the tests have been made.

2. In many instances local authorities have, contrary to their powers under the *Public Health Act*, permitted cesspools to discharge via drains to soakaway pits or sub-irrigation systems or otherwise, at a risk of causing nuisance or endangering public health.

3. Local authorities generally have not attempted to empty cesspools weekly or otherwise as may be necessary to take away the entire contents of sewage, and where local authority arrangements exist for cesspool-emptying the most common routine is to take away a load or so of solids content from each cesspool once every three months.

Cesspool drainage is thus, in the main, a mere pretence at sanitation, and the work that is done on cesspool-emptying is usually no more than a gesture to conceal the general contravention of the *Public Health Act*.

In the author's opinion, the law should be altered and installation of cesspools *should not be permitted unless it is demonstrated that proper provision for emptying can be, and will be made*. Only in such circumstances is cesspool-drainage a satisfactory method.

SEWAGE DISPOSAL FOR ISOLATED BUILDINGS

There are some circumstances in which a cesspool may be considered the proper solution of the sanitary problem. For example, there are buildings which are not normally inhabited but which are occasionally visited by staff and at which a water-closet and lavatory basin or sink may need to be provided. In such a case the amount of water used and discharged as sewage can be very little indeed, and it then becomes possible for truly watertight and non-overflowing cesspools to store the whole of the sewage discharged over a period of, say, three months.

Construction of cesspools. Cesspools are usually constructed circular on plan, the floor being of concrete and the walls of circular brickwork in header bond; or pre-cast concrete pipes surrounded with concrete. The work is rendered watertight in accordance with *Building Regulations*. Ventilation is essential. Earlier cesspools had corbelled or arched roofs; reinforced concrete slabs are now usual. Access is by (nominally) airtight manhole cover, which should not be of too light a type.

At one time it was usual for cesspools to be provided with chain pumps for emptying their contents on to the land—a most offensive process. Now chain pumps are seldom provided, cesspool emptying being effected by suction under vacuum into the tank of a cesspool-emptying vehicle.

A cesspool should be sufficiently large to store the whole of the sewage likely to accumulate in the interval between emptying times, and the capacity should be an exact multiple of the capacity of the local cesspool-emptying vehicle, for the reasons given in the next section.

FIG. 144
Chain pump
By courtesy of Burn Bros. (London), Ltd.

The regulations that cesspools shall be of such a depth to enable complete emptying and shall have a capacity of not less than 4000 gallons virtually determines the design of those of minimum size. As (as is mentioned later) the maximum depth below road-level is 17 feet and if the inlet pipe comes in at about 4 feet below ground-level, the working depth of the cesspool is about 13 feet. This calls for a tank of 8-foot diameter.

Tank vehicles and their influence on the design of cesspools and septic tanks. The tanks of most cesspool-emptying vehicles have capacities in the region of 750–800 gallons. For this reason every cesspool or septic tank that is to be emptied by vehicle should have a capacity which is a multiple of 750 gallons (3·4 cubic metres) or some other figure according to the capacity of the local tank vehicles, otherwise the vehicle will make its final journey, on each occasion of emptying, only partially full, which may mean that the householder has to pay the price for removal of a full load for the last few gallons removed.

Cesspool emptiers carry long lengths of suction hose: some are normally equipped with 200 feet of flexible suction pipe. For this reason it is not necessary for cesspools to be placed near the public highway, or for access roads to be laid to the cesspools; but obviously, the cesspool must be sufficiently near the nearest point of access for the length of hose normally carried to reach from the vehicle to the *bottom* of the cesspool.

Another point to be kept in mind is that the contents of cesspools and septic tanks are sucked out by a vacuum created in the cesspool emptier and therefore there is a limit to the total lift from bottom of cesspool or septic tank to the top of the tank of the cesspool emptier. This is usually stated as being a maximum of 25 feet. As the cesspool emptier stands about 8 feet high to top of tank, the bottom of cesspool should not be lower than 17 feet below road (say 5 metres).

Small treatment works. Before deciding on the construction of new treatment works, the designer should first ascertain if new works are, in fact, necessary. There have been so many instances where works were proposed and even constructed, but where, in fact, connection could have been made to a public sewer at lower overall cost. The cost of a small treatment works is equal to that of a considerable length of small-diameter drain or sewer, and therefore the designer has to look not merely for the existence of public sewers in the immediate vicinity of the site, but ascertain whether or not it is possible to connect to a sewer 1000 or more feet away. Even if it is slightly more costly in the first instance, the connection to a public sewer does away with all worry about maintenance, and often means that surface water and soil-sewage can be discharged via the same pipe. There is also the possibility of delivering sewage to the public sewers by pumping. In more than one instance this has been possible and has been adopted as an alternative to reconstruction of existing treatment works.

The methods that have been used for treating sewage from isolated houses, schools and institutions are:

1. Treatment by settlement tank and percolating filter.
2. Treatment by settlement and on land.
3. Treatment by settlement and in small activated sludge works.

Disposal to cesspools should be condemned as incompatible with sanitation, for the reasons already given. Disposal by sub-irrigation or other means of soakage into the ground after settlement in a septic tank is also to be condemned in this country, although it may have its legitimate uses overseas in sparsely populated districts. Its effects are exactly the same as those of the overflowing or leaking cesspool, for the septic tank receives little or no attention, and the effluent, similar in character to crude sewage, goes where it may be a danger to public health.

Far too often local authorities do nothing about the danger or nuisance that is involved. In one instance the writer inspected a septic tank which had been described as "satisfactory;" in fact, it was discharging crude sewage, including unbroken faeces and paper, to a stream which, apart from the sewage flow, was dry in dry weather. In another instance where a septic tank serving a large institution was discharging, without any treatment, to a stream, the local authority officer said that these works "had been a running sore for years;" yet he had never done anything about them. These are typical examples taken from many at random.

The methods of treatment that can be used with satisfaction in practically all circumstances are settlement in a septic tank followed by aeration on a percolating filter, and/or land treatment. However, land treatment as an alternative to percolating filter treatment is not generally to be recommended for two reasons: it is liable to produce offensive conditions near to premises, and it requires more intelligent and interested attention than can be expected, other than at large premises, where a good handyman (perhaps) can be trusted to do the work properly. The activated-sludge method is limited to those larger establishments where a competent handyman can attend to the plant. Thus the problem is simplified by being in most instances restricted to the proper design and construction of septic tanks and percolating filters.

Exclusion of rain-water. One of the axioms of the design of new small treatment works is the total exclusion of rain-water from the flow to be treated. If separate drains are laid for soil sewage and surface water, the rate of flow of soil sewage will not vary much with rainfall and the surface water can be discharged without treatment to the nearest watercourse or soakaway, according to the nature of the subsoil. If, on the other hand, storm water is combined with soil sewage, the sewage works will receive very intense flows during severe storms of short duration, which will tend to disturb the sludge in the septic tank, washing it into and through the percolating filter and producing a bad effluent.

There are, however, many instances where there are existing treatment works receiving combined drainage, or places where works have to be provided for elaborate combined drainage systems which cannot easily or at low cost be converted to separate systems. The problem then arises as to what shall be done—whether new separate surface water and soil drains should be laid (which is usually best) or whether the whole of the flow should be pushed through the treatment works to receive such treatment as can be given in the circumstances.

In such an instance, over-statement of the axiom should be avoided. Septic tanks normally have to accommodate very varied flows, and will tolerate some degree of overload. Percolating filters are not upset by occasionally receiving an excess of suspended solids, provided it is not too often, and perform at their maximum efficiency when loaded up to a rate of flow of 1000 gallons per day per superficial yard of bed. Therefore, it appears reasonable that where the run-off from a small area of roofed or paved area is discharged to the soil drains, no serious trouble need be expected.

This suggestion is borne out by the observation that two small sewage works that give excellent final effluents in dry weather receive a certain amount of rain-water. In these instances the only effect of wet weather, according to the effluent analyses, is to dilute the effluent, thereby *improving* the sample.

It is not easy to decide just where to draw the line in any particular instance, but the writer has no hesitation in suggesting that in most instances no serious harm would be done, provided that the surface area of roof or pavement drained to the works does not exceed 1 superficial foot per gallon of septic tank capacity.

Where, however, a large area, such as the whole of the drained surface area, is connected to combined drains and a flow likely to upset the septic tank is to be expected, it is advisable to reconstruct the drainage system, which usually means the laying of a new system of separate surface-water drains.

Septic tanks. Septic tanks have been wrongly constructed and operated time and time again because of erroneous beliefs as to their function. The design was originated as a means of settlement of municipal sewage and solution of the sludge problem by "digestion" of the sludge, which was permitted to accumulate over long periods in the bottom of the tank. However, experience proved that the theory did not work well in practice, for the amount of digestion was much less than was expected, with the result that municipal septic tanks became choked with heavy sludge and scum and had to be cleansed, which was a difficult and dirty process (see Chapter XXXI). This was the reason that, over forty years ago, septic tanks had already gone out of fashion as far as the treatment of municipal sewage was concerned. Their place was taken by sedimentation tanks designed to be sludged at least once a week under hydrostatic head by the process of "bleeding off" the sludge— i.e. drawing it off at a slow rate or intermittently, taking care not to draw off sewage with it.

The reason why the septic tank has remained the most favoured type of tank for isolated premises is that at the private house, the small group of houses, or the comparatively small institution, no one is willing to spend a considerable amount of time in carefully drawing off sludge from a settlement tank, and disposing it on to drying beds, from which he will have to remove it after it has dried. It is so much easier to store the sludge for long periods and then get rid of it by running out the whole quantity in one go at the end of several months. It is still easier and better for the occupier of premises if this sludging can be done by the local authority or a contractor with the aid of a cesspool-emptying vehicle, which sucks out the contents of the tank and takes them right away.

Thus, the modern septic tank is a settlement tank, made of adequate proportions to have the capacity necessary not only for settlement to take place, but also to store several months' accumulation of sludge.

However, the name and history of the septic tank have been cause for misconceptions resulting in improper design and operation. Many people concerned with design, construction and operation of sanitary works still believe that septic tanks digest a large percentage of the sludge which accumulates in them and also that they produce effluents very different in character from those of other settlement tanks.

Dealing first with the amount of digestion, the theoretical maximum that can take place is in the region of one-third of the total sludge, for usually more or less half of the solids are inorganic, and of the remaining organic matter not all is capable of being digested. According to G. B. Kershaw, Engineer to the Royal Commission on Sewage Disposal, the amount of

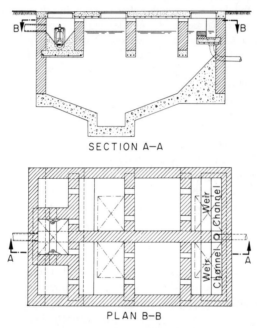

Fig. 145—Typical septic tank

digestion observed by Harrison at Leeds was 30 per cent, at Manchester by Dr. Fowler 25 per cent, at Sheffield by Haworth 32·9 per cent and at Birmingham by Watson and O'Shaughnessy 10 per cent of the suspended matter entering the tank. These were figures for municipal works; in the case of septic tanks for small works, much lower digestion percentages are to be expected if the tanks are frequently sludged, as is necessary if satisfactory effluents are to be produced.

Recently some so-called research figures were published to show that, if septic tanks were not sludged except at very infrequent intervals, a very small proportion of the original suspended matter was to be found in them. But examination of these figures shows that if the loss of solids is entirely due to digestion, not only must the whole of the organic matter have been digested but also sand and silt and similar indestructible material. It is therefore obvious that the solids must have been lost by some means other than by digestion, and in this connection it is significant that the research workers admitted that they had not been able to make a complete "solids balance" between incoming solids, solids in sludge and solids going out with the effluent: an error had crept in somewhere. Here again one can go right back to the findings of the Royal Commission as summarised by their Engineer:

> If the sludging question alone had to be considered with septic tanks, it would generally be preferable to sludge at infrequent intervals; but there are several objections to this, the chief one being that after a certain time the suspended solids issuing in the tank liquor increase largely, and cleaning out becomes a necessity, whilst the sludge in the tank becomes denser and more difficult to manipulate.
>
> It would therefore appear that some figure should be decided upon as regards the limit of suspended solids permissible in the tank liquor, and when this figure is exceeded the tanks should be cleaned out. If this is not done, the heavy increase in suspended solids will rapidly choke filters of medium-sized material or land.

This finding has been amply confirmed by the writer's observations. Of a considerable number of small sewage works with which he has had to deal, all, with rare exceptions, were found to be completely unsatisfactory when first inspected owing to neglect and improper design. In many instances, because of infrequent sludging and the use of too fine a grade of medium, the percolating filters were choked and inoperative. But after the medium had been replaced with new and coarser material, and arrangements made for the regular sludging of septic tanks at three-monthly intervals, not one instance of filter chokage or even local ponding was observed. It was also found that well-designed septic tanks, provided they were sludged as aforementioned, gave effluents lower in suspended solids than is usually claimed for municipal sedimentation tanks. If, however, a tank was not sludged according to programme, the effluent eventually deteriorated because sludge was escaping with it.

The reasonable interpretation of the foregoing is that while some digestion may take place in septic tanks, the reason for there being little sludge left in a tank that is left for a long time without sludging (and, incidentally, in tanks of insufficient size or of poor design) is that the sludge is escaping with the effluent, to the detriment of treatment. It follows that septic tanks should be designed so as to be able to store the whole of the sludge likely to be settled in the interim between sludgings, and that they should not be left unsludged for longer than the design periods.

The second (in this case partly) erroneous belief is that septic tanks give some form of treatment to the sewage other than separation of suspended solids. According to some, the change is one which renders the sewage more easy to treat on percolating filters; others say that septic-tank effluent is offensive, in fact acid and septic, and more difficult to treat than the effluent from a frequently sludged sedimentation tank. These opinions are diametrically opposed, and therefore one must be wrong, perhaps both. And on this matter no one can speak with absolute authority, for there are no side-by-side tests of the two types of tank treating the same effluent. It is certain that the septic tanks must have some effect other than settlement, for such little digestion as may take place effects oxidation by taking oxygen from the water but also produces from the solids content a strong liquor which will pass out with the effluent. Septic tanks are generally of larger relative capacity than sedimentation tanks and, for this reason, if sludging is not neglected, and also provided they are properly designed, they effect more efficient settlement than the average settlement tank.

As to actual observation of the performance, it should be borne in mind that, after consideration of all the evidence submitted, the Royal Commission recommended the same loading rates in gallons per day per cubic yard of filter medium for filters treating septic-tank liquors as for filters treating the effluents of continuous-flow sedimentation tanks of large capacity—i.e. 15 hours' dry-weather flow. On the basis of this, it would appear that at the present time, when sedimentation tanks are made of small capacity and septic tanks can be designed and operated to higher standards of efficiency than formerly, septic-tank effluent should be as good as sedimentation-tank effluent. Here again, the writer, like all other writers on the subject, cannot quote direct comparisons made under test conditions, for such do not exist, but he can say that small sewage works consisting of septic tanks of good design and percolating filters of normal proportions, together with efficient humus removal, can secure final effluents better than Royal Commission standard and equal to the best at municipal works.

As to septicity of septic-tank effluent, a number of chemical analyses have been examined for tanks varying in capacity from 36 to 56 hours' dry-weather flow (actual). In no case was there any nuisance, and in no case was an acid effluent found, in all circumstances the effluent being virtually neutral or slightly alkaline, and generally of almost exactly the same pH value as the crude sewage. This should finally dispel the belief in septicity of effluents from tanks of large capacity dealing with sewages from isolated premises.

A septic tank, to be efficient for sedimentation purposes, should be of sufficient capacity. It should have satisfactory inlets and outlets. It should be so proportioned that sludge can settle at the bottom and scum accumulate at the surface, while space is left for the sewage to flow through without unduly disturbing sludge and scum. In order that this space is not taken up by sludge and scum, the tank should be of sufficient capacity for the sludge and scum to occupy not more than, say, one-half of the total capacity at the end of the storage period.

Dealing first with capacity, analysis of crude sewage from purely domestic premises shows that the amount of solids likely to be brought into the tank is almost exactly 1 gallon of dry matter per head of population per three months. If a settlement efficiency of 70 per cent is assumed, digestion of 25 per cent of the solids and an average moisture content of the settled sludge 95 per cent, the quantity of sludge likely to be stored in three months will

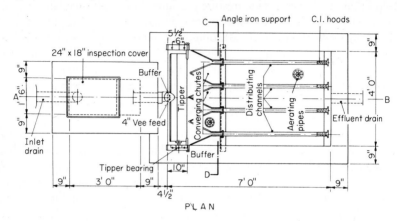

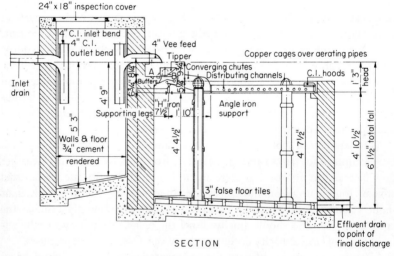

FIG. 146—Septic tank and percolating filter
By courtesy of William E. Farrer, Ltd.

amount to no less than about 10 gallons per person, which means that the septic tank should have a capacity of not less than 20 gallons per head of population served if sludging is at three-monthly intervals, 40 gallons if sludging is at six-monthly intervals, or 80 gallons if sludging is at twelve-monthly intervals; any lower capacity must tend towards deterioration of effluent towards the end of the storage period.

The Royal Commission recommended that septic tanks should have capacities of 24 hours' dry-weather flow on the basis of a sewage flow of 40 gallons per head of population per day. This Royal Commission recommendation should just about serve in such circumstances and on the assumption that sludging is not less frequent than at six-monthly intervals. But it was based on municipal conditions and higher rates of flow of sewage per head of population than are usually discharged from isolated premises. More recently, septic tanks have been designed on the basis of 24 hours' dry-weather flow, but at lower rates of flow per head of population per day, and sludging periods of six months or over. Simple arithmetic shows that this cannot give good results.

It is for these reasons that the writer has recommended that septic tanks should be sized not in terms of hours' dry-weather flow, but according to the population and the length of the sludging period. The figures already quoted as based on analyses showed that a septic tank of *at least* $\frac{2}{9}$ gallon per head of population per day of storage period is essential. But having regard to the common experience that sludging is seldom carried out with absolute regularity, because the occupier of the premises forgets to give the order, the owner of the tank-vehicle is unable to carry out the service just when needed or for some other reason, the writer suggests, as a general rule, that the capacities of septic tanks should be greater than this figure. A figure which has been used with absolute satisfaction and, for good work, is not too lavish is to allow for a tank capacity of $\frac{2}{5}$ gallon per head of population per day of sludge storage. This figure has been used for some years, and as a result of observations and tests it can now be said that satisfactory septic tank settlement can be expected if the tank is made to hold $\frac{2}{7}$ gallon per head of population; any capacity much smaller than this is liable to produce an inferior effluent.

In addition to the size of tank calculated on the basis of storage of sludge, there is a minimum capacity which will give satisfactory settlement. The rate of flow from isolated premises is irregular owing to the discharge of water-closets, baths, etc., and the contents of a very small tank are liable to be stirred up by these discharges to the detriment of tank effluent. Putting aside overseas practice, which is not applicable to English conditions, the minimum capacities that have been recommended by different authorities range between 1000 and 500 gallons for the smallest premises and *Building Regulations* now require 600 gallons. But these figures are based on the merest guess. As all engineers who have made a scientific study of the theory of sedimentation know, there is no critical capacity at which a marked change occurs in settlement efficiency as capacity is changed, for the curve of detention period related to suspended solids settled is virtually a straight line when plotted to exponential scales. It follows that it is unsound to talk about a strictly scientific approach to an exact size of tank from this point of view, and therefore the writer prefers a purely practical approach and recommends the acceptance of any figure between the limits of 1000 and 500 gallons, and suggests that, if the tank is to be sludged by cesspool-emptying vehicle, it

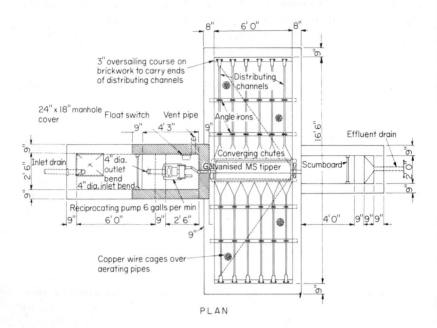
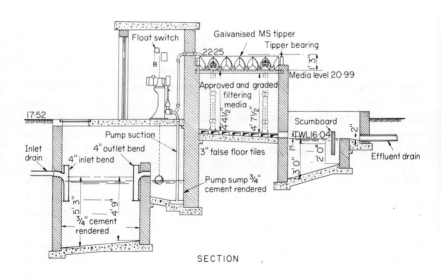

Fig. 147—Septic tank, automatic pump and percolating filter
By courtesy of William E. Farrer, Ltd.

SEWAGE DISPOSAL FOR ISOLATED BUILDINGS 451

should be of the same capacity as the vehicle or a multiple thereof. This is for two reasons: first, the contractors charge the same for full or partial loads, and if the tank does not contain an exact number of loads, the householder pays for sludge that is not taken away. Second, when the tank is equal to the capacity of the vehicle it is possible to check that the tank has been actually emptied: scamping of the work of tank-emptying is very prevalent.

The inlets of septic tanks should be so designed that the velocity of inflow *never* exceeds 8 inches per second. To make sure that this condition is ensured, the incoming drain should enter a chamber or baffle-box with out-

TABLE 59

APPROX. HEAD OF POPULATION SERVED BY SEPTIC TANKS

(Based on $\frac{2}{5}$ gallon per head per day)

Periods between times of sludging, months	Capacity of septic tank, gallons			
	750	1500	2250	3000
6	10	20	30	40
4	15	30	45	60
3*	20	40	60	80
2	30	60	90	120

TABLE 60

PROPORTIONS OF SEPTIC TANKS

Total approximate capacity		Number of tanks	Dimensions of each tank in metres		
gallons	cu metres		length	breadth	depth
750	3·4	1	2·4	0·8	1·75
1500	6·8	2	2·4	0·8	1·75
2250	10·2	2	2·86	1·03	1·75
3000	13·6	2	3·44	1·14	1·75

lets below, and certainly not above, top water-level of tank. On no account should the incoming pipe discharge direct into the tank above top water-level. Dip-pipe inlets can be used, but these should connect from a chamber with their inverts below top water-level of tank.

Dip-pipes will also serve as outlets for small septic tanks, but weirs, with scum-boards carried to 2 feet below top water-level of tank, are to be preferred, provided that the wall of the tank slopes from the weir steeply down under the scum-board in such a manner as to prevent local settlement of sludge that may eventually rise and pass over the weir (see Fig. 145).

To facilitate sludging, the floor of the tank should fall to a low point or sump as near the centre as possible.

* Usual period

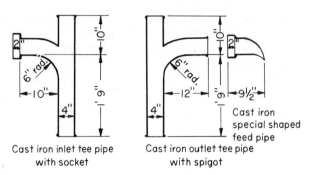

FIG. 148—Cast-iron dip-pipes for septic tank
By courtesy of William E. Farrer, Ltd.

It is usual to cover septic tanks to prevent people from falling into them and to conceal their contents. But covered tanks should be well ventilated to prevent accumulation of explosive gas, and access manholes should be provided to expose inlets and outlets and to permit the dropping-in of suction pipes for sludging.

Sludging. There is a fact in the sludging of sewage tanks which is of primary importance. When sludge is drawn off from the bottom of the tank, at first the small quantity of sludge in the immediate vicinity of the outlet or suction pipe comes away, after which *sewage* is drawn off because the sludge, being only slightly heavier but much more viscous than the sewage, lies in banks away from the point of outlet and the scum remains floating on the surface. With continued draw-off more sewage is removed, until finally only sludge and scum remain in the tank. These come off last, and then only if there is sufficient slope on the floor of the tank for them to gravitate to the outlet. This is the reason for the slow bleeding-off of sludge from steep-bottomed pyramidal sedimentation tanks and for the need for the sludging of septic tanks by *complete emptying*. If septic tanks are sludged by partial removal only of the contents, they become more and more filled with sludge and scum, and the quality of the effluent falls off.

Frequently it is found that virtually no sludge settles in the septic tank but all the suspended solids collect as a thick blanket of scum at the surface. In such a case it is obvious that the tank cannot be sludged by drawing off part of the contents from the bottom: it must be emptied of sewage before the scum can be finally withdrawn.

The sludging of septic tanks and comparatively flat-bottomed settlement tanks under hydrostatic head by means of a sludge pipe which collects from the lowest point in the tank but discharges at a higher level, so that the tank cannot be emptied, is far too common and very unsatisfactory. It has been perpetuated by illustrations in various publications, including the catalogues of manufacturers. Such an arrangement should never be installed.

By far the most satisfactory means of sludging a septic tank is total removal of contents by tank-vehicle. This is particularly satisfactory to the occupier of premises if the local authority is under obligation, under the *Public Health Act*, 1936, to empty cesspools, for then the authority must empty septic tanks when called upon to do so, for septic tanks are legally cesspools. In such circumstances the reasonable procedure of the designer of works is to ascertain the routine of the local authority—for many authorities work to a three- or four-month sludging period—and determine the size of the tank accordingly. If the authority is not under an obligation, the tank can be emptied by contractor.

If a septic tank is to be emptied by vehicle, the tank must, as mentioned earlier in this chapter, be located in such a position that the vehicle can get to it. The septic tank must be placed near to a hard road that will bear the weight of a heavy tank-vehicle, or a road must be constructed for the purpose. There have been many instances where this has been overlooked and tanks placed in positions where they cannot be sludged at all, or at best only during dry weather when vehicles can run over soft ground.

At all except the smallest works with the minimum size of tank equal to that of the emptying vehicle, there should be two tanks so that one may be emptied while the other is left in operation. It is a serious drawback at comparatively large works if there are not two tanks that can be isolated in turn, for the rate of flow of incoming sewage is sometimes so great that the tank-vehicle cannot get the sludge away fast enough to be able completely to empty the tank. If emptying is by contract, this means that the occupier has to pay for more loads than would otherwise be necessary. Where two tanks are provided, their inlets should be controlled by penstock set partly below top water-level, and their outlets should be weir outlets, carefully adjusted to equal level after the tanks have been filled with water, for if one weir is lower than the other, one tank will do all the work. One way of adjusting weirs is to set them level and then grind them down with carborundum stone or rub out the mortar joints until both weirs appear to be discharging equally.

Sewage works are, of course, placed downhill of the premises served whenever possible, in order to avoid pumping. But there is no need for septic tank and filter to be close together: the septic tank can be near to the road and the filter much farther downhill, away from the premises.

When emptying by vehicle is out of question, where the premises are away from other properties, and if there is enough land for the sewage works to be well away from the buildings and for sludge to be disposed of locally, the sludge can be discharged to lagoons for drying prior to being dug out and used as fertiliser. Sludge will dry reasonably well if discharged into shallow lagoons excavated in porous subsoil, into trenches or into the furrows of a ploughed field. But the capacity of the lagoons or trenches must be *at least equal to the total capacity of the septic tank*, otherwise the tank cannot be emptied;

and of course the lagoon or trenches must be dug out and dried sludge removed before the next sludging.

If a septic tank cannot be emptied to lagoons, trenches or on to the land by gravity it has to be pumped out. It should hardly be necessary to mention that where there is sufficient fall for gravitation of tank effluent through a percolating filter, there is usually also sufficient fall for gravity sludging and therefore, in most instances, gravity sludging is possible. Where, however, it is necessary to pump settled effluent to the percolating filter, sludge may have to be pumped to the lagoon or trenches, and as sewage pumping usually means electric pumping, it follows that an electric pump should be installed if current is available. In far too many instances, septic tanks have been fitted with hand-operated pumps, and as the emptying of a large septic tank involves some hours of hard labour, the inevitable result is neglect of sludging, with consequent choking of filter and the production of a bad effluent.

Not all pumps will handle sludge, because large solids tend to interfere with the working of the valves. Some reciprocating pumps are quite useful, including the contractor's diaphragm pump. But by far the most common installation is the chain pump, which is a permanent fixture built into the septic tank and worked by rotating a handle or by power. Chain pumps consist of an endless chain carrying discs which are drawn up a vertical tube and bring with them liquid and solids (see Fig. 144). They are unchokable and very reliable. The base of the pump is mounted on the concrete slab that covers the septic tank, and the outlet spout is usually about 3 feet above this. From this spout the sludge can be delivered to a near-by lagoon by means of timber troughing, or the sludge may be discharged into a portable tank. Table 61 gives the approximate deliveries of chain pumps.

TABLE 61

DELIVERY OF CHAIN PUMPS

Diameter of barrel (inches)	2	$2\frac{1}{2}$	3	$3\frac{1}{2}$	4	$4\frac{1}{2}$	6
Gallons raised per hour at 60 revs per minute (approx.)	780	1200	1750	2400	3100	3950	7000

Converting cesspools. By slight modifications an existing cesspool can be converted into a septic tank for use with a new percolating filter. If there is sufficient fall for the percolating filter to be fed by gravity and drained by gravity to the point of outfall, all that is necessary to convert the cesspool is to reconstruct the inlet so as to prevent disturbance of the contents and, on the opposite side from the inlet, insert a dip-pipe, or break out the wall and construct a weir protected by a scum-board and discharging to a channel or a chamber, from which a drain can lead to the filter.

If, owing to the lie of the land, it is necessary to pump the tank effluent, the wall opposite the inlet can be broken out, a weir constructed protected by a scum-board carefully arranged so that scum cannot rise underneath it, and a small chamber built in which the tank-effluent can collect. In this chamber a small automatic electric pump can be fixed and protected by a wooden or brick house. (Suitable types of pump are described later.)

Percolating filters. Percolating filters used for small sewage works are similar in general principle to those for municipal works, differing only according to the scale of the works and the circumstances in which they are used.

The Royal Commission on Sewage Disposal recommended capacities of filter medium for various strengths of sewage and assuming the condition that treatment works would deal with flows up to three times dry-weather flow. Small domestic sewage works do not have to deal with prolonged discharges at rates up to three times dry-weather flow, and although they may have to cope with hour-to-hour variation of flow and strength, the use of large-capacity septic tanks evens this out. On the other hand, small private works have to withstand occasional severe differences of strength and character of sewage, for such contingencies as the heavy use of disinfectants during a case of illness, or the throwing away of some unwanted household fluid or of waste food, can put a very heavy load on the percolating filter or militate against its operation. Again, the distributors used for the smallest types of percolating filter do not always give even distribution over the surface, so that part of the medium is not effectively dosed. In some instances the author found that two-thirds only of the medium was receiving flow. Furthermore, filters for small works are often constructed to too shallow a depth to be efficient.

Practice for many years has been to construct very small percolating filters on the basis of $\frac{1}{2}$–1 cubic yard of medium per person served, according to the need for a high-quality effluent; this compares with the Royal Commission recommendations, which average about 0·533 cubic yard per person.

For many years media of 1–1$\frac{1}{2}$-inch grade were generally accepted for normal, straight-run percolating filters either at municipal or small domestic works. Smaller media were considered to be more efficient for treating weak sewages because of the larger surface area of particle exposed to the air: the Royal Commission found that coarser medium gave better results with strong sewages, presumably because of the better ventilation of the mass of the bed that the larger particles permitted.

In examining a considerable number of existing small works, most of which were performing very badly, the writer found that one of the most common faults was a choked condition of the beds. This was, in part, due to the common practice of leaving septic tanks unsludged for too long a period, but the trouble was augmented by the small particle size of the media. Also, in many of the works which were inoperative or giving very inferior

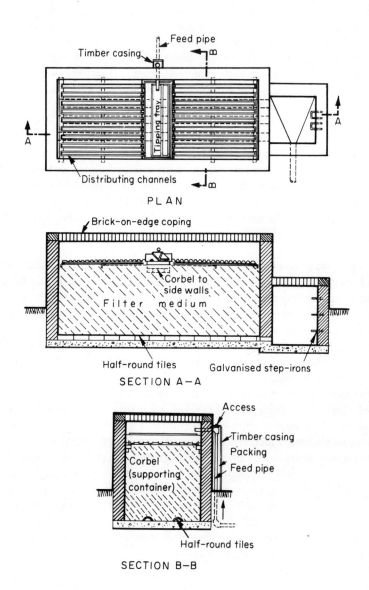

Fig. 149—Small rectangular percolating filter with Tuke and Bell tipping-tray mechanism

effluents, the beds were graded from comparatively coarse at the bottom to fine at the surface, so that the percolating filters were, in fact, *filters* tending to separate out suspended solids, producing ponding at the surface and chokage some inches below. He therefore decided to adopt a standard particle size in the region of $2\frac{1}{2}$ inches, to take special care that no dust was included with the medium and to allow a minimum of $\frac{2}{3}$ cubic yard per person.

Accordingly, wherever existing works were sufficiently well designed or of sufficient size to be capable of being made satisfactory by modification, the media were removed and replaced with new material consisting of particles of at least 2-inch size and of even size from the top to within 9 inches of the bottom, where a layer of large stones or clinkers was provided to assist ventilation. Where works were not capable of modification, entirely new installations were built and in these, the same types and grades of media were used.

The result of this policy was to improve the performance of existing works even when the quantity of medium was below standard. Some of the new works gave excellent results. In not one instance has ponding or chokage been observed at any of the works constructed or modified in this way.

A further cause of inferior performance of existing small sewage works is improper design of the filter. Many percolating filters are built without sufficient ventilation at the bottom and are provided with cover slabs or boards to hide the mechanism of the distributor. In the course of reconstruction, it was found necessary to provide air shafts to take air down to the bottoms of existing filters, to lay additional agricultural tiles underneath the medium, and to remove any cover slabs that had been provided.

From the foregoing observations the following recommendations are made. Small percolating filters should have capacities not less than as given in Table 62. The medium should be of at least 2-inch size particle to as much as 3-inch, of even size throughout the bed (apart from the large material in the bottom 9 inches) and absolutely free from dust. (There is no need to specify medium in accordance with the existing British Standard, for the tests included rule out some quite satisfactory material, and in any case the laboratory tests are far more elaborate than the quantity of material used justifies.) Generally, the material should be hard gasworks clinker, or broken stone of cubical—not flaky—shape, free from all soft material. It is impracticable to call for this material to be free from dust as delivered to the site, but essential that all dust should be removed on the site by passing the medium over wire sieves, for even small percentages of dust tend to accumulate in layers which interfere with the ventilation of the bed.

The percolating filter should have a concrete floor, and on this either a false floor of filter tiles or bricks should be constructed, or a layer of large stones with lines of agricultural drains at not more than 3-foot centres. The medium should be handled with forks, not shovels, and not walked over any more

458 PUBLIC HEALTH ENGINEERING PRACTICE

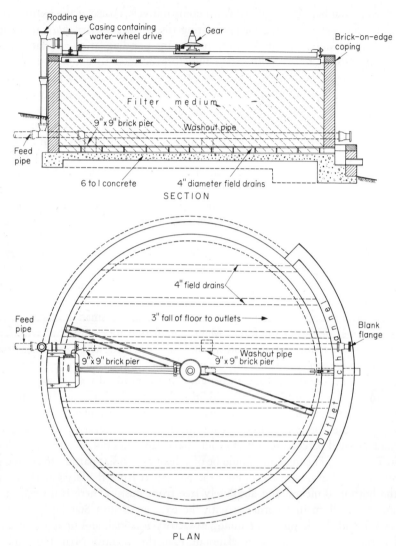

Fig. 150—Small circular percolating filter with Ames Crosta Mills rotating distributor

(Works consisting of the septic tank shown in Fig. 145, the above design of filter and a small irrigation area have given superb effluents)

than is absolutely necessary. The walls of the filter should be of solid (not honeycomb) brickwork and carried up to the level of the top of the medium, or higher if the type of distributor mechanism permits this. The ends of the under-drains should discharge to a chamber or channel giving ample ventilation, or connected to vertical pipes brought up to above ground-level outside (not inside) the bed.

SEWAGE DISPOSAL FOR ISOLATED BUILDINGS 459

The depth of bed should not be less than 1¼ metres or so great as to make maintenance difficult. There is, however, no danger of making a bed of coarse medium too deep to work properly, for no one has yet made a deep filter which was not more efficient than one of shallow depth. On no account should percolating filters be covered over to conceal the mechanism or to protect the bed, except by fine wire netting to keep out leaves where the beds are unavoidably constructed under trees (see Fig. 153).

The type of distributor mechanism determines the shape of the bed. For very small beds, fixed distributors are available which consist of cast-iron distributor channels with V-notch outlets that are dosed by automatic

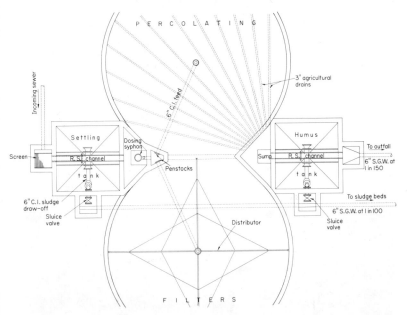

FIG. 151—Small sewage-treatment works suitable for an institution, village or camp

tipping-tray mechanisms. This dosing is essential to all fixed distributors except those receiving flow by pumping, for without dosing the small rate of flow through the distributor channel results in the greater part of the flow escaping by one of the V-notches only, the lowest. There is one disadvantage of tipping-tray mechanisms: although usually fitted with rubber buffers, they are not absolutely silent, and cause annoyance at night, particularly when the buffers fall out.

Where distributors of this kind are used, the percolating filters are rectangular, and the dimensions of length and breadth have to be chosen according to the dimensions of the proprietary mechanism selected.

When a sufficient quantity of medium is required to make possible the construction of a circular bed of about 10 feet diameter, or larger, it costs less

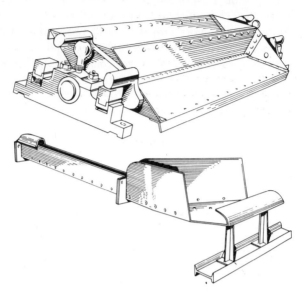

Fig. 152—Tipping-tray mechanism and one of the distribution channels to which tipping-tray discharges

By courtesy of William E. Farrer, Ltd.

Fig. 153—Wire-netting cover to percolating filter to keep out falling leaves

By courtesy of William E. Farrer, Ltd.

to use a rotating sparge or distributor. There are several mechanisms of this kind, most of which are made to serve beds from 10 to 20 feet in diameter; Hartleys (Stoke-on-Trent), Ltd., make distributors for beds as small as 4 feet 6 inches diameter. Above 20 feet diameter, the type of mechanism is similar to that used at municipal works, and has perforated sparge pipes which need cleaning out more frequently than is always practicable at private premises. However, many of the mechanisms made for beds up to 20 feet diameter have open-trough sparges with V-notch outlets that do not tend to choke.

TABLE 62

MINIMUM CAPACITIES OF SMALL PERCOLATING FILTERS

Minimum depth in metres	Minimum cubic metres per person served	
	Revolving distributor	Tipping tray
1·85	0·4	0·6
1·65	0·45	0·675
1·50	0·5	0·75
1·35	0·55	0·825
1·25	0·6	0·9

Rotating distributors are rotated by the flow of sewage and need in the region of 15–20 inches head of water for their operation. Whereas sparges of the municipal type for large beds are mostly rotated by reaction of the jets and, in order that they shall not stand still and dribble when flow is slack, have dosing tanks and siphons to operate them, smaller sparges have dosing tanks and siphons built into the mechanism at the centre of the bed. Rotating open-trough channels are operated by water-wheels and similar devices. Distributors are either over-fed—i.e. the tank effluent is brought over the bed and discharged into the central vessel of the distributor; or under-fed, in which case the tank effluent is brought to the distributor via a pipe carried through the medium and connected to the central column up which the tank effluent rises.

All distributing mechanisms need maintenance. Generally, a weekly inspection, oiling and clearing of any chokage suffices. The beds require to be weeded and cleared of fallen leaves.

Humus removal. An effluent of Royal Commission standard cannot be assured on all occasions unless something is done to remove the suspended solids that come out of the percolating filter. Samples from percolating filter effluents from which humus has not been removed are sometimes virtually free from suspended solids; others may be heavily loaded. It is a matter of chance whether or not a clot of humus happens to be coming away at the time that the sample is taken. And when there is much humus in the sample, not only is the analysis bad in having a high suspended-solids figure, but the B.O.D. value also is high because of the oxygen demand of the humus.

Humus is usually removed at small sewage works either by settlement in humus tanks or settlement by irrigation over ploughed land or through scrubby vegetation. Humus tanks, which at one time were considered a normal component of any sewage works, fell into disfavour, as far as small private installations were concerned, because if badly neglected they made the effluent poorer than it would have been had they not been there, and such tanks were often badly neglected. But, as in other aspects of this study, over-statement should be avoided, and humus tanks should not be condemned out of hand. For the fact remains that the humus tank of a small installation can give excellent results if cleaned out once a week, and may even perform reasonably well with less attention: it is excessive neglect that causes trouble. As an instance, a plant which was visited only once a week and had its humus tank sludged even less frequently gave, on the average, an effluent well inside the Royal Commission standard, and frequently the suspended-solids figure was recorded as nil.

Humus tanks *should* be provided whenever septic-tank effluent has to be lifted any considerable height to the filter, for in such circumstances the humus tank can be 2 or 3 feet above the septic tank, and then sludging under hydrostatic head is simple, for the tank can be sludged merely by opening a valve that connects from the bottom of a pyramidal humus tank and discharging the humus to the inlet manhole of the septic tank. Such easy humus removal is not so likely to be neglected as is sludging by pumping out or, as has often been necessary, baling out.

There are no exact data available as to the capacity and surface area of humus tanks, and practice has tended to change over the years, so that at present it is usual to provide, for municipal works, tanks with a capacity of one-sixth of a day's flow. This is probably excessive; but the figure should be satisfactory for application to small private works where any tank of smaller capacity might be of negligible size.

Many existing humus tanks are flat-bottomed and rectangular, and have no proper arrangement for sludging. A tank in the form of an inverted pyramid with sides sloping at 60° to the horizontal and provided with a properly baffled inlet is much better. The sludge should be drawn off from the bottom of the hopper under gravity or by pumping and discharged to trenches for drying. As it is a good manure, it should be preserved for use on the land.

There is a danger of humus gassing and rising to the surface, then passing over the outlet weirs. For this reason weir outlets should be protected by scum-boards. These need not be carried as deep into the tank as they are in septic tanks—a few inches' submersion is adequate. Alternatively, a dip-pipe may be used at very small installations, but preferably not one of less than 6 inches internal diameter.

Removal of humus by irrigation of the effluent over grass land is very effective indeed. On this matter again, no accurate data are available; but the writer can say from first-hand experience that in all instances where the total area of land made available for the purpose has been 4 superficial yards per head of population served, the results have been excellent. A smaller

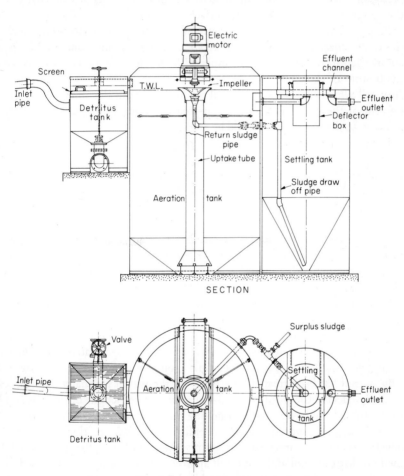

FIG. 154—Simplex aeration plant for an institution
By courtesy of Ames Crosta Mills & Co., Ltd.

area would probably be satisfactory (see "Effluent polishing," Chapter XXIX).

The filter effluent should be distributed over the land by means of a ditch dug along a contour and provided with controllable outlets, such as agricultural tiles, set to equal levels and partly plugged with clay so as to adjust the flow. If the irrigation area is ploughed, the furrows should run truly along the contours. This work should be carefully watched by the engineer or architect responsible, for ploughmen are seldom used to contour ploughing. But by far the greatest trouble is the man who thinks he knows better than the designer and ploughs straight down the hill, so that the water can get away easily. This is just what is not wanted: the purpose of the contour ploughing is to hold up the effluent as long as possible so that it will settle-out the humus and not scour away the surface of the ground.

Effluent from irrigation areas sometimes contains a little top-soil which spoils the chemical analysis. To avoid this, a sampling point should be constructed in the form of a manhole with an inlet pipe raised a bit above the bottom of the ditch which collects the effluent from the irrigation area. Any soil which is washed away will tend to collect in this ditch, but if the bottom of the ditch rises, so as to approach the invert of the pipe which connects to the manhole, it should be dug again.

Land treatment. The land treatment methods of broad irrigation and land filtration are not favoured for use in connection with isolated buildings, because strong septic-tank liquor is particularly liable to be offensive when treated on the surface of the ground. Also land treatment necessitates intelligent attention that is seldom given where a full-time operative is not employed.

The reasons for adopting land treatment at small private works at all are that it does away with the necessity for pumping when there is little or no fall to play with, and it avoids the cost of construction of a filter.

For details, reference should be made to Chapter XXII.

Sub-irrigation. Sub-irrigation is not to be confused with land treatment, for it is soaking-away of tank effluent below the surface of the ground through a system of field drains. Even if carried out extensively, the amount of treatment that it can give to the sewage is very speculative, and often negligible, and, therefore, it is legitimate only as a means of disposal of effluent that has been treated on a filter. In the way it is usually applied, it certainly effects little purification, because no greater length of drain is laid than is thought to be necessary for getting rid of the tank effluent by soakage.

Sub-irrigation appeals to the unenlightened and to those property developers who are interested in getting away with the cheapest possible method that the local authority will approve. That is why it is still used in this country in those localities where the local government officers are not well informed on sewage treatment. It is no doubt a legitimate means of disposal of sewage in undeveloped countries that are sparsely populated and where the soakage of untreated sewage is not likely to be a nuisance, cause of complaint or danger to health. But in England the method should not be used or accepted, for although in many cases it does not create an obvious nuisance, it is usually a potential danger to health.

Activated sludge methods. Activated sludge methods are rarely applied to small flows from isolated buildings, because they require intelligent attention. A portable plant is manufactured by Messrs. Ames Crosta Mills & Co., Ltd., by which the "Simplex" method can be applied to small sewage flows and there are some non-portable plants of this make (see Figs. 154, 155).

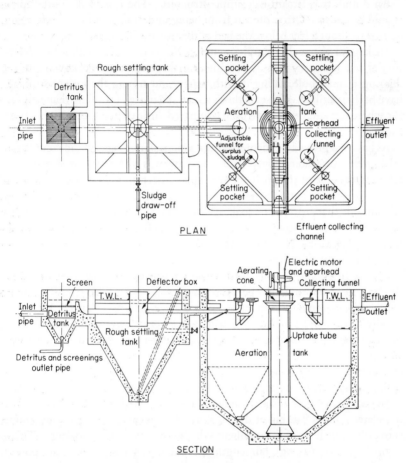

FIG. 155—Simplex aeration plant for an institution
By courtesy of Ames Crosta Mills & Co., Ltd.

Of recent years a number of proprietary "packaged plants" have come on the market. These treatment works, usually with iron aeration and sedimentation tanks, can be installed with the minimum of building work. The principle of treatment involves extended aeration by activated sludge treatment continued at high rates so as to effect aerobic digestion of organic solids.

Final discharge. A very simple rule can be used with, of course, discretion, in deciding on the outfall. If there are watercourses, the finally treated effluent should be discharged to them: if there are none, the fact can be taken as evidence that the nature of the subsoil is such that all surface-water soaks away; and in this circumstance some form of soakaway may be the only practicable means of disposal of the effluent.

An outfall to a stream is a simple structure. The end of the outfall pipe should be concreted to protect it from being undermined, the concrete being carried to about 2 feet below the bed of the stream.

Soakage is not such a simple matter. Of recent years, methods of determining the necessary area of bottom of soakage trench or soakaway pit, by digging a small hole, filling it with water, and timing the rate of soakage, have been described. These methods tend to be more misleading than helpful, for where test and result have been compared, little or no agreement has been found.

As has been proved by theoretical and practical investigations in connection with wells for water supply (and a soakaway is virtually a well acting in reverse), the size of the pit has a very small affect on the rate of soakage. It is the local geology and the level of the water-table, as well as the porosity of the subsoil, that matter. In soaking away sewage effluent, there is another factor to be taken into account—the choking effect of the suspended solids that escape with the effluent. If there is much humus in the effluent, a large soak pit may quickly choke.

For the foregoing reasons, no satisfactory rule can be recommended as to the size of soakaways except for surface water (see Chapter VI). There are, however, some rules of design which are worth considering. First, it is usually a waste of money to construct an elaborate system of parallel drains or "herring-bone" drains, because a single ring-main surrounding the same area of land is just as effective (when first put in) as the most elaborate and costly grid system that it could enclose. The reason for this is that when a soakaway system has been in operation for some time, the level of the water-table rises to the level of the drainage system. Throughout the irrigation area the water-table is flat, but at the outskirts it slopes away at a gradient, and it is the possible slope of this gradient which limits the rate of soakage. There is some virtue in having a fair length of drain other than that necessary purely for enclosing the areas, as a long drain will take longer to choke than a short one.

Dumb-wells or soakaway pits are very useful for dealing with run-off of rainfall, but can choke quickly when used for sewage effluent. They have the one virtue of being easy to dig out and enlarge. But there is a limit to this activity, and there have been instances where, when one soakaway has been choked, another has been constructed farther downhill, and so on, until the property owner had no more land on to which he could extend.

In chalk areas it is not advisable to spend any great amount of money on soakaway arrangements until the works have been in operation for some time and it has been ascertained whether or not the effluent can find a natural channel. In places where the chalk is naturally shattered there are crevices in the subsoil, and if the effluent is discharged merely on to the surface of the ground, it will most probably find one of these and make its way down. In fact, this has happened in *all* except one of the cases which the writer has observed, even where large soakaway pits have choked.

Pumping. A percolating filter scheme, for easy construction, requires a minimum fall of about 8 feet between invert of incoming drain and highest water-level in the stream to which the outfall discharges, or an absolute minimum of something more than 6 feet. The greater part of this head is taken up by the depth of the percolating-filter medium, about 2 feet for the fall between top water-level of septic tank and top of medium, including the head which operates the distributor mechanism and the remainder in various losses such as through humus tank or over land, and in channels and pipes. Where there is not sufficient head for gravity through the system, pumping becomes necessary.

Electricity is available in many parts of the country even where public water supplies have not extended. Electric pumping is most convenient at small sewage works, for it makes possible automatic starting and stopping of pumps, thereby doing away with the need for storage of sewage, hand-starting of pumping plant and other technical difficulties (which will be discussed).

A mistake that has occasionally been made is to provide a pumping station to which the crude sewage is drained and whence it is pumped to the septic tanks. This should not be done unless absolutely unavoidable, for the pumping of crude sewage necessitates the installation of large pumps which can handle solids without choking. Such pumps are expensive, and a complete station, consisting of electric pumps in duplicate, costs well over £1500. The main technical disadvantage of the arrangement is that the high rate of discharge of the pump stirs up the septic tank to such an extent that even when the tank is large, as, for example, when installed to serve a population of 100 persons, sedimentation efficiency may be reduced to as low as about 50 per cent.

By far the more usual and correct arrangement is to place the septic tank in a convenient position to which the drains can gravitate, to collect the tank effluent to a small suction-well and pump to the percolating filter, which can then be arranged at any convenient level. Efficiently settled sewage is easily handled by the small vertical-spindle automatic electric pumps, sometimes known as "cellar-emptying" pumps, which can deliver upwards of 10 gallons a minute. These installations are very low in cost. They consist of a centrifugal pump which is submerged below the level of the sewage in the suction-well, a vertical spindle connecting to an electric motor supported over the suction-well, a float and switch, the whole embodied in one unit which is easily installed. The rising main of such a pump can consist of about $1\frac{1}{2}$-inch internal diameter galvanised-iron barrel or larger according to the rate of flow: as a rule, the diameter of the pipe in inches can be taken as the square root of the pumping rate in cubic feet per minute. A small house needs to be built over the pump (or duplicate pumps, for it is always advisable to have a stand-by) to protect the motor and switchgear from the weather.

It is of great importance that, when tank effluent is pumped to a filter, the quantity pumped at one time is not too great. If the suction-well is too large, or the float gear badly arranged, the equivalent of *several hours'* flow may be pumped away in a matter of *minutes*, so that the settled sewage floods through the filter, receiving little or no treatment. Neglect of this factor has been the cause of unsatisfactory effluents in a number of instances. Wherever practicable, the suction well should be so sized, and the cut-in and cut-out levels of the float gear so arranged, that the quantity pumped at any one time does not exceed 2 gallons for every superficial yard of filter-bed surface.

The delivery from the automatic electric pump can discharge to the dosing mechanism of the filter distributor, if this has been provided. But often, when sewage is pumped, the pumping in itself provides the necessary dosing, and therefore in many instances dosing mechanism can be omitted, with a small saving of cost.

Rising mains and all pumping plant should be given adequate protection against frost.

When electricity is not available, pumps driven by internal-combustion engines have to be used, and consequently hand-starting and stopping becomes necessary, because automatic control of small petrol and oil engines is not a practical proposition. In order that it will not be necessary to attend to the plant more than once a day, the flow should be collected into a tank capable of taking a whole day's flow. By far the best arrangement in these circumstances is to pump fully treated effluent that has passed through the percolating filter: the storage tank or suction-well, if properly designed, can serve as humus tank. This may involve some minor difficulties or cost in the arrangement of the percolating filter somewhat deep in the ground, but it overcomes other technical difficulties, for the pumping of a large quantity at a time, after treatment, can have no adverse effect on treatment. Moreover, when effluent is discharged to a stream in a large dose instead of steadily throughout the day, the effect on the stream, should the effluent contain suspended matter, is much less noticeable—a practical point from the householder's point of view.

The alternative to pumping after complete treatment is pumping settled tank effluent to an overhead storage tank from which the settled sewage is fed at a steady slow rate of flow to the percolating filter. In practice, it is not too satisfactory, for modules for controlling very small flows of liquid not altogether free from suspended solids are liable to choke or otherwise malfunction, and need constant attention. Also, some of them freeze up very easily in cold weather. If sewage is pumped to storage prior to sedimentation, the difficulties are similar to, but greater than, those encountered when settled tank effluent is pumped.

Example calculation. The following is a calculation for what is at present a very common case—the use of a large isolated country house as a small boarding-school or institution.

Particulars
Premises: Boarding-school
Population: 68, including 25 staff
Quarterly water usage as measured by meter: 136,500 gallons
= 1495 gallons per day
= 22 gallons per head per day
Nearest sewer: 1½ miles distant and cannot be reached by gravity

Local authority have cesspool-emptying service and are obliged, under the *Public Health Act*, 1936, to empty cesspools. Their usual routine is to empty all cesspools once every three months.
Capacity of cesspool-emptier, 750 gallons.

The house is built below the road on falling land, at the bottom of which there is a stream on the premises 537 feet from the house. On the bank of the stream and extending towards the premises for a distance of 196 feet is a level area below a terrace. This is just above flood-level, and should serve as an irrigation area.

Invert-level of outgoing 4-inch drain of house . . . . . 105·48 N.D.
Ground-level below terrace . . . . 94·12 N.D.
Therefore available fall . . . . 11·36 feet
Length from house to bottom of terrace . . . . . 341 feet
Allowing a 4-inch drain at the minimum gradient of 1 in 90, loss of head in
  drain . . . . $= \dfrac{341}{90} =$ 3.79 feet
Loss through septic tank outlet channels, say . . . . 0·50 foot
Head required to operate percolating filter distributor mechanism . . 1·75 feet
Loss through outlet channels, etc., of percolating filters . . . . 0·50 foot

Total loss of head, excluding fall through filter medium . . . 6·54 feet
Therefore depth available for filter medium . . . . 11·36 − 6·54
                                                              = 4·82, feet say, 1·5 metres

Septic tank. As a free cesspool-emptying service is available, advantage should be taken of this. As the usual emptying period is three months, a septic tank should be capable of storing three months' sludge in one-half its capacity. As the size of the cesspool-emptier is known to be 750 gallons, total capacity of septic tanks should be an exact multiple of this capacity so as to ensure efficient and complete emptying.

The amount of sludge likely to be accumulated is between $\frac{1}{7}$ and $\frac{1}{5}$ gallon per person per day. Then, capacity of septic tank falls between $\frac{2}{7}$ and $\frac{2}{5}$ gallon per person per day, which in the case of a population of 68 persons and a storage period of three months is 1773–2482 gallons. The multiple of 750 gallons which falls between these figures is 2250 gallons (say 10 cubic metres), and this capacity should be selected. As the total capacity is greater than the capacity of the tank vehicle, two septic tanks should be provided in parallel, each having a capacity of 1125 gallons (say 5 cu. metres). Satisfactory proportions for each of these (to brick dimensions) are: length, 9 feet 4½ inches; breadth, 3 feet 4½ inches; depth, 5 feet 9 inches (or 2·86, 1·03 and 1·75 metres respectively).

Percolating filter. Allowing 0·50 cubic metre of medium per head of population (see Table 62), required capacity of bed = 34 cubic metres. At a depth of $1\frac{1}{2}$ metres this gives an area of 22·7 square metres, and therefore a diameter of about 5·36 metres, if a circular bed is used. As this is in the economic range of diameter, a circular bed could be designed and fitted with a good-quality rotating distributor.

Irrigation area for humus removal. Distance from percolating to stream, about 196 feet.

Available width for irrigation area, about 100 feet.

Therefore available superficial area, 2178 square yards = 32 square yards per head of population.

As 4 square yards per head of population is ample, the available irrigation area, if not wanted for other purposes, could be used without any special preparation: alternatively, part only could be used.

APPENDIX I

DEFINITIONS AND EQUIVALENTS OF BRITISH AND METRIC UNITS

Acre. An area of 43,560 square feet, 4046·86 square metres or 0·404686 hectare.

Atmosphere. A pressure of 760 millimetres of mercury, 29·92 inches of mercury, 33·9 feet of water, 1·01325 bars, 14·6959 pounds per square inch, 1·03323 kilograms per square centimetre or 1,013,250 dynes per square centimetre.

Bar. A pressure of 14·5038 pounds per square inch, 1·01972 kilogrammes per square centimetre, 100,000 newtons per square metre, 29·53 inches or 750·062 millimetres of mercury.

Board of Trade Unit (B.T.u.). The legal name for the kilowatt-hour. It is the equivalent of 3415 British thermal units.

British Thermal Unit (B.th.u.). The amount of heat required to raise the temperature of one pound of water by 1 °F. It is the equivalent of 0·252 kilogramme-calorie or 778·169 foot-pounds.

Calorie. On the matter of the value of the calorie there is every possibility of confusion when dealing with international literature or even the literature of different sciences in the same country. To British engineers and scientists a calorie is the amount of heat required to raise the temperature of 1 gramme of water by 1 °C. The terms "standard calorie," "normal calorie" and "mean calorie" are very nearly, but not quite, the same because the value varies with initial temperature. The Calorie (spelt with a capital C, although most writers forget to do this) is the unit used for quoting the energy values of foods. This has the value of 1000 calories as above defined. It is also known to most British scientists and engineers as the "kilogramme-calorie."★

The position, however, appears to be different in France. According to the "Petit Larousse Illustré Nouveau Dictionnaire Encyclopédique" the French use the word "calorie" for the equivalent of the British kilogramme-calorie and have the term "petit calorie" for the British calorie.

According to Webster's "New International Dictionary," the calorie, as first defined herein, is called specifically the "small calorie" (cal) and the kilogramme-calorie the "large (or great) calorie" (Cal).

As the calorie varies with temperature, the international table calorie was defined as a fixed unit in 1956 as the equivalent of 4·1868 joules exactly.

One kilogramme-calorie is the equivalent of 3·96832 British thermal units, 426·935 kilogramme-metres, 3088·03 foot-pounds and 4186·8 joules.

Centigrade heat unit. The amount of heat required to raise the temperature of one pound of water through 1 °C. It is the equivalent of 1400·7 foot-pounds, 1·8 British thermal units, 0·453592 kilogramme-calorie, 193·654 kilogramme-metres and 1899·1 joules.

Cubic foot. Measure of capacity which is the equivalent of 0·0283168 cubic metre, 6·22883 Imperial gallons, 7·4805 United States gallons and 28·3168 litres.

★ Better called kilocalorie.

Cubic feet per minute. Rate of discharge which is approximately 9000 Imperial gallons per day.

Cubic metre. Measure of capacity which is the equivalent of 35·3147 cubic feet, 219·975 Imperial gallons, 264·2 United States gallons and 1000 litres.

Cubic metre per minute. The equivalent of 1,440,000 litres per day.

Dyne. The force which, acting upon 1 gramme, will impart an acceleration of 1 centimetre per second per second. 100,000 dynes is the equivalent of one newton.

Erg. The work done by the force of one dyne acting through one centimetre.

Foot. The equivalent of 0·3048 metre.

Foot-pound. The work done by the force of one pound weight (force) acting through one foot.

Foot-poundal. The work done by the force of one poundal acting through one foot.

Gallons, Imperial. Measure of capacity which is almost exactly the equivalent of 1·2 United States gallons, 0·160544 cubic foot, 4·54596 litres and weighs 10 pounds if of pure water.

Gallons, United States. Measure of capacity which is almost the equivalent of 0·833 Imperial gallon, 0·1337 cubic foot, 3·785 litres and weighs about 8·33 pounds if of pure water.

Gramme equals one thousandth of a kilogramme.

Grammes per litre equals 1000 parts per million.

Hectare. Measure of area which is the equivalent of 10,000 square metres, 107,639 square feet or 2·471 acres.

Horse-power. British measure of rate of doing work. It is the equivalent of 33,000 foot-pounds per minute, 42·41 British thermal units per minute, 0·7457 kilowatt, 1·014 metric horse-power and about 10·7 kilogramme-calories per minute.

Horse-power (metric). The equivalent of 75 kilogramme-metres per second.

Inch. The exact equivalent of 25·4 millimetres.

Joule. Work done by one ampere of electric current through the resistance of one ohm in one second. It is the equivalent of 0·737562 foot-pound.

Kilogramme. The equivalent of 2·2046 pounds. The weight of one litre of water.

Kilogramme-calorie. The equivalent of 3·96832 British thermal units, 3088·03 foot-pounds, 0·001558 horse-power hour, 426·935 kilogramme-metres, 0·001162 kilowatt-hour and 4186·8 joules.

Kilogramme-calorie per minute. The equivalent of 51·43 foot-pounds per second, 0·09351 horse-power (British) and 0·06972 kilowatt.

Kilogramme per square metre. The equivalent of 0·2048 pound per square foot.

Kilogrammes per square centimetre. The equivalent of 14·223 pounds per square inch.

Kilowatt. The equivalent of 1000 watts, 56·92 British thermal units per minute, 44,250 foot-pounds per minute, 1·341 horse-power (British), 14·34 kilogramme-calories per minute and 6118·3 kilogramme metres per minute.

Kilowatt-hour. The equivalent of 3415 British thermal units, 2,655,000 foot-pounds, 1·341 horse-power hours, 860·5 kilogramme-calories and 367,100 kilogramme-metres.

Litre. The volume of 1 kilogramme of distilled water at 4 °C. and at a pressure of 760 millimetres of mercury. It has a capacity of 1000 cubic centimetres, 0·03531 cubic foot, 0·22 Imperial gallon and 0·2642 United States gallon.

Megawatt. The equivalent of 1,000,000 watts.

Metre. The equivalent of 3·28084 feet and 39·37 inches.
Module. Flow of 100 litres per second.
Newton. The force which, acting upon one kilogramme, will impart an acceleration of one metre per second per second. It is the equivalent of 100,000 dynes, 0·224809 pound force or 7·23301 poundals.
Pound. The equivalent of 0·45359237 kilogramme.
Pound per square inch. Pressure that is the equivalent of 0·070307 kilogramme per square centimetre.
Pound per square foot. Pressure that is the equivalent of 4·88243 kilogrammes per square metre.
Pound weight (Pound force). 32·174 poundals or 4·44822 newtons.
Poundal. The force which, acting upon a mass of one pound, will impart an acceleration of 1 foot per second per second.
Square metre. The equivalent of 10·7639 square feet.
Therm. The equivalent of 100,000 British thermal units.
Watt. Unit of power equivalent to one ampere at a potential of one volt. One-thousandth of a kilowatt.

APPENDIX II
NOMINAL EQUIVALENT IMPERIAL AND METRIC PIPE DIAMETERS

For many years the nominal diameters of pipes in some materials have varied from the actual because this has been convenient for manufacture. These variations have been as much as 9 per cent below and $4\frac{1}{2}$ per cent above the actual. It is, therefore, not unreasonable that, with the adoption of the metric system, nominal metric diameters should be applied to equivalent diameters in inches, seeing that this can be done with a variation from the actual which never exceeds 2 per cent less. At the time of writing the appropriate British standards have not been ratified but the figures in the table opposite are in agreement with the majority of those published by manufacturers in their trade brochures.

APPENDIX II

METRIC EQUIVALENTS OF NOMINAL PIPE DIAMETERS

	Internal diameter	
Inches	Actual millimetre equivalent	Nominal millimetres
$1\frac{1}{2}$	38·1	38
2	50·8	50
$2\frac{1}{2}$	63·5	60
3	76·2	75
4	101·6	100
5	127·0	125
6	152·4	150
7	177·8	175
8	203·2	200
9	228·6	225
10	254·0	250
12	304·8	300
14	355·6	350
15	381·0	375
16	406·4	400
18	457·2	450
20	508·0	500
21	533·4	525
22	558·8	550
24	609·6	600
26	660·4	650
27	685·8	675
28	711·2	700
30	762·0	750
32	812·8	800
33	838·2	825
36	914·4	900
38	965·2	950
39	990·6	975
40	1016·0	1000
42	1066·8	1050
44	1117·6	1100
45	1143·0	1125
46	1168·4	1150
48	1219·2	1200

APPENDIX III
FLOW IN PARTLY-FILLED SEWERS

This material has been added at a very late stage of publication and is therefore included as an appendix.

Experiments on partly-filled pipes have been made by T. R. Camp, *Sewage Works Journ.*, **18**, 1, 3 (January 1946); D. L. Yarnell and S. M. Woodward, Bull. No. 854, U.S. Dept. of Agriculture (1920); E. R. Wilcox, Bull. No. 27, Univ. of Washington Engineering Experiment Station (1924); K. W. Cosens, *Sewage and Industrial Wastes*, **26**, 1, 42 (January 1954); L. G. Straub and H. M. Morris, Tech. Paper No. 4, Ser. B, St. Anthony Falls Hydraulic Laboratory, Univ. of Minnesota; R. D. Pomeroy, *Journal Water Pollution Control Federation* (September 1967). Also the author has analysed the results of many tests made by the London County Council Chief Engineer's Department on twenty-eight large sewers.

Analysis of certain of these data showed that, in average conditions, flows in partly-filled sewers or open channels are best calculated by adding to the wetted perimeter in the flow formula one-half of the free-surface top-water width of flow. The following table compares the data used with the figures so calculated (using Escritt's formula given in Vol. I, Table 10).

PROPORTIONAL VELOCITIES AND DISCHARGES IN CIRCULAR
SEWERS RUNNING PARTLY FULL

Prop. depth	Empiric proportional velocity*				Unweighted average	Calc. prop. velocity	Prop. area	Calc. prop. discharge
	Camp	Yarnell and Woodward	Cosens	Wilcox				
1·0	1·0000	1·0000	1·0000	1·0000	1·0000	1·0000	1·0000	1·0000
0·9	1·0400	1·0400	1·0200	1·0333	1·0333	1·0394	0·9480	0·9854
0·8	1·0133	1·0400	1·0133	1·0400	1·0267	1·0189	0·8576	0·8738
0·7	0·9567	1·0000	0·9800	1·0100	0·9867	0·9765	0·7477	0·7302
0·6	0·8900	0·9166	0·9166	0·9600	0·9208	0·9173	0·6265	0·5747
0·5	0·8066	0·8233	0·8500	0·8900	0·8425	0·8425	0·5000	0·4213
0·4	0·7166	0·7166	0·7600	0·8100	0·7508	0·7432	0·3735	0·2776
0·3	0·6100	0·5867	0·6568	0·6934	0·6367	0·6427	0·2523	0·1621
0·2	0·4900	—	0·5400	—	0·5150	0·5102	0·1424	0·0727
0·1	0·3270	—	—	—	0·3270	0·3373	0·0520	0·0175

* Scaled from Fig. 4 of Pomeroy's paper.

BIBLIOGRAPHY

1. H. C. Adams. "Sewerage of Seacoast Towns." 1911.
2. S. H. Adams. "The Ministry of Health 'Requirements,' etc." 1921.
3. S. H. Adams. "Modern Sewage Disposal and Hygienics." 1930.
4. W. E. Adeney. "The Dilution Method of Sewage Disposal." 1928.
5. C. C. Andersen. "Sludge Digestion Tanks: Condition after Seven Years' Service." *Sewage Works Journal*, 1938.
6. Norval E. Anderson. "Design of Final Settling Tanks for Activated Sludge." *Sewage Works Journal*, January 1946.
7. F. V. Appleby. "Impervious Factors." *Journal Institution of Municipal and County Engineers*, Vol. LXIII, No. 16, p. 1077.
8. E. Ardern and C. Jepson. "Short Period Mesophilic Sludge Digestion." *Proceedings Institute of Sewage Purification*, 1938, Part 1.
9. E. Ardern, C. Jepson and L. Klein. "Short Period Mesophilic Sludge Digestion at Daveyhulme." *Proceedings Institute of Sewage Purification*, 1940.
10. E. H. Arrowsmith. "Some Practical Experiments in Flocculation without Mechanical Means." *Proceedings Institute of Sewage Purification*, 1942.
11. John Austin. "The Hygienic Treatment and Disposal of Offal and By-Products in Abbatoirs." *Journal of the Royal Sanitary Institute*, March 1939.
12. H. E. Babbitt. "Sewerage and Sewage Treatment." 1953.
13. H. E. Babbitt and J. J. Doland. "Water Supply Engineering." 1939.
14. J. L. Bannister. "The Significance of the Time of Concentration Factor in the Design of Storm Water Sewers." *Journal Institution of Municipal and County Engineers*, Vol. LXVI, No. 7, p. 321.
15. A. A. Barnes. "Hydraulic Flow Reviewed." 1916.
16. D. H. Barraclough. "Recent Contributions to the Study of Industrial Waste Treatment." *Proceedings Institute of Sewage Purification*, 1941.
17. N. W. Barritt. "The Respiration of Activated Sludge." *Proceedings Institute of Sewage Purification*, 1939, Part 1.
18. J. R. Barron. "The Evils of Cesspool Drainage." *Journal Royal Sanitary Institute*, Vol. LIX, No. 2.
19. E. O. Baxter. "Rainfall Intensities." *Journal Institution of Municipal and County Engineers*, Vol. LXII, No. 10, p. 529; No. 19, p. 1026; No. 25, p. 1425.
20. J. L. Beckett. "Partial Activated Sludge Treatment of Sewage: Notes on Experimental Work." *Journal Royal Sanitary Institute*, Vol. LX, No. 2.
21. D. B. Bevan. "Recent Sewerage and River Works in Birmingham." *Journal Institution of Municipal and County Engineers*, Vol. LVIII, No. 2, p. 217.
22. E. J. Bevan and B. T. Rees. "Sewers: Theory, Design, Specification and Construction." 1938.
23. E. G. Bilham. "Classification of Heavy Falls in Short Periods." *British Rainfall*, 1935.
24. E. G. Bilham. "Climate of the British Isles." 1938.

25. Sir Alexander R. Binnie. "Rainfall, Reservoirs and Water Supply." Second Impression, 1925.
26. G. M. Binnie. "Model Experiments on Bellmouth and Siphon-bellmouth Overflow Spillways." *Journal Institution of Civil Engineers*, No. 1, 1938–9.
27. E. H. Blake. "Drainage and Sanitation." Sixth Edition revised by W. R. Jenkins. 1942.
28. C. D. C. Braine. "Draw-down and Other Factors Relating to the Design of Storm-water Overflows on Sewers." *Journal Institution of Civil Engineers*, No. 6, April 1946–7.
29. C. D. C. Braine. "The Adaptation of Small Siphons for Sewerage and Sewage Disposal." *Surveyor* (London), 2 April 1937.
30. British Rainfall Organisation. *British Rainfall*. Published annually.
31. J. A. R. Bromage. "The Sewage Disposal of Delhi." *Journal Institution of Civil Engineers*, No. 6, 1939–40.
32. C. E. N. Bromehead. "Fluorosis in England." *The Lancet*, May 1941.
33. C. E. N. Bromehead. "The Early History of Water Supply." *The Geographical Journal*, Vol. XCIX, Nos. 3, 4, March and April 1942.
34. F. Buckley. "River Pollution Prevention as Affecting the Municipal Engineer." *Journal Institution of Public Health Engineers*, Vol. LXV. Part 4, October 1966.
35. H. J. Bunker. "The Corrosion of Mains in Clay Soils." *Journal Institution of Sanitary Engineers*, No. 5, 1944.
36. T. R. Camp. "A Study of the Rational Design of Settling Tanks." *Sewage Works Journal*, September 1936.
37. C. H. Capen, Jr. *Eng. News-Record*, **99**, 883, 1927.
38. George Chamier. "Capacities Required for Culvert and Flood-openings." *Proceedings Institution of Civil Engineers*, Vol. 134, 1897–8.
39. E. Sherman Chase. "Trickling Filters—Past, Present and Future." *Sewage Works Journal*, September 1945.
40. H. H. Clay. "Camp Sanitation: Modern Practice in Treatment and Disposal of Sullage Water." *Journal Royal Sanitary Institute*, Vol. LXII, No. 1.
41. C. H. J. Clayton. "Land Drainage from Field to Sea." 1919.
42. W. Clifford and M. E. Windridge. "Experiments with Model Tanks." *Journal Institute of Sewage Purification*, 1935, Part 1.
43. C. F. Colebrook. "Turbulent Flow in Pipes with Particular Reference to the Transition Region between the Smooth- and Rough-pipe Laws." *Journal Institution of Civil Engineers*, No. 4, 1938–9.
44. C. F. Colebrook and C. M. White. "The Reduction of Carrying Capacity of Pipes with Age." *Journal Institution of Civil Engineers*, No. 1, 1937–8.
45. G. S. Coleman. "The Estimation of Storm-water Run-off from Inhabited Areas." Selected Engineering Paper, No. 4, Institution of Civil Engineers. 1923.
46. G. S. Coleman. "Storm-water Run-off". *Journal Institution of Municipal and County Engineers*, Vol. L, No. 17, p. 771.
47. G. S. Coleman and A. Johnson. "Rainfall Run-off." *Journal Institution of Municipal and County Engineers*, Vol. LVIII, No. 18, p. 1403.
48. B. A. Copas. "Storm Water Storage Calculation." *Journal Institution of Public Health Engineers*, Vol. LVI, Part 3, July 1957.

BIBLIOGRAPHY

49. R. J. Cornish. "The Analysis of Flow in Networks of Pipes." *Journal Institution of Civil Engineers*, No. 2, 1939–40.
50. G. T. Cotterell. "The Problem of Sewage Disposal in Rural Areas." *Journal Royal Sanitary Institute*, Vol. LXIV, No. 1.
51. G. T. Cotterell. "The Effect of Grease on the Operation of Sewage Disposal Works and Steps taken for its Elimination." *Journal Institution of Sanitary Engineers*, No. 5, January 1944.
52. H. T. Cranfield. "The Manurial Value of Sewage Sludge." *Proceedings Institute of Sewage Purification*, 1941.
53. W. Santo Crimp and C. Ernest Bruges. "Crimp and Bruges Tables and Diagrams for Designing Sewers and Water Mains." Revised W. E. Bruges. 1936.
54. Professor Hardy Cross. "Analysis of Flow in Networks of Conduits or Conductors." University of Illinois Engineering Experiment Station, Bulletin No. 286. November 1936.
55. E. M. Crowther and A. H. Bunting. "The Manurial Value of Sewage Sludges." *Proceedings Institute of Sewage Purification*, 1942.
56. E. P. Currell. "On Stoneware Pipes and Sewers." Selected Papers, Institution of Civil Engineers.
57. P. Davies. "The Laws of Syphon Flow." *Proceedings Institution of Civil Engineers*, Vol. 235.
58. J. R. Daymond. "The Hydraulic Problem Concerning the Design of Sewage Storage Tanks and Sea-outfalls." *Journal Institution of Civil Engineers*, No. 3, 1939–40.
59. Don and Chisholm. "Modern Methods of Water Purification." 1911.
60. T. Donkin. "The Effect of the Form of Cross-section on the Capacity and Cost of Trunk Sewers." *Journal Institution of Civil Engineers*, No. 2, 1937–8.
61. Downing, Boon and Bayley. "Aeration and Biological Oxidation in the Activated Sludge Process." *Journal Institute of Sewage Purification*, 1962, Part I, pp. 66–70.
62. A. L. Downing and J. D. Swanwick. "Treatment and Disposal of Sewage Sludge." *Journal Institution of Municipal Engineers*, 1967, 94, 83.
63. Dunbar and Calvert. "Principles of Sewage Treatment."
64. C. B. Egolf and W. L. McCabe. "Rate of Sedimentation of Flocculated Particles" *Transactions of the American Institution of Chemical Engineers*, **33**, 620, 1937.
65. E. F. Eldridge. "Industrial Waste Treatment Practice." 1942.
66. L. B. Escritt. "Sewerage Engineering." 1939.
67. L. B. Escritt. "Surface Drainage." 1944.
68. L. B. Escritt. "Sewerage Design and Specification." 1947.
69. L. B. Escritt. "Sewage Treatment: Design and Specification." 1950.
70. L. B. Escritt. "A Code for Sewerage Practice." 1950.
71. L. B. Escritt. "Surface-water Sewerage." 1950.
72. L. B. Escritt. "Building Sanitation." 1953.
73. L. B. Escritt. "Sewerage and Sewage Disposal." 1956.
74. L. B. Escritt. "Pumping Station Equipment and Design." 1962.
75. L. B. Escritt. "Design of Surface-water Sewers." 1964.
76. L. B. Escritt. "Sewers and Sewage Works with Metric Calculations and Formulae." 1971.

77. L. B. and V. P. Escritt "Escritts' Tables of Metric Hydraulic Flow." 1971.
78. Professor J. A. Ewing. "Sewerage." Encyclopaedia Britannica 1886.

79. Gordon Maskew Fair and John Charles Geyer. "Elements of Water Supply and Waste-water Disposal." 1958.
80. Federation of Sewage Works Associations. "Occupational Hazards in the Operation of Sewage Works." Manual of Practice, No. 1, U.S.A. 1944.
81. Federation of Sewage Works Associations. "Utilisation of Sewage Sludge as Fertiliser." Manual of Practice, No. 2, U.S.A. 1946.
82. J. Finch. "The Possibilities of, and Economics Relating to, Organic Manures as Applied to Air-dried Sewage Sludge." *Proceedings Institute of Sewage Purification*, 1941.
83. H. Lorain Folkes. "Some Recent Developments in Settling Tanks." *Proceedings Institute of Sewerage Purification*, 1938, Part 2.
84. A. P. Folwell. "Sewerage." 1936.
85. T. P. Francis. "Modern Sewage Treatment." 1931.
86. Dr. Ing Otto Franzuis. "Waterway-Engineering." Trans. L. G. Straub. Technology Press, Cambridge, Mass., 1936.

87. E. T. M. Garlick. "Sedimentation and Anaerobic Digestion of Sewage-sludge at Colac, Echuca and Mildura, Australia." *Journal Institution of Civil Engineers*, No. 4, 1935–6.
88. J. F. Garner. "The Law of Sewers and Drains." Third Edition, 1962.
89. J. H. Garner. "Notes on Sewage Disposal and Rivers Pollution Prevention." *Proceedings Institute of Sewage Purification*, 1938, Part 2.
90. R. C. Gibbs. "Mechanical Flocculation of Water and Sewage." *Institution of Sanitary Engineers*, 11 November 1938.
91. A. H. Gibson. "Tidal and River Models." *Supplement, Journal Institution of Civil Engineers*, No. 8, October 1936.
92. J. Glasspoole. "The Areas Covered by Intense and Widespread Falls of Rain." *Proceedings Institution of Civil Engineers*, Vol. 229, Part I, 1929–30.
93. H. H. Goldthorpe. "Experimental Rapid Filtration at Huddersfield." *Proceedings Institute of Sewage Purification*, 1938, Part 1.
94. H. H. Goldthorpe. "Some Comments on Recent Experiments on Activated Sludge." *Proceedings Institute of Sewage Purification*, 1939, Part 2.
95. H. H. Goldthorpe. "A Report upon the Treatment of Sewage on Percolating Filters at Huddersfield." *Proceedings Institute of Sewage Purification*, 1942.
96. W. M. Griffith. "Discharge by Surface Floats." *Journal Institution of Civil Engineers*, No. 4, 1940–41.
97. W. M. Griffith. "A Theory of Silt and Scour." *Proceedings Institution of Civil Engineers*, Vol. 223.
98. W. M. Griffith. "Silt Transportation and its Relation to Regime Channel Sections." *Journal Institution of Civil Engineers*, April 1944. (Abstracted Paper.)
99. E. Dixon Grubb. "Impermeability with Special Reference to Gaugings." *Surveyor* (London), 14 April 1939.

100. Walter H. Haile. "Land Drainage in England and Holland." *Journal Institution of Municipal and County Engineers*, Vol. LX, No. 14, p. 1024.

101. E. J. Hamlin and H. Wilson. "Investigations on Percolating Filters." *Proceedings Institute of Sewage Purification*, 1938, Part I and 1939, Part I.
102. C. A. Hart. "Correspondence on Rainfall and Run-off." *Journal Institution of Municipal and County Engineers*, Vol. LIX, No. 18, p. 978.
103. A. Hazen. "Flood Flows." 1930.
104. A. Hazen. *Transactions of the American Society of Civil Engineers*, 1904, **53**, 45.
105. G. R. Hearn. "The Effect of Shape of Catchment on Flood Discharge." *Proceedings Institution of Civil Engineers*, Vol. CCXVII, 1924.
106. Augustine Henry. "Forests, Woods and Trees in Relation to Hygiene." 1919.
107. R. Hicks. "The Effect of Pressure on the Rate of Oxidation of Sewage in the Presence of Activated Sludge." *Proceedings Institute of Sewage Purification*, 1938, Part 2.
108. J. Hirst. "Experimental Work to Improve the Performance of a Bio-aeration Plant." *Proceedings Institute of Sewage Purification*, 1942.
109. H. J. N. Hodgson. "Sewage and Trade Wastes Treatment." 1938.
110. G. D. Holmes and W. P. Gyatt. "Syracuse Sewage Treatment Works, Summary of Four Years' Operating Experiences." *Sewage Works Journal*, **1**, 318, 1929.
111. A. Holroyd. "The Operation of the Dagenham Pruss Tank: A Study of Radial-flow Sedimentation." *Journal Institute of Sewage Purification*, 1937.
112. Brian D. Horan. "The Design of Storm Water Sewers." *Journal Institution of Municipal and County Engineers*, Vol. LX, No. 6, p. 409.
113. Neil Hoskins. "The Rational System of Calculations of Sewer Discharges and the Design of Modern Sewage Pumping-stations." *Journal Institution of Municipal and County Engineers*, Vol. LVIII, No. 6, p. 494.
114. H. J. Huber. "House Laterals and Connections." *Sewage Works Journal*, July 1942. See also *Journal Institution of Sanitary Engineers*, No. 2, April 1943.
115. Sir George W. Humphreys. "Main Drainage of London. Descriptive Account, with Accompanying Tables, Plans, Diagrams and Illustrations."
116. Alexander Hunter and T. Cockburn. "Operation of an Enclosed Aeration Filter at Dalmarnock Sewage Works." *Journal Institute of Sewage Purification*, 1944.
117. J. Hurley. "The Mechanical Flocculation of Sewage." *Surveyor* (London), 9 January 1942.
118. J. Hurley. "A Critical Review of Recent Work on Sewage Filtration." *Proceedings Institute of Sewage Purification*, 1942.
119. J. Hurley. "Post-War Possibilities in Sewage Works Design." Paper read at Midland Branch of Sewage Purification, 1944.
120. J. Hurley, J. McNicholas and G. B. O. Jones. "The Pre-treatment of Trade Effluent." *Proceedings Institute of Sewage Purification*, 1940.
121. J. Hurley, J. McNicholas, H. Myatt and A. J. Clifford. "Symposium on Practical Problems of Sewage Works Management." *Proceedings Institute of Sewage Purification*, 1941.
122. J. Hurley and M. Lovett. "The Purification of Sewage on Covered Filters." *Proceedings Institute of Sewage Purification*, 1938, Part I.
123. J. Hurley and M. E. D. Windridge. "Enclosed Aerated Filters." *Proceedings Institute of Sewage Purification*, 1938, Part I.

124. Karl Imhoff. "The Arithmetic of Sewage Treatment Works."
125. Imhoff and Fair. "Sewage Treatment." 1956
126. Karl Imhoff, W. J. Muller and D. K. B. Thistlewayte. "Disposal of Sewage and Other Water-borne Wastes." 1956.
127. Institute of Sewage Purification. "Brief Directory of Methods in Use at Sewage Disposal Works." *Proceedings Institute of Sewage Purification*, 1938, Part 1.
128. Institute of Sewage Purification. "Directory of Sewage Works." 1963.
129. Institution of Civil Engineers. "Interim Report of the Committee on Floods in Relation to Reservoir Practice." 1933.
130. Institution of Civil Engineers. "Location of Underground Services." Report of Joint Committee of Institution of Civil Engineers and Institution of Municipal and County Engineers.
131. Ippen and Carver. "Basic Factors of Oxygen Transfer in Aeration Systems." *Sewage and Industrial Wastes*, July 1954.

132. S. H. Jenkins. "Progress in the Methods of Treatment and Disposal of Sewage Sludge." *Proceedings Institute of Sewage Purification*, 1939, Part 2.
133. S. H. Jenkins and C. H. Hewitt. "The Effect of Chromium Compounds on the Purification of Sewage by the Activated Sludge Process." *Proceedings Institute of Sewage Purification*, 1942.
134. A. Johnson. "Rainfall Run-off." *Journal Institution of Municipal and County Engineers*, Vol. LXI, No. 16, p. 830.
135. F. Johnstone-Taylor. "River Engineering." 1938.
136. C. C. Judson. "Difficulties in the Design and Construction of a Surface-water Conduit." *Journal Institution of Municipal and County Engineers*, Vol. LVIII, No. 10, p. 787.
137. C. C. Judson. "Run-off Calculations: A New Method." *Journal Institution of Municipal and County Engineers*, Vol. LIX, No. 15, p. 861.
138. C. C. Judson. "Rainfall as Affecting Flow in Sewerage Systems." *Surveyor* (London), 11 August 1933.

139. C. E. Keefer. "Sewage-treatment Works. Administration and Operation." 1940.
140. R. G. Kennedy. "The Prevention of Silting in Irrigation Canals." *Proceedings Institution of Civil Engineers*, Vol. CXIX, 1895.
141. C. G. Kent. "The Construction of Main Sewers." 1937.
142. G. B. Kershaw. "Modern Methods of Sewage Disposal." 1911.
143. G. B. Kershaw. "Sewage Purification and Disposal." 1925.
144. J. B. C. Kershaw. "The Recovery and Use of Industrial and Other Wastes." 1928.
145. M. A. Kershaw. "The Inspection and Control of Trade Waste Effluent Discharges." Paper read at a meeting of the Institution of Sanitary Engineers, 27 February 1948.
146. H. R. King. "Oxygen Absorption." Discussion, *Sewage and Industrial Wastes*. May 1955.
147. H. R. King. "Mechanics of Oxygen Absorption in Spiral-flow Aeration Tanks." *Sewage and Industrial Wastes*. August 1955.
148. L. P. Kinnicutt, C. E. A. Winslow and R. W. Pratt. "Sewage Disposal." 1919.

BIBLIOGRAPHY 483

149. B. G. Kirk. "Description of the Luxborough Main Outfall Works, Chigwell." *Journal Institute of Sewage Purification*, 1942.
150. L. Klein. "The Strength of Sewage: Some Comparative Results." *Proceedings Institute of Sewage Purification*, 1941.
151. E. R. Knight. "The Equipment and Maintenance of Automatic Sewage Pumping Stations." *Journal Institution of Civil Engineers*, No. 1, 1943-4. (Abstracted Paper.)
152. E. Kuichling. Report to the Common Council of Rochester, U.S.A. 1889.
153. E. Kuichling. "The Relation between the Rainfall and the Discharge of Sewers in Populous Districts." *Transactions of the American Society of Civil Engineers*, 1889, **20**, 1.

154. Gerald Lacey. "Uniform Flow in Alluvial Rivers and Canals." *Proceedings Institution of Civil Engineers*, Vol. 237, Part 1, 1933-4.
155. Gerald Lacey. "Stable Channels in Alluvium." *Proceedings Institution of Civil Engineers*, Vol. 229, 1930.
156. Ernest Latham. "Marine Works." 1922.
157. J. P. Lawrie. "Chemicals from Methane." Science Services, Ltd. 1947.
158. F. C. Lea. "Hydraulics." Fourth Edition, 1923.
159. F. C. Lea. "Rainfall and Run-off into Sewers." *Journal Institution of Municipal and County Engineers*, Vol. LIV, No. 19, p. 1097.
160. G. E. Lillie. "Discharge from Catchment Areas in India, as affecting the Waterways of Bridges." *Proceedings Institution of Civil Engineers*, Vol. CCXVII, 1924.
161. H. M. Limb. "Sewerage Systems with Special Reference to Run-off of Surface Waters." *Journal Institution of Municipal and County Engineers*, Vol. LIV, No. 13, p. 895.
162. L. Lloyd, J. F. Graham and T. B. Reynoldson. "Material for a Study in Animal Competition. The Fauna of the Sewage Bacteria Beds." *The Annals of Applied Biology*, Vol. XXVIII, No. 1, February 1940.
163. D. E. Lloyd Davies. "The Elimination of Storm-water from Sewerage Systems." *Proceedings Institution of Civil Engineers*, Vol. CLXIV, 1906.
164. London County Council. Report, 31 March 1913.
165. London County Council. "London Main Drainage System: General Information." October 1959.
166. A. S. Lowe and C. D. Bottomley. "Recent Developments in Sewage Sludge Dewatering and Incineration." *Proceedings Institute of Sewage Purification*, 1938, Part 1.
167. C. Lumb. "Heat Treatment as an Aid to Sludge Dewatering—10 Months Fullscale Operation." *Proceedings Institute of Sewage Purification*, 1940.
168. C. Lumb. "Suspended Solids Balance in Purification of Sewage by Activated Sludge. Some Experiments on Precipitation of a Trade Waste Sewage." *Proceedings Institute of Sewage Purification*, 1940.
169. C. Lumb. "Some Investigations into the Digestion and Drying of Humus and Activated Sludges." *Proceedings Institute of Sewage Purification*, 1941.

170. R. Manning. "On the Flow of Water in Open Channels and Pipes." *Transactions Institution of Civil Engineers of Ireland*, Vol. XX, 1891.
171. A. J. Martin. "The Activated Sludge Process."

172. A. J. Martin. "The Bio-aeration of Sewage." *Proceedings Institution of Civil Engineers*, Vol. CCXVII.
173. D. K. McKenzie. "The Design of Combined Sewerage Systems." *Journal Institution of Municipal and County Engineers*, Vol. LVIII, No. 11, p. 857.
174. J. McNicholas. "The Neutralisation of Acid Trade Effluent." *Proceedings Institute of Sewage Purification*, 1939, Part 1.
175. J. A. McPherson. "Waterworks Distribution." 1907.
176. J. B. L. Meek. "Sewerage, with Special Reference to Rainfall." Engineering Conference of the Institution of Civil Engineers, 1928.
177. L. Metcalf and H. P. Eddy. "American Sewerage Practice." 1928, 3 vols.
178. L. Metcalf and H. P. Eddy. "Sewerage and Sewage Disposal." 1930.
179. E. V. Mills. "The Treatment of Settled Sewage in Percolating Filters in Series with Periodic Changes in Order of the Filters: Results of Operation of the Experimental Plant at Minworth, Birmingham, 1940 to 1944." *Journal Institute of Sewage Purification*, 1945, Part 2.
180. E. V. Mills. "The Treatment of Settled Sewage by Continuous Filtration with Recirculation of Part of the Effluent." *Journal Institute of Sewage Purification*, 1945, Part 2.
181. R. C. R. Minikin. "Practical River and Canal Engineering." 1920.
182. Ministry of Health. Advisory Committee on Water Report on the Measures for the Protection of Underground Water. 1925.
183. Ministry of Health. "Accidents in Sewers: Report on the Precautions Necessary for the Safety of Persons Entering Sewers and Sewage Tanks." 1934.
184. Ministry of Health Departmental Committee on Rainfall and Run-off. *Journal Institution of Municipal and County Engineers*, Vol. LVI, No. 22, p. 1172.
185. Ministry of Health. "Methods of Chemical Analysis as Applied to Sewage and Sewage Effluents." Reprinted 1946. (See also 187.)
186. Ministry of Health. "The Bacteriological Examination of Water Supplies." Reprinted 1946.
187. Ministry of Housing and Local Government. "Methods of Chemical Analysis as Applied to Sewage and Sewage Effluents." 1956.
188. Ministry of Housing and Local Government. "Pollution of the Tidal Thames." (Pippard Report.) 1961.
189. G. E. Mitchell. "Sanitation, Drainage and Water Supply." Fifth Edition, 1954.
190. Morgan. "Clogging Studies of Fine Bubble Diffuser Media." *Sewage and Industrial Wastes*, February 1959, pp. 153–63.
191. K. A. Murray. "The Control of Filter Flies on Percolating Filters." *Proceedings Institute of Sewage Purification*, 1939, Part 2.

192. H. C. Nicholson. "The Principles of Field Drainage." 1942.

193. J. J. O'Kelly. "The Employment of Unit Hydrographs to Determine the Flows of Irish Arterial Drainage Channels." *Proceedings Institution of Civil Engineers*, Part III, August, 1955.
194. M. T. M. Ormsby. "Rainfall and Run-off Calculations." *Journal Institution of Municipal and County Engineers*, Vol. LIX, No. 16, p. 889.

195. A. Parker. "Treatment of Water for Domestic and Industrial Requirements: Some Problems and Methods." *Journal Institution of Civil Engineers*, No. 8, October 1941–2.
196. A. Parker. "The Treatment and Disposal of Trade Waste Waters." *Journal Institution of Civil Engineers*, No. 5, 1937–8.
197. H. W. Parkinson and H. D. Bell. "Insect Life on Sewage Filters." 1919.
198. W. K. Parry and W. E. Adeney. "Discharge of Sewers into a Tidal Estuary." *Proceedings Institution of Civil Engineers*, Vol. CXLVII, Part I, 1901–2.
199. Dr. A. Pasveer. "Efficiency of the Diffused Air System." *Sewage and Industrial Wastes*, January 1956.
200. V. G. Pickering. "Practical Sewerage Problems." *Journal Institution of Municipal and County Engineers*, Vol. LX, No. 17, p. 1239.

201. W. B. Redfern. *Sewage Works Journal*, January 1939.
202. John Reid. "The Estimation of Storm-water Discharge." *Journal Institution of Municipal and County Engineers*, Vol. LIII, No. 23, p. 997.
203. T. B. Reynoldson. "The Role of Macro-organism in Bacteria Beds." *Journal Institute of Sewage Purification*, 1939, Part I.
204. T. B. Reynoldson. "Enchtræid Worms and the Bacteria Bed Method of Sewage Treatment." *The Annals of Applied Biology*, Vol. XXVI, No. 1, February 1939.
205. T. B. Reynoldson. "On the Life-history and Ecology of *Lumbricillus Lineatus* Mull. (Oligochaetal)." *The Annals of Applied Biology*, Vol. XXVI, No. 4, November 1939.
206. T. B. Reynoldson. "The Biology of Macro-fauna of a High-rate Double Filtration Plant at Huddersfield." *Proceedings Institute of Sewage Purification*, 1941.
207. T. B. Reynoldson. "Further Studies on the Biology of a Double Filtration Plant at Huddersfield." *Proceedings Institute of Sewage Purification*, 1942.
208. B. D. Richards. "Flood-hydrographs." *Journal Institution of Civil Engineers*, No. 5, 1936–7.
209. B. D. Richards. "Further Note on Flood-hydrographs." *Journal Institution of Civil Engineers*, No. 6, 1938–9.
210. B. D. Richards. "Flood Estimation and Control." 1944.
211. Samuel Rideal. "Water and its Purification." Second Edition, 1902.
212. D. Wearing Riley. "Notes on Calculating the Flow in Surface Water Sewers." *Journal Institution of Municipal and County Engineers*, Vol. LVIII, No. 20, p. 1483.
213. Leslie Roseveare. "Run-off as Affecting the Flow in Sewers." *Journal Institution of Municipal and County Engineers*, Vol. LVI, No. 22, p. 1177.
214. W. Rudolfs. "Principles of Sewage Treatment." 1935.
215. W. Rudolfs. "Industrial Wastes, Their Disposal and Treatment." 1953.
216. H. F. Rutter. "Kempe's Engineers Year Book." 1935.
217. William J. Ryan. "Water Treatment and Purification." 1937.

218. Dr. J. C. Samuel. "Some Aspects of Flocculation." *Chemistry and Industry*, 28 August 1936.
219. H. Schraeder. "Economic Aspect of the Design of Sewage Treatment and Disposal Works." 1934.

220. A. Sciver and E. H. M. Badger. "The Effect of Spent Gas Liquor on Cold Sludge Digestion." *Journal Institution of Sanitary Engineers*, No. 3, July 1943.
221. W. Scott and W. Foster. "The Treatment of Trade Waste Waters at Traders' Premises." *Proceedings Institute of Sewage Purification*, 1942.
222. W. Scott, H. Turner, S. M. Sykes and A. R. Ward. Symposium on "The Manufacture and Utilisation of Sewage Sludge as a Fertiliser." *Proceedings Institute of Sewage Purification*, 1942.
223. Scouller and Watson. "The Solution of Oxygen from Air Bubbles." *Journal Institute of Sewage Purification*, 1934, Part 1.
224. E. J. Silcock. "Kempe's Engineers Year Book." 1935.
225. Sir John Simon. "English Sanitary Institutions." Second Edition, 1897.
226. Dr. B. A. Southgate. "Treatment and Disposal of Industrial Waste Waters." 1948.
227. George S. Standley and Bernard J. Withers. "Current Activities in Coulsdon and Purley." *Journal Institution of Municipal and County Engineers*, Vol. LIX, No. 19, p. 997.
228. Stationery Office. The Reports of the Royal Commission on Sewage Disposal.
229. Stationery Office. "Effect of Discharge of Crude Sewage into the Estuary of the River Mersey and the Amount and Hardness of the Deposit in the Estuary." Water Pollution Research Technical Paper, No. 7.
230. Stationery Office. "Effects of Polluting Discharges on the Thames Estuary." 1964. Water Pollution Research Technical Paper, No. 11.
231. E. W. Steel. "Water Supply and Sewerage." Revised Edition, 1947.

232. G. M. C. Taylor. "Sea Outfalls in the Borough of Bangor on Belfast Lough." Belfast and District Association of the Institution of Civil Engineers. April 1938.
233. D. H. Thatcher. "Rectangular Tanks—Efficiency and Adaptation to Desludging Mechanism." *Journal Institute of Sewage Purification*, 1940.
234. J. T. Thompson. "High Rate Dosing of Gravel Percolating Beds." *Proceedings Institute of Sewage Purification*, 1942.
235. J. T. Thompson and J. W. Proctor. "Filter Pressing of Sludge." *Proceedings Institute of Sewage Purification*, 1938, Part 1.
236. Thresh, Beale and Suckling. "The Examination of Waters and Water Supplies." Fourth Edition, 1933.
237. C. B. Townend. "The Elimination of the Detritus Dump." Public Works, Roads and Transport Congress. 1937.
238. C. B. Townend. "West Middlesex Main Drainage: Ten Years' Operation." *Journal Institution of Civil Engineers*, No. 4, 1946–7.
239. J. H. T. Tudsbery and A. W. Brightmore. "The Principles of Waterworks Engineering."
240. Turneaure and Russell. "Public Water Supplies." (New York.) Third Edition, 1924.

241. T. H. P. Veal. "The Disposal of Sewage." 1939.
242. F. C. Vokes. "The Design and Operation of the Coleshill Sewage-disposal Works of the Birmingham, Tame and Rea District Drainage Board." *Journal Institution of Civil Engineers*, No. 3, 1937–8.

243. F. C. Vokes and C. B. Townend. "Power Gas from Sewage-sludge at the Works of the Birmingham, Tame and Rea District Drainage Board." *Proceedings Institution of Civil Engineers*, Vol. 226, Part 2, 1927-8.
244. F. C. Vokes and S. H. Jenkins. "The Yardley Sewage-disposal Works of the Birmingham, Tame and Rea District Drainage Board." *Journal Institution of Civil Engineers*, No. 2, 1942-3.
245. F. C. Vokes and S. H. Jenkins. "Experiments with a Circular Sedimentation Tank." *Journal Institution of Civil Engineers*, No. 2, 1942-3.

246. R. D. Walker. "The Principles of Underdrainage." 1929.
247. John H. Walton. "The Determination of Fill and Surcharge Loads on Glazed Vitrified Clay Pipelines." Parts 1, 2 and 3.
248. A. R. Ward. "The Kessener Brush Aeration Process at the Stockport Sewage Disposal Works." *Proceedings Institute of Sewage Purification*, 1939, Part 1.
249. D. M. Watson. "West Middlesex Main Drainage." *Journal Institution of Civil Engineers*, No. 6, 1936-7.
250. D. M. Watson. "Modern Sanitation in Great Britain." *Journal Institution of Civil Engineers*, No. 2, December 1939-40.
251. W. Watson. "Purification of Strong Sewage by Re-circulation through a Percolating Filter." *Proceedings Institute of Sewage Purification*, 1941.
252. P. R. Welsh. "Garbage Digested with Sewage Sludge." *Sewage Works Journal*, 1938.
253. E. G. White. "Attempted Sterilisation of Sewage Works Effluents by Means of Chlorine." *Proceedings Institute of Sewage Purification*, 1939, Part 2.
254. H. C. Whitehead. "The Design of Sewage Purification Works." *Journal Institution of Civil Engineers*, No. 5, March 1940-41.
255. H. C. Whitehead and F. R. O'Shaughnessy. "The Treatment of Sewage Sludge by Bacterial Digestion." *Proceedings Institution of Civil Engineers*, Vol. 233, Part 1, 1931-2.
256. C. Bransby Williams. "Flood Discharge and the Dimensions of Spillways in India." *The Engineer*, Vol. 134, 1922.
257. J. Williamson. "Considerations on Flow in Large Pipes, Conduits, Tunnels, Bends and Syphons." *Journal Institution of Civil Engineers*, No. 6, April 1938-9.
258. H. M. Wilson and H. T. Calvert. "A Textbook on Trade Waste Waters: Their Nature and Disposal." 1913.
259. M. E. D. Windridge. "Some Views and Experiences on the Treatment of Trade Wastes." *Proceedings Institute of Sewage Purification*, 1941.
260. J. M. Wishart, R. Wilkinson and W. C. Tomlinson. "Purification of Settled Sewage in Percolating Filters in Series, with Periodic Change in the Order of the Filters." *Proceedings Institute of Sewage Purification*, 1941.
261. Horace B. Woodward. "Soils and Subsoils from a Sanitary Point of View: with Especial Reference to London and its Neighbourhood." Memoirs of the Geological Survey, England and Wales. 1906.
262. Horace B. Woodward. "The Geology of Water Supply." 1910.
263. J. E. Worth and W. S. Crimp. "The Main Drainage of London." *Proceedings Institution of Civil Engineers*, Vol. 129, 1897.
264. H. C. Wyatt. "Sewers and Sewerage." 1941.

INDEX

Accidental injury, 210
Accidents in Sewers. Report on the Precautions Necessary for the Safety of Persons Entering Sewers and Sewage Tanks, 133
Acidity, 221
Activated Sludge Ltd., 301, 390, 406
Activated-sludge methods, comparison of, 411
Activated-sludge plant operation, 400
Activated-sludge plant, performance formula, 414
Activated sludge, surplus, 324, 405
Activated sludge, surplus, dewatering of, 339
Activated sludge, surplus, quantity of, (fig.) 406
Activated-sludge systems, 390
Activated-sludge treatment for small works, 465
Aeration channels, proportions of, 395
Aeration, methods of, 366
Aeration, requirements of, 365
Aeration-tank capacity, 393
Aeration tank, flat-bottomed, (fig.) 394
Aeration, volume of air required, 396
Aerodromes, drainage of, 193
Agricultural drainage, 190
Agricultural tiles, 191
Air-actuated devices, 142
Air blowers, efficiencies of, (table) 397
Air compression, 168
Air, filtration of, 397
Air-lifts, 142
Air mains, 398
Air mains, loss of pressure in, (table) 170, (table) 171, 172
Air mains, velocities in, (table) 397
Air required for sewage ejectors, 167
Air supply, 396
Alga, 367
Alkalinity, 221
Alternating double filtration, 372, 386
Ames Crosta Mills & Co. Ltd., 339, 349, 390, 405, 409
Ames Crosta Mills grit extractor, (fig.) 270
Ames Crosta Mills primary digestion tanks (fig.) 357
Ames Crosta Mills "Simplex" digestion tank, (fig.) 351
"Amphistoma" pump, (fig.) 144
Area occupied by sewage works, 239
Artificial lakes, purification in, 232
Asbestos-cement pipes, 112
Asphyxiation, 209
Asplit, 432
Automatic pumping, 138, 149
Automatic sampling, 225
Automatic sewage sampler, (figs.) 227
Automatic starting of diesel engines, 159

Backdrop, (fig.) 128, 133
Bacteriological qualities of effluents, 230

Balancing pumping flow, 285
Balancing tanks, 286
Bank-full flow, 204
"Barrington" pump, (fig.) 161, (fig.) 163
Bazalgette, Sir Joseph William, 5
Beet-sugar wastes, 426
Bellmouth chambers, 136
Benching, (fig.) 108, 129
Bilham, E. G., 31
Bilham's formula, 31
Bio-aeration, 422, 423
Biochemical oxygen demand (B.O.D.), 220, (table) 223
Biofiltration (Dorr process), 384, (fig.) 385
Birmingham formula, 30
Bleeding-off sludge, 307
B.O.D.—*see* Biochemical oxygen demand
Brick culverts, 120
Brick manholes, 128
Bricks for sewerage, 120
Broad irrigation, 235
Brush aeration system, 411
Brush drainage, 192
Building Regulations, 435, 436, 449
Building sites, under-drainage of, 189
Bulking, 401

Calculations for soil sewers, (table) 22
Camber, road, 99
Camp., T. R., 304, 305
Capen, C. H., 292, 303
Carbon dioxide, 211
Carbon monoxide, 211
Cascades, 133
Cast-iron pipes, 110
Cast-iron segmental sewer, (fig.) 118
Cast-iron segmental sewers, proportions of, (table) 117
Cast-iron segments, 116
Cast-iron tubbing details, (fig.) 118
Catchment, estimation of run-off from, 196
Catchment, rainfall allowance on, 199
Catchpits, 192
Cattle and sewage effluents, 230
"Cement Prodor", 431
Centrifugal pumps, 142, 143
Cesspool drainage, 439
Cesspool emptier, 438
Cesspools, 435
Cesspools, construction of, 441
Cesspools, conversion of, 454
Chain pump, (fig.) 441
Chain pumps, delivery of, (table) 454
Chamier, George, 198
Channels, inlet and outlet, 320
Chemical closets, 436, (fig.) 438
Chemical-mechanical treatment, 422
Chemical-resisting construction, 429
Chemistry of sewage, 217, 219
Chlorine, 220

INDEX

Circular sedimentation tanks, 310
Cladothrix, 401
Clifford, W., 295, 301, 303, 305
Clowes, F., 5
Coastal towns, 173
Cockburn, T., 374
"Coilfilter" installation, (fig.) 333, 363
Cold digestion, 336
Coleman and Johnson method, 56, 58
Combined system, 7
"Comminutors," 251, 256, 260, (fig.) 261
Compressed air, work in, 213
Compression of air, 168
Compression of air, power required for, (table) 169
Concrete pipes, 111
Concrete pipes, joints for, 111
Concrete pipe sewer, (fig.) 122
Concrete surround, (fig.) 108
Conductivities, thermal, (table) 345
Conservancy methods, 434
Constant-velocity detritus channel, 263, (fig.) 268, (fig.) 269, (fig.) 272, (fig.) 273
Construction of large sewers, 115
Construction of manholes, 127
Construction of sea outfalls, 182
Construction of small sewers, 104
Contact beds, 418
Control room, (fig.) 228
Copas, B. A., 69
Cost of large sewers in heading, 119
Cost of sewage works, 238
Cost of sludge disposal, 329
Crimp's formula for sludge dewatering, 339
Crops, 236
Crossness sewage-treatment works, (fig.) 243
Culvert, brickwork, 120
Culvert, mass concrete, (fig.) 121
Curves of settlement, 298
Cut-in and cut-out levels, 153, (table) 156

Dangerous machinery, 210
Danson, P. S. S., 415
Daymond, J. R., 180
Delivery pipes, sizes of, 148
Department of the Environment, 21, 24
Detention period, 297
Detergents, 221, 415
Detritus, analysis of, 274
Detritus channels, aerated, 263
Detritus channels, constant velocity, 263, (fig.) 268, (fig.) 269, (fig.) 272, (figs.) 273
Detritus channels, parabolic, 270, (fig.) 271
Detritus tanks, 262
Dewatering surplus activated sludge, 339
Diameters, minimum and maximum, 15
Dibdin slate beds, 418
Dickens' formula, 197
Diesel engines, automatic starting of, 159
Differential float control, 252, 256
Diffused-air plant, (fig.) 391, (fig.) 399
Diffused-air system, 392
Diffusers, 395
Diffusion, methods of, 395
Dilution, disposal of sewage by, 231
Dip-pipes, (fig.) 452

Direct-oxidation process, 420
Disintegration of screenings, 256, 259
Disposal in sealed capsules in deep sea, 427
Dissolved oxygen, 221
Ditches, 193
Dome diffusers, (fig.) 393, (fig.) 394, 396
Dorr Biofiltration process, 384, (fig.) 385
Dorr "Clarifier" type "A," (fig.) 315
Dorr "Clarifier" type "SC," (fig.) 317
Dorr "Detritor," 263, (fig.) 264, 265, (fig.) 266
Dorr digester, (fig.) 336
Dorr-Oliver "B.G.H." digesters, (fig.) 341, (fig.) 344, 349
Dorr-Oliver Co. Ltd., 333, 349, 363, 406
Dorr type "T" screen and disintegrator, (fig.) 252
Dortmund tank, (fig.) 302, 303
Dosing, 375
Dosing tank and percolating filter, (fig.) 377
Drainage of aerodromes, 193
Drainage of playing fields, 193
Drainage to stabilise earthworks, 194
Drawings, preparation of, 23
Drowning, 209
Dry-weather flow defined, 218
Dry-well pumping station, (fig.) 160, 161
Dumb-wells, 466

Earth-closets, 435, 436
Earthworks, drainage to stabilise, 194
Economic size of rising main, 144
Eddies, using, 301
Eddy, H. P., 293, 296, 299, 303
Edmonson, J. H., 423
Effluent polishing, 388
Effluent, Royal Commission, 229
Effluents, bacteriological qualities of, 230
Effluents discharged to sea, standards of, 173
Effluents, standards of, 228
Effluents treated on grass, 389
Egg-shaped sewer, 121, (fig.) 122
Egolf, C. B., 301
Ejectors, 142, 165, 166, 167
Ejectors, air required for, 167
Ejectors, emergency, 167
Electrical treatment, 419
Electrostatic air filter, 398
"Elsan" chemical closet, (fig.) 437
Embankment development, (fig.) 245
Embankment plan, 244
Embankments, river, 206
Emergency ejectors, 167
Emscher tank, 358
Enclosed filtration, 372
Enclosed percolating filters, 387
Environment, Department of the, 21, 24
Example calculation for small sewage works, 468
Example pumping calculation, 154
Excavation plan, 243
Excess-gas burner, 354
Exclusion of rainwater, 443
Experiments, model, 207, 305
Extended aeration, 465

Fair, Professor G. M., 255, 293, 305, 336

490 INDEX

Falling speeds of particles, (table) 265
Farrer, Wm. E., Ltd., 390
Feed-pipes, progressive bifurcation of, (fig.) 321
Fieldhouse tank, 418
Filling trenches, 113
Filter pressing, 332
Filtration, enclosed, 372
Filtration, high-rate, 383
Filtration of air, 397
Final discharge, 465
Final sedimentation tanks, 401
Fisher, A. J., 294
Fittings, frictional resistance of, (table) 155
Flexibility of operation, 246
Flexible clay pipe joints, approved, 105
Float experiments, 175
Flocculation, 293
Flood prevention, 204
Flow, horizontal, velocity of, 296
Flow of sewage, hourly variation of, (fig.) 20
Flow of sewage, rate of, 19, 218
Flow records, 202
Flow regulating devices, 287
Flow regulation of sewage, 288
Flushing of sewers, 92
Flushing siphon, (fig.) 93
Flushing tank, (fig.) 94
Foaming, 416
Foyn, Dr. Ernst, 420
Freezing of sludge, 363
Frictional resistance of fittings, (table) 155
"FS Disposal System," 363
Fuel store, 163

Gas, discharge of through orifices, (table) 356
Gas liquor, 426
Gas, sludge—*see* Sludge gas
Gases and vapours, 211
Gasholder, 350, 356
Gehm, H. W., 294
Geyer, J. C., 255
Gibson, Professor A. H., 207
Glasspoole, J., 199
Goodrich, S. R., 423
Gradients for land drains, (table) 191
Gradients for sludge mains, (table) 364
Gradients, maximum, for sewers, 13
Gradients, minimum, for sewers, (table) 12
Gradients, optimum, for rising mains, (table) 12
Gradients, self-cleansing, 9
Grass plots, effluent treated on, 389
Grease balls, 277
Grease removal, 276, 277
Groynes, 187
Gullies, 96
Gully spacing, (table) 97

Hartley sludging mechanism, (fig.) 319
Haunching, (fig.) 108
Haworth, John, 422, 423
Hazen's theory of sedimentation, 263, 265
Headings, 113
Heat losses from sludge digestion tanks, 343
"Heatamix" system, 348

Heating sludge digestion tanks, 349
Hermite process, 419
High-rate filtration, 383
Historical methods of sewage treatment, 417
Holroyd, A., 295
Horse-power calculations, 155
Horse-power calculations, metric, 156
Horseshoe-shaped sewers, (fig.) 121
House connection, (fig.) 17
Houston, Sir Alexander, 236
Houston, H. C., 5
Humus, 326
Humus, average quantity of, 324
Humus removal, 461
Humus tanks, 382
Hunter, Alexander, 374
Hurley, J., 294
Hydrogen, 212
Hydrogen sulphide, 211
Hydrograph, 63
Hydrograph method, 64, (table) 65, (table) 66, 200
Hydrolytic tank, 358, 418
Hypothetical storm, 57, 61, (table) 62
Hypothetical storm curve, (fig.) 61

Imhoff, Karl, 292, 293, 305, 358
Imhoff tanks, 358, 417
Impeller for use with sludge, (fig.) 144
Impermeability, change of during rainfall, 40, (fig.) 41
Impermeability factor, 26, (table) 27
Impermeability of surfaces, 26
Impervious area, 26
Impervious area, allowance for irregular distribution of, 49
Incineration of sludge, 362
Incremental feeding, 398
Industrial wastes, 424
Infection, 209, 214
Injury, 209
INKA system, 406
Inlet channels, 320
Inlet effect, 294
Instrument panel, 227, 358
Intercepting sewers, 79
Inverted siphon, 85, (fig.) 87, (fig.) 88
Inverts, 129
Irrigation over ploughed land, 235
Isolated buildings, sewage disposal for, 433

Jenkins, S. H., 415
Johnson Coupling, 148
Johnstone-Taylor, F., 197
Joints for clay pipes, 105
Joints for concrete pipes, 111
Jones & Attwood Comminutor, 251, 256, 260, (fig.) 261

Keefer, C. E., 294
Kennedy, R. G., 203, 296
Kershaw, G. B., 236, 297, 445
Kershaw, M. A., 428
Kessener system, 390, 410
Kinnear Moodie pre-cast concrete segments, 119

INDEX

Kirk, B. G., 275
Komline-Sanderson "Coilfilter," (fig.) 333
Komline-Sanderson Dissolved Air Flotation Thickening Process, 339

Labyrinth, (fig.) 428, 429
Lacey, G., 203, 296
Ladders, galvanised steel, 137
Lakes, artificial, purification in, 232
Land drains, 191
Land drains, gradients for, (table) 191
Land filtration, 235, 389
Land treatment, 235, 464
Land treatment, area for, 236, (table) 237
Landreth "Direct-oxidation" process, 420
Laughlin process, 422
Lay-out of mains, 246
Lay-out of sewage works, 238, 241
Level survey, 16
"Lift and Force" ejectors, 166
Lime-mixing plant, (fig.) 427
Lloyd-Davies calculation, example, 37
Lloyd-Davies formula, 30
Lloyd-Davies method, 25, 29, 36
Lloyd-Davies theory, proof of, 35
Location of sewers, 15
Lunar clock, 185

Mains, lay-out of, 246
Manhole, brick, 128
Manhole chambers and shafts, 130
Manhole covers, 136
Manhole ironwork, 136
Manhole, mass concrete, 128
Manhole, pre-cast concrete, 127, (fig.) 128
Manhole, pressure, 135
Manhole, shallow, (fig.) 129
Manhole, side-entrance, (fig.) 128
Manhole walls, thickness of, (fig.) 131
Manholes, construction of, 127
Manholes on large sewers, 132
Manurial value of sludges, 361
Marston formula, 123
Mass concrete culvert, 120, (fig.) 121
Mass concrete manhole, 128
Maximum diameters, 15
Maximum gradients, 13
McCabe, W. L., 301
McGowan formula, 222
Mechanical agitation, 407
Mechanical screens, (fig.) 254
Mesophilic digestion, 336
Metcalf, L., 293, 296, 299, 303
Methane, 211
Micro-straining, 388
Midday peak, smoothing out, 287
"Mieder" Scraper, (fig.) 320
Miles' acid process, 421
Miles, George W., 421
Milk wastes, 426
Mills, E. V., 373
Minimum diameters, 15
Ministry of Health Departmental Committee on Rainfall and Run-off, 30
Ministry of Health standard curve, 30
Ministry requirements, 14

Mistakes in pumping-station design, 140
Mobrey magnetic level switch, 150
Model Byelaws, 15
Model experiments, 207, 305
Module, 73, 288, (fig.) 289
Mogden formula, 429
Mohlman sludge-volume index, 400
Mole drainage, 193
Momentary impermeability, 41

Nicholls-Herreshoff "Multi-hearth" incinerator, 362
Night-soil content of sewage, 224
Nitrogen, 220
Nixon, Marshall, 203
Noise, reduction of, 163
Nylon, 432

Occupational Hazards in the Operation of Sewage Works. Manual of Practice No. 1, U.S.A., 209
Organic carbon, 220
Ormsby and Hart method, (fig.) 51, 56, 60, 61, (calculation) 63
Outfall calculation, stage-by-stage method, 180
Outfall, sea, discharge of, 177
Outfall, sea, plan of, (fig.) 185
Outfall, sea, position of, 175
Outfall, steel, flotation of, (fig.) 183
Outlet channels, 320
Oxygen deficiency, 212
Oxygen, dissolved, 221

Packaged plants, 465
Parabolic detritus channel, 270, (fig.) 271
Partially-separate system, 7
Particles, falling speeds of, (table) 265
Percolating filter and sedimentation tank, (fig.) 376
Percolating filter, circular, (fig.) 458
Percolating-filter distributors, 379
Percolating-filter medium, effect of size of, (table) 375
Percolating-filter medium, required properties of, 367
Percolating-filter medium, retention of settled sewage in, (table) 373
Percolating filter with tipping-tray, (fig.) 456
Percolating filters, 367
Percolating filters, capacity of, 369
Percolating filters, construction of, 380
Percolating filters, depth of, 374
Percolating filters, enclosed, 387
Percolating filters, feed-pipe system to, 379
Percolating filters for small works, 455
Percolating filters, loading of, (table) 372
Percolating filters, operation and maintenance of, 387
Percolating filters, rectangular, 381
Percolating filters, shapes of, 374
Percolating filters, small, capacity of, (table) 461
Petrol vapour, 212
Phormidium, 367
Pipe-laying, 106

Pipe-sewers, stoneware, 104
Pipes, asbestos-cement, 112
Pipes, buried, pressures on, 123
Pipes, cast-iron, 110
Pipes, concrete, 111
Pipes, concrete, joints for, 111
Pipes, pre-stressed concrete, 111
Pipes supported on piers, 83
Pipework of pumping station, 147
Playing fields, drainage of, 193
Pneumatic ejectors, 165
Poisoning, 209
Pollution, organic, 219
Polyethylene, 432
Polytetrafluoroethylene, 432
Polyvinyl chloride, 432
Ponding, 383
Ponds, storage, 72
Porteous method of sludge dewatering, 361
Porteous plant, (fig.) 362
Pre-cast concrete manhole, 127, (fig.) 128, 130
Pre-cast concrete segments, 119
Precautions, 212
Preliminary treatment, 249
Pressure manholes, 135
Pressures on buried pipes, 123
Pre-stressed concrete pipes, 111
Primary sedimentation tanks, (fig.) 410
Primary sludge, 325
Priming, automatic, 160
"Prodorcote," 432
"Prodorfilm," 432
"Prodor-glas," 432
Prodorite Ltd., 431
"Prodorkitt," 431
"Prodorphalte," 431
Proportional flow in partly full pipes, 11, 476
Proportional-flow plate weir, (fig.) 267
Protozoa, 401
Prüss, Dr., 337
Pumping at small sewage works, 467
Pumping, automatic, 138, 149
Pumping, example calculation, 154
Pumping flow, balancing of, 285
Pumping from excavation, 113
Pumping machinery, arrangement of, 145
Pumping of sewage, 138
Pumping station, dry-well, (fig.) 160, 161
Pumping station head losses, 154
Pumping station lay-out, 148
Pumping station, mistakes in design of, 140
Pumping station pipework, 147
Pumping station ventilation, 162
Pumping station without suction-well, 157
Pumps, centrifugal, 142, 143
Pumps, number of, 146
Pumps, reciprocating, 142, 143
Pumps, starting times of, (table) 153
Pumps, types of, 142
Purging plant, 358
Pyramidal sedimentation tanks, 308, (fig.) 311
Pyramidal sedimentation tanks, construction of, 322

Radioactive contamination, 214

Rainfall allowance on river catchment, 199
Rainfall and run-off coefficients for rational method, (table) 45
Rainfall intensities, 30, (table) 32
Rainwater, exclusion of, 443
Ramp, (fig.) 134
Rational method, 25, 44
Rational method, example calculation, 46
Rational method, run-off and rainfall coefficients, (table) 45
Raymond drier, 362
Reciprocating pumps, 142, 143
Recirculation, 372
Records of flow, 202
Rectangular percolating filters, 381
Rectangular sedimentation tank, 313
Rectangular sedimentation tank with travelling scraper, (fig.) 318
Reflux valves, loss of head through, (fig.) 154
Regionalisation, 239
Reinforced concrete, 120
Reinforced-concrete sewer, (fig.) 121
Reinstatement of surfaces, 113
Requirements, Ministry, 14
Reynoldson, T. B., 367
Riley, D. Wearing, 54
Rising mains, economic diameter of, 145
Rising mains, economic size of, 144
Rising mains, optimum gradients for, (table) 12
River embankments, 206
River maintenance, 195
Rivers, self-purification of, 231
Road camber, 99
Road channel levels, difference of, (table) 101
Road drainage, 96
Road drainage at intersections, 101
Roof slabs, 130
Royal Commission, 1882, 5
Royal Commission effluent, 228, 370
Royal Commission on Sewage Disposal, 7, 74, 369, 446
Rudolfs, Dr. W., 337
Run-off and rainfall coefficients for rational method, (table) 45
Run-off from catchment, estimation of, 196

Safety, 209
Sampler, automatic, (figs.) 227
Sampling, 224
Sampling, automatic, 225
Screenings, 255
Screenings, disintegration of, 259
Screenings, quantities of, (table) 255
Screens, 161, 249
Screens, calculating sizes of, 257
Screens, hand-raked, 250
Screens, mechanical, 251, (fig.) 254
Screens, mechanical, proportions of, 252
Screens, number of, 258
Scum-boards, 319, 320
Scum removal, 319
Sea outfall, 233
Sea outfall, construction of, 182
Sea outfall, discharge of, 177
Sea outfall plan, (fig.) 185
Sea outfall, position of, 175

INDEX 493

Sea outfall, submission of plans of, 184
Sea, protection from, 186
Sea walls, 187
Sedimentation experiments, 304
Sedimentation involving digestion, 358
Sedimentation tank and percolating filter, (fig.) 376
Sedimentation tank, bridge-type, (fig.) 311
Sedimentation tank, circular, 310
Sedimentation tank, final, 401
Sedimentation tank, flow in, 294
Sedimentation tank, pyramidal, 308, (fig.) 310
Sedimentation tank, pyramidal, construction of, 322
Sedimentation tank, rectangular, 313
Sedimentation tank, rectangular, with travelling scraper, (fig.) 318
Sedimentation tank to percolating filter, loss of head, (fig.) 378
Sedimentation tank, types of, 307
Sedimentation, two-stage, 303
Segmental sewer, cast-iron, (fig.) 118
Segmental sewer, cast-iron, proportions of, (table) 117
Segments, cast-iron, 116
Segments, pre-cast concrete, 119
Self-cleansing gradients, 9
Self-purification of rivers, 231
"Semtex," 431
Separate system, 7
Separating weirs, 279
Septic tank, 358, 417, 444, (fig.) 445
Septic tank, automatic pump and percolating filter, (fig.) 450
Septic tanks, population served by, (table) 451
Septic tanks, proportions of, (table) 451
Septic tanks, sludging of, 452
Settlement curves, 298
Settlement of activated sludge, (fig.) 300
Settlement of particles, 291
Settlement of river mud, (fig.) 300
Settlement produced at various detention periods, (fig.) 299
Sewage, average, 224
Sewage, chemistry of, 217, 219
Sewage, hourly variations in flow and strength, (fig.) 20
Sewage, night-soil content of, 224
Sewage, rate of flow of, 19
Sewage, regulation of flow and strength, 288
Sewage sampler, automatic, (figs.) 227
Sewage sickness, 237
Sewage, storage of, 177
Sewage, strength of, 221, (table) 223
Sewage, treatment of on land, 235
Sewage-treatment works, area occupied by, 239
Sewage-treatment works, cost of, 238
Sewage-treatment works, lay-out of, 238
Sewage-treatment works, siting of, 238
Sewage-treatment works, small, 442, (fig.) 459
Sewage-treatment works, small, example calculation, 468
Sewer, concrete pipe, (fig.) 122
Sewer, egg-shaped, 121, (fig.) 122
Sewer, horseshoe-shaped, (fig.) 121

Sewer plan and section, (fig.) 18
Sewer, reinforced-concrete, (fig.) 121
Sewer sections, preparation of, 17
Sewer, steel, (fig.) 125
Sewer, steel, on piers, (fig.) 86
Sewer, U-shaped, (fig.) 122
Sewerage, history of, 3
Sewerage systems, 7
Sewers above ground-level, 83
Sewers, cross-sections of, 115
Sewers in heading, cost of, 119
Sewers, intercepting, 79
Sewers, large, construction of, 115
Sewers, location of, 15
Sewers, minimum gradients for, (table) 12
Sewers, small, construction of, 104
Sewers, thickness of walls of, (table) 126
Sewers, ventilation and flushing of, 92
Sheffield system, 422
"Shone" ejector, (fig.) 165
Side-entrance manhole, (fig.) 128, (fig.) 132
"Simplex" aeration tanks, (fig.) 411
"Simplex" high-intensity aerating cone, (fig.) 412
"Simplex" plant for an institution, (fig.) 463, (fig.) 465
"Simplex" system, 390, 407, (fig.) 408
Siphon, inverted, 85, (fig.) 87, (fig.) 88
Siphon, multiple, 88
Siphon, precautions against silting, 89
Site of building defined, 435
Siting of sewage works, 238
Sizes of suction and delivery pipes, 148
Sludge as fertiliser, 359
Sludge, bleeding-off, 307
Sludge calculating quantities of, 325
Sludge, dewatering, 338
Sludge, dewatering, Porteous method, 361
Sludge digestion, 335
Sludge digestion, cold, mesophilic and thermophilic, 336
Sludge digestion, heat requirements of, 342
Sludge digestion period, 337
Sludge digestion tanks, capacity of, 337
Sludge digestion tanks, feeding, 347
Sludge digestion tanks, hazards of operating, 347
Sludge digestion tanks, heat losses from, 343
Sludge digestion tanks, methods of heating, 349
Sludge digestion tanks, shapes of, 340
Sludge digestion, two-stage, 337
Sludge disposal, 328
Sludge disposal, comparative costs of, 329
Sludge dry solids, average quantity of, 324
Sludge drying, 329
Sludge drying beds, 330
Sludge drying beds, area of, 330
Sludge, flow of in pipes, 363
Sludge freezing, 363
Sludge gas, content and calorific value of, 351
Sludge gas produced according to detention period, (table) 353
Sludge gas, quantity of, 340
Sludge gas, recirculation of, 348
Sludge gas, uses for, 354

Sludge gas, utilisation of, 351, (fig.) 359
Sludge incineration, 362
Sludge mains, gradients for, (table) 364
Sludge, manurial value of, 361
Sludge, primary, 325
Sludge produced at different velocities, (table) 296
Sludge, quantities of, 324
Sludge recirculation, 398
Sludge removal, 316
Sludge seeding and mixing, 347
Sludge separators, 332
Sludge, specific heat of, (table) 343
Sludge-volume index, 400
Sludging mechanism, (fig.) 319
Sludging mechanism for circular tank, (fig.) 312
Soakaway, (fig.) 82
Soakaway, capacity of, 81
Soakaway pits, 466
Soakaway systems, 80
Soil sewerage, 16
Soil sewers, calculations for, (table) 22
Soil sewers, normal requirements in design, 13
"Solids Diverter," 144
Stability, 221
Stable channels, 203
Stage-by-stage calculation for storage tank, 180
Standards of effluents, 228
Steel outfall, flotation of, (fig.) 183
Steel sewer, (fig.) 125
Steel sewer on piers, (fig.) 86
Steel tubes, 111
Step aeration, 398
Step-irons, 137
Stokes' Law, 265, 291, 292, 306
Stoneware-pipe sewers, 104
Storage of sewage, 177
Storage of storm water, 159
Storage of surface water, 64
Storage ponds, 72
Storage tanks, 186
Storm, hypothetical, 57, 61, (table) 62
Storm-overflow weir, (fig.) 78, (fig.) 280
Storm-relief sewers, 79
Storm-relief works, 74
Storm-tank policy, 76
Storm tanks, 281
Storm tanks away from sewage works, 77
Storm tanks, details of, 284
Storm tanks, purpose and capacity of, 282
Storm water, 219
Storm-water separation, 279
Storm water, storage of, 159
Storms, areas covered by, 199
Strength of sewage, hourly variation of, (fig.) 20
Sub-irrigation, 464
Suction pipes, sizes of, 148
Suction-well, capacity of, 155
Suction-well, size of, 150
Sulphur compound, 432
Sulzer disintegrator, 256
Supernatant water, removal of, 350
Surface aeration, 390, 407

Surface-water sewerage, 24
Surface-water sewerage practice, early, 28
Surface-water sewerage theory, 25
Surface-water, storage of, 64
Surge in incoming sewer, 158
Surplus activated sludge, 324, 405
Surplus activated sludge, quantity of, (fig.) 406
Survey, level, 16
Systems of sewerage, 7

Tangent curves, (fig.) 56, (fig.) 57
Tangent method, 52, (fig.) 54
Tangent method, revised or modified, 54
Tank sewers, 184
Tank vehicles, 441
Tapered aeration, 398
Testing lines of pipes, 114
Thermal conductivities, (table) 345
Thermophilic digestion, 336
Three-point problem, (fig.) 176
Tidal experiments, 174, 175
Tidal reaches, 207
Timbering, 113
Time-area graph, 50
Tipping-tray mechanism, (fig.) 460
Trade wastes, charges for treatment of, 429
Trade wastes classified, 425
Trade wastes, control of at point of origin, 427
Travis hydrolytic tank, 418
Travis, W. Owen, 418
Treatment, choice of method of, 241
Turbulence, effect of, 294
Two-stage sedimentation, 303

Under-drainage of building sites, 189
Unit-hydrograph method, 201
Unit-hydrograph method, application of, (table) 202
Upward-flow principle, 302
U-shaped sewer, (fig.) 122
Utilisation of Sewage Sludge as Fertiliser: Manual of Practice No. 2, U.S.A., 360

Vacuum filters, 332
Vapours, 211
Velocities, self-cleansing, 24
Velocity of scrapers, peripheral, 314
Ventilation of pumping station, 162
Ventilation of sewers, 92

Walker, R. D., 197
Wash-outs, 316
Water-cushions, 133, 134
Water-hammer, 148
Wave action, 186
Weil's disease, 214
Williams, G. Bransby, 198
Windridge, M. E. D., 295, 301, 305
Wire-netting cover to percolating filter, (fig.) 460
Woolf, Albert E., 419
Worms, 367

Zeolite treatment, 421
Zimmerman process, 363